Stereochemistry and Reactivity of Systems Containing π Electrons

Methods in Stereochemical Analysis

Volume 3

Series Editor
Alan P. Marchand
North Texas State University
Department of Chemistry
Denton, Texas 76203

Stereochemistry and Reactivity of Systems Containing π Electrons

Proceedings of the Symposium "Stereochemistry and Reactivity in Pi Systems" held May 19-22, 1982 in recognition of Professor Paul Bartlett's many years of leadership in physical organic chemistry.

Edited by

William H. Watson

Verlag Chemie International
Deerfield Beach, Florida

William H. Watson, Ph.D.
Chairman, Department of Chemistry
Texas Christian University
Fort Worth, Texas 76129

Library of Congress Cataloging in Publication Data
Main entry under title:

Stereochemistry and reactivity of systems containing [Greek letter pi] electrons.

(Methods in stereochemical analysis ; v. 3)
Includes bibliographies and index.
1. Stereochemistry — Congresses. 2. Reactivity (Chemistry) — Congresses.
I. Watson, William H., 1931- . II. Bartlett, Paul D. III. Series.
QD481.S758 1983 541.2'23 83-14760
ISBN 0-89573-117-7

Printed in the United States of America.

ISBN 0-89573-117-7 Verlag Chemie International, Deerfield Beach
ISBN 3-527-26115-X Verlag Chemie GmbH, Weinheim

FOREWORD

The hope for a fertile conference rests on fortunate timing, fitting, and phasing of diverse approaches to a common theme. At this point we can see that the subject of the present conference was a happy choice. It brought in some of the leading current thinking on the theoretical treatment of systems and definitive studies of the role of stereochemistry in some quite varied cases. To the great pleasure of the hosts, contributors came from long distances to present their unique findings, and it seems that the discussions greatly accelerated the usually slow process of forming a consensus. Hearty thanks to those who made these valuable contributions.

Paul D. Bartlett

PREFACE

In undergraduate courses a carbon-carbon double bond is described in terms of sp^2 hybridized carbon atoms which form three coplanar σ bonds. These bonds exhibit valence angles of 120°. The remaining p orbitals are perpendicular to the σ plane and overlap to form a π bond. The π bond is symmetric with respect to the σ plane. Although distortions of this idealized geometry have been known for years, their effects upon stereochemistry and reactivity are of renewed interest. In small ring alkenes the valence angles differ considerably from 120°, and the resulting effects are discussed in terms of bond angle deformations or strain. When the ends of a carbon-carbon double bond are twisted in opposite directions this may reduce p-orbital overlap, and the effects are discussed in terms of torsional strain. Both of these effects are emphasized in the chapters on bridgehead alkenes.

Although scattered references to "pyramidalization" of sp^2 hybridized carbon atoms can be found in the literature, little consideration was given this topic until the late 1970's. Theoretical calculations and experimental studies of rates and stereochemistries indicated the properties of molecules such as norbornene might be due to an asymmetry of the π-electron system and a "pyramidalization" of the carbon atoms forming the π bonds. There was little direct experimental evidence to confirm these predictions until the large distortions in the syn-sesquinorbornenes and related molecules were reported. The 16° to 20° deviations from planarity provided evidence that "pyramidalization" was a real and significant effect. Several chapters in this volume are concerned with "pyramidalization" and asymmetric electron distributions. The active discussion following each presentation indicates the evolving nature of the field. The origin of the asymmetry and the factors responsible for the observed stereochemistries and accelerated rates will continue to be debated.

The symposium was organized to bring together active workers in the field. A few were unable to attend, and we missed their insight and critiques. The symposium also was organized to honor Professor Paul Bartlett who continues to be in the forefront of research in physical organic chemistry. His research group is active in the investigation of sesquinorbornene systems and some unexpected reactions of isodicyclopentadiene.

It was a pleasure to organize and host the conference, and I am indebted to all the participants who contributed to the excitement of the meeting. We hope the presentations and discussions will stimulate the reader as much as they did those in attendance.

William H. Watson

CONTRIBUTORS

WALDEMAR Adam, Institut fur Organische Chemie, Universitat Wurzburg, Am Hubland, D-8700 Wurzburg, F.R.G.

PAUL D. BARTLETT, Department of Chemistry, Texas Christian University, Fort Worth, Texas, 76129 U.S.A.

ANDREW J. BLAKENEY, Department of Chemistry, Texas Christian University, Fort Worth, Texas. 76129 U.S.A.

MICHAEL BOHM, Organisch-Chemisches Institut der Universitat, Im Neuenheimer Feld 270, D-6900 Heidelberg, F.R.G.

JACK D. BOYD, Department of Chemistry and Biochemistry, University of California, Los Angeles, California 90024 U.S.A.

NESTOR CARBALLEIRA, Institut fur Organische Chemie, Universitat Wurzburg, Am Hubland, D-8700 Wurzburg, F.R.G.

GERALD L. COMBS, Department of Chemistry, Texas Christian University, Fort Worth, Texas 76129 U.S.A.

OTTORINO DE LUCCHI, Institut fur Organische Chemie. Universitat Wurzburg, Am Hubland, D-8700 Wurzburg, F.R.G.

CHRISTOPHER S. FOOTE, Department of Chemistry and Biochemistry, University of California, Los Angeles, California 90024 U.S.A.

JEAN GALLOY, FASTBIOS Laboratory. Department of Chemistry, Texas Christian University, Fort Worth, Texas 76129 U.S.A.

ROLF GLEITER, Organisch-Chemisches Institut der Universitat, Im Neuenheimer Feld 270, D-6900 Heidelberg, F.R.G.

FREDERICK D. GREENE, Department of Chemistry Massachusetts Institute of Technology, Cambridge, Massachusetts, 02139 U.S.A.

CHEE-LIANG GU, Department of Chemistry and Biochemistry, University of California, Los Angeles, California 90024 U.S.A.

KARLHEINZ HILL, Institut fur Organische Chemie, Universitat Wurzburg, Am Hubland, D-8700 Wurzburg, F.R.G.

KENDALL N. HOUK, Department of Chemistry, University of Pittsburgh, Pittsburgh. Pennsylvania 15260 U.S.A.

HERBERT O. HOUSE, Department of Chemistry, Georgia Institute of Technology. Atlanta, Georgia 30332 U.S.A.

MARK L. KACHER, Department of Chemistry and Biochemistry, University of California, Los Angeles, California 90024 U.S.A.

RICHARD KANNER, Department of Chemistry and Biochemistry, University of California, Los Angeles, California 90024 U.S.A.

MICHAEL KRAMER, Department of Chemistry and Biochemistry, University of California. Los Angeles, California 90024 U.S.A.

JANG-JENG LIANG, Department of Chemistry and Biochemistry, University of California, Los Angeles, California 90024 U.S.A.

LEWIS E. MANRING, Department of Chemistry and Biochemistry, University of California, Los Angeles, California 90024 U.S.A.

WAYLAND E. NOLAND, School of Chemistry, University of Minnesota, 207 Pleasant St. SE, Minneapolis, Minnesota 55455 U.S.A.

PETER R. OGILBY, Department of Chemistry and Biochemistry, University of California, Los Angeles, California 90024 U.S.A.

Leo A. PAQUETTE, Evans Chemical Laboratory, The Ohio State University, Columbus, Ohio 43210 U.S.A.

ANTONIUS A. M. ROOF, Department of Chemistry, Texas Christian University. Fort Worth, Texas 76129 U.S.A.

KENNETH J. SHEA, Department of Chemistry. University of California. Irvine, California 92717 U.S.A.

RAVI SUBRAMANYAN, Department of Chemistry Texas Christian University, Fort Worth, Texas 76129 U.S.A.

PIERRE VOGEL, Institut de Chimie Organique, 2 rue de la Barre, CH 1005 Lausanne, Switzerland

PETER J. WAGNER, Department of Chemistry Michigan State University, E. Lansing, Michigan 48824 U.S.A.

WILLIAM H. WATSON, FASTBIOS Laboratory, Department of Chemistry, Texas Christian University, Fort Worth, Texas 76129 U.S.A.

WILLIAM J. WINTER, Department of Chemistry, Texas Christian University, Fort Worth, Texas 76129 U.S.A.

CHENGJIU WU, Department of Chemistry, Texas Christian University, Fort Worth, Texas 76129 U.S.A.

CONTENTS

1. THEORETICAL STUDIES OF ALKENE PYRAMIDALIZATIONS AND ADDITION STEREOSELECTIVITIES

Kendall N. Houk

Department of Chemistry, University of Pittsburgh, Pittsburgh, Pennsylvania 15260

Today I will talk about a subject which is not unfamiliar to all of you. You have probably seen this molecule before!

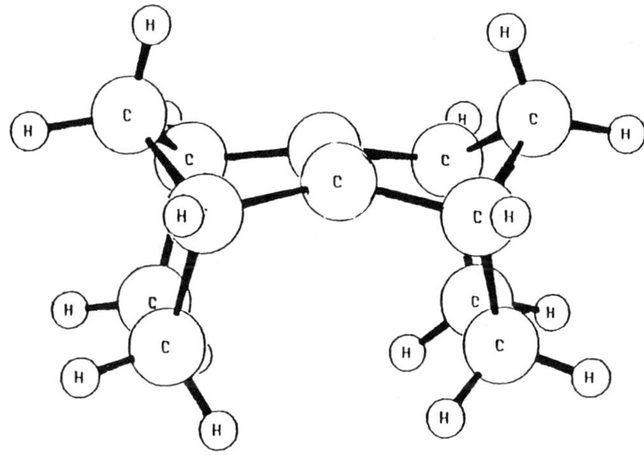

In many ways it is responsible for giving us the opportunity to gather in Fort Worth. *Syn*-sesquinorbornene has attracted much attention because of its unusual properties, which in some ways remind us of the nonclassical era of physical organic chemistry. Perhaps we might represent the molecule with a variety of dashed lines to indicate the fact that it has been suggested that there are special and significant interactions between σ and π bonds or even between hydrogen atoms and π bonds[1].

Before beginning, I must acknowledge a number of people to whom I am indebted for inspiration for our work. These people happen to be most of the speakers on the program today and part of tomorrow. Professors Watson, Bartlett, Paquette, Vogel and Gleiter have accumulated some extremely interesting experimental results on sesquinorbornenes and related molecules, and our work is highly dependent on their discoveries.

I am going to talk today about our ideas about the origins of the spectacular pyramidalizations of the alkene carbons in syn-sesquinorbornene, and the reasons for some of the peculiar reactivity and stereoselectivity phenomena that have been observed for this and related molecules. I will not try to review what others have concluded about this molecule, because these gentlemen will have their chance later. Let me describe briefly what I will cover in this lecture. In his introduction, Bill Watson mentioned our predictions about alkene pyramidalizations. I will briefly review our predictive work and that of Morokuma and Wipff in that area. I will describe the experimental realization of these pyramidalization effects, and then talk about the origin of pyramidalization. I will discuss the consequences of pyramidalization on the electronic features of π bonds, and then I will make a few additional predictions. Other molecules that may be highly pyramidalized follow directly once the reasons for pyramidalization in these systems is understood. Finally, I will discuss reactions of pyramidalized alkenes, that is, alkenes in which the alkene carbons do not lie in a molecular symmetry plane. Such molecules have two different kinds of π faces, which are attacked at unequal rates. First I will describe the thermal reactions of such species, and then our studies of the Bartlett experiments on the triplet photoreductions of syn-sesquinorbornene. Our ongoing experimental and theoretical studies of the stereoselectivity of cycloadditions to 7-substituted norbornadienes will form the last part of my talk.

In 1980, we developed some inkling that some norbornadienes would not have strictly planar alkene carbon systems [1]. This came from model calculations designed to determine whether the force constants for bending of hydrogen atoms exo or endo in norbornadiene were different. We found that indeed the force constants were different, and a nonplanar minimum was predicted [1]. Later, we carried out optimizations using minimal basis set *ab initio* calculations on norbornene, norbornadiene and some related bicyclic molecules shown here [2].

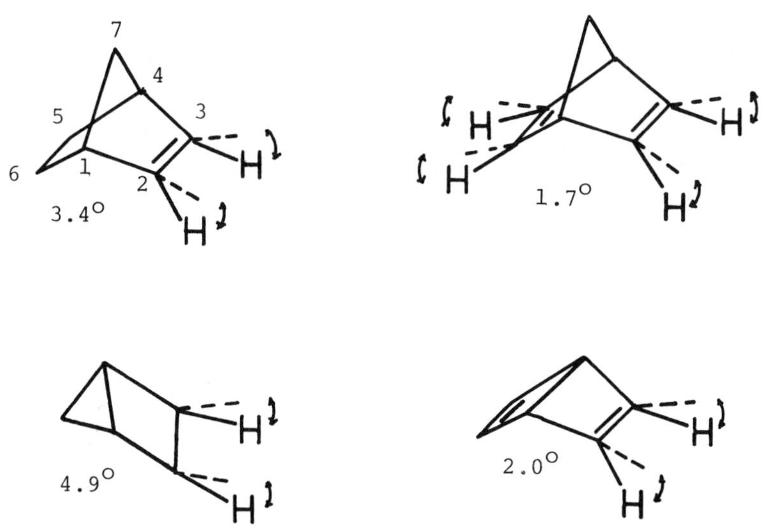

Somewhat before our publication, Wipff and Morokuma published similar optimizations of norbornadiene, norbornene, and hydroxy-substituted derivatives which showed the same effect [3].

Now I suppose that throughout this symposium there may be some confusion on how nonplanarity, or alkene pyramidalization, is described. In the figure above, 3.4° is the endo bending of the hydrogens of norbornene out of the C1C2C3C4-plane. More frequently, pyramidalization is described in terms of the dihedral angle X-C=C-X (trans), which is, in general, larger than the out-of-plane angle.

Norbornene is pyramidalized slightly in the endo direction. Norbornadiene is bent to a smaller extent, and the largest pyramidalization for molecules in the simple bicyclic series is observed in bicyclopentene. A partial optimization of Dewar benzene was already in the literature [4]. A full optimization indicates there is about a 2° out-of-plane bending of the hydrogens in the endo direction.

We wanted to determine whether there was some special feature of these strained molecules which caused bending, or whether the arrangement of allylic bonds with respect to the π bond causes the pyramidalization. We did model optimizations which have been published on propene in which one allylic C-H was constrained to a dihedral angle of 90° with respect to the three carbons of propene [2]. On the left below are indicated the various distortions that occur at alkene carbons when one allylic C-H bond is fixed at a 90° dihedral. The hydrogen

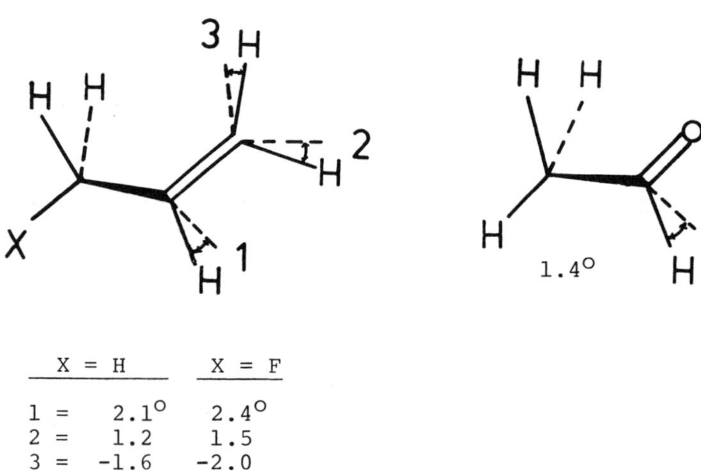

	X = H	X = F
1 =	2.1°	2.4°
2 =	1.2	1.5
3 =	-1.6	-2.0

at C-2 moves down toward the hydrogen fixed at a 90° dihedral, and the out-of-plane angle is 2.1°. At the same time the terminal CH_2 group essentially rotates to maintain overlap of the p orbital on C-1 with the p orbital on C-2 which no longer lies strictly perpendicular to the plane of the heavy atoms.

We have done computations on a number of other molecules as well. I won't review all of the results, but some of them are indicated above. Allyl fluoride also shows pyramidalization, when the C-F bond is fixed perpendicular to the plane of the heavier atoms. Acetaldehyde is bent to a lesser extent. It has a downward bending of the carbonyl hydrogen by 1.4°.

More recently, we have used force-field or molecular mechanics techniques rather than quantum mechanical techniques to investigate this phenomenon. This was done because our hypothesis was that purely torsional interactions caused the pyramidalization, and molecular mechanics treats such interactions rather well. Shown below are Newman projections of the Allinger MM2 optimized [5] structures of propene and acetaldehyde. One C-H bond was fixed with a dihedral angle of 90° with respect to the heavy atom plane.

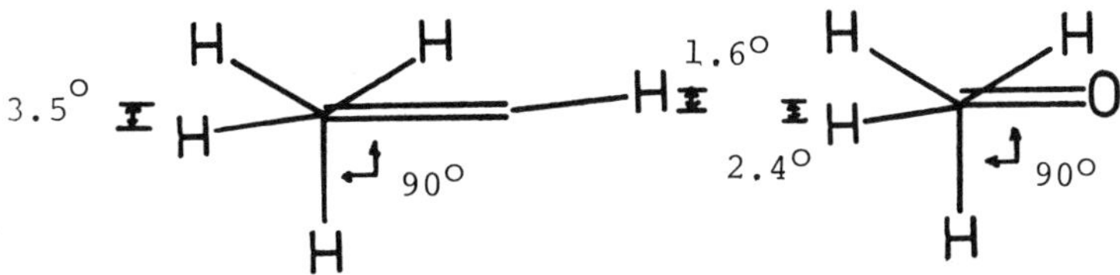

We see that although the numbers are not exactly the same, the same distortions and trends are observed as in the ab initio calculations. The propene C-2 hydrogen moves down by 3.5°, and there is the same kind of rotation of the terminal CH_2. The pyramidalization in the case of acetaldehyde is smaller as was also predicted by the ab initio calculation. I should also mention that Ermer [6] did force-field calculations on norbornene and norbornadiene which shows the same kind of pyramidalizations predicted by the ab initio calculations. These force-field calculations came several years before the ab initio results. Burkert has published in this area, also [7]. There are even experimental suggestions of pyramidalization in norbornadiene from NMR studies in liquid crystal phases [8]. In a moment I will return to the significance of the success of force-field calculations in reproducing the pyramidalization.

In molecules such as norbornene, looking down a double barreled Newman projection, or in the simpler propene molecule, where one CH dihedral angle is fixed at 90°, the direction of pyramidalization is such that the bonds to the alkene carbon become slightly staggered with respect to the allylic bond. After analyzing the orbital interactions in detail, and considering a number of fancy orbital explanations, we concluded that the origin of this pyramidalization is the relief of various torsional interactions that occur in the planar alkenes.

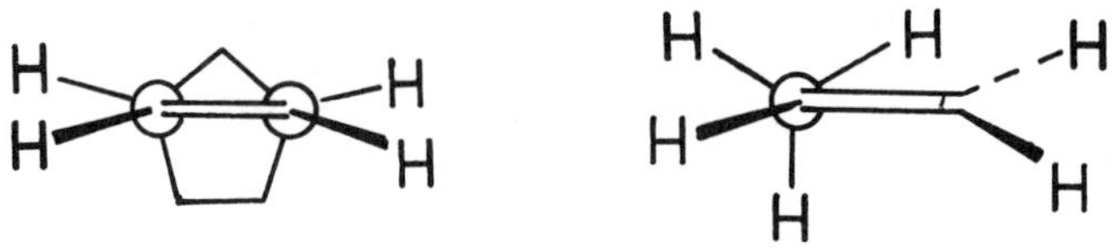

Interactions between the alkene C-H bond, the alkene C-C bond and perhaps the π orbital as well and the three bonds to the allylic carbon are all partially eclipsed in the planar species, but are relieved in the nonplanar. I will try to show you through various computational experiments that in fact these simple torsional interactions are the origin of alkene pyramidalizations.

Now let me say something about orbitals. Lately, a common question at the end of my talks is "What happened to the orbitals?" Nevertheless, they are still there controlling everything, but it is not always necessary to look at them to understand things. Nevertheless, I have to put a few orbitals in here. I want to slightly digress and talk about the barrier to rotation in ethane, because we believe the same thing that causes ethane to be staggered produces the pyramidalization in the molecules that we have discussed. There are a number of elegant quantitative [9] as well as qualitative [10] investigations of this phenomenon, and let me summarize briefly the results of these investigations. The pictures below show first of all two filled σ C-H orbitals of ethane in an eclipsed conformation. On the right I have shown one pair of anti CH orbitals in ethane in a staggered conformation. Of course there are three such interactions. It is clear that the overlap of these filled CH orbitals in the syn arrangement (eclipsed), is better than the overlap of the two filled obitals in the anti arrangement (staggered). This interaction is a repulsive one because both of these are filled, so four electron interactions cause the staggered conformation in ethane to be favored over the eclipsed. Of course this is a well known phenomenon, [9,10] and it is called closed-shell repulsion, exclusion repulsion, or exchange repulsion. It favors the staggered conformation of ethane. Although a second effect, described in the lower half of the figure, is not

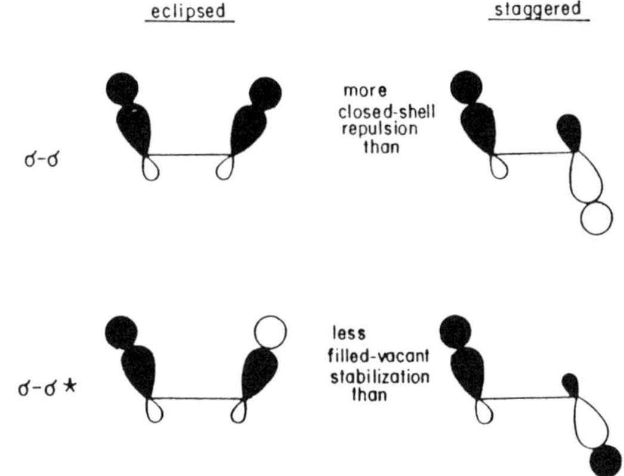

very significant energetically in ethane itself, as revealed by calculations [9], the interaction of filled orbitals on one fragment with vacant orbitals on the other fragment also will favor the staggered conformation. At the bottom of the figure are shown a filled C-H σ orbital on one carbon and a vacant C-H σ* orbital on the other. The overlap in this case is larger for the anti (staggered) arrangement than it is for the syn (eclipsed) arrangement. The fact that there is no node on the back tail of the σ* orbital, whereas there is a node right through the middle of the front side of this orbital causes the overlap between these to be very small in the syn arrangement and larger in the anti. Since this is a two-electron interaction, it is stabilizing and more so in the anti (staggered) case. Although this is again not very significant for ethane, it may be significant for cases where relatively polar bonds or strain may lower the energy of the σ* orbital.

We feel that these kinds of interactions, namely, minimization of closed-shell repulsion between vicinal bonds and perhaps to a smaller extent the maximization of filled-vacant interactions between vicinal bonds, cause ethane to be staggered and also cause the alkene carbons of the bicyclic molecules we have described to be pyramidalized, and I will try to prove it later.

About the same time as some of these computations were done, a number of experimental results were reported which confirm, often in spectacular fashion, the predictions of alkene pyramidalizations. I am sure a number of others will be reported today as well.

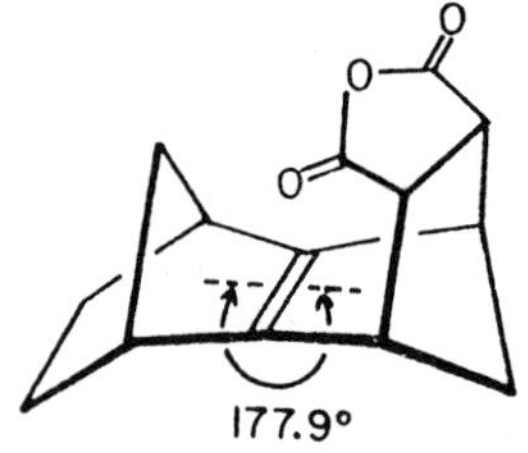

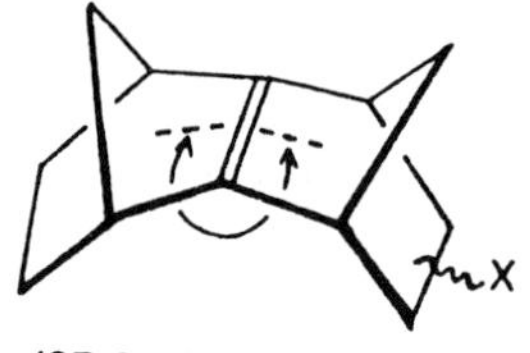

163.6° (exo-anhydride)
162.0° (exo-SO_2 Ph)

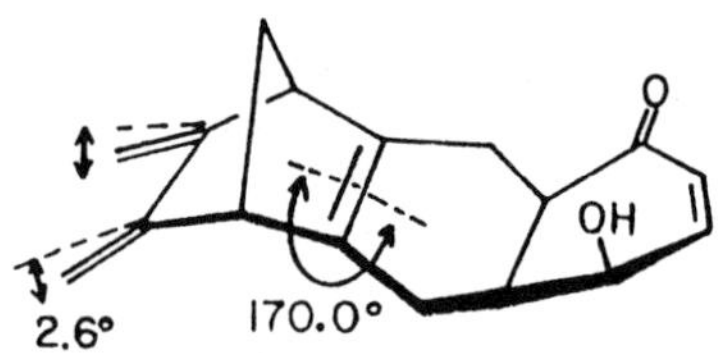

At the very top is a sketch of the low-temperature neutron diffraction structure of acetamide done by George Jeffrey and co-workers [11]. Because of crystal forces the methyl group in acetamide adopts an unusual conformation, one that is not present in the gas phase. The C-H bond is arranged essentially at a 90° dihedral angle with respect to the plane of the heavy atoms. As a consequence, there is a slight pyramidalization of the carbonyl carbon. The oxygen moves down out of the plane of the other three heavy atoms by 1.6°. More recently, Jeffrey has studied other amino acids and peptides by neutron diffraction and has carefully analyzed other literature results as well [12]. The direction of pyramidalization found in acetamide, toward a staggered arrangement, is uniformly present in such molecules when they have this unusual arrangement of the allylic bond. When the molecule adopts the conformation it prefers in the gas phase, namely with one of these allylic bonds eclipsed with the carbonyl group, then the molecule is flat, or the experimental deviation from planarity is less than 0.5°.

Other molecules show much more spectacular pyramidalizations. Watson and Bartlett reported that the _anti_-sesquinorbornene derivative shown is essentially flat, but the two _syn_-sesquinorbornene derivatives have significant folding around the double bond. The alkene carbon atoms pyramidalize so that the dihedral angles are 164-166° [13]. That is, bending away from planarity by 16 to 18° is observed. Paquette and Blount also reported the crystal structure of a _syn_-sesquinorbornene derivative which shows a similar degree of pyramidalization [14]. Vogel reported a crystal structure of a molecule with significant bending around the norbornene double bond, and smaller (he calls it insignificant, but I like to think it is significant) bending of 2.6° for the exocyclic double bonds [15]. Can it be possible that these kinds of pyramidalizations are caused merely by minimization of torsional effects between bonds to the alkene carbon and allylic bonds? The results I have shown you are consistent with this idea, and I have mentioned before that we have also investigated some of these molecules using molecular mechanics calculations [5]. Other force-fields give pyramidalization as well. Any time you do a force-field calculation it is possible by adjustment of parameters to obtain almost any result you wish. In a sense our calculations suffer from this defect as well. If normal parameters are used for _syn_- and _anti_-sesquinorbornene, both the molecules are significantly pyramidalized. However, if the small ring parameters are used, which include stiffer bending potentials about the alkene carbons, then more reasonable results are obtained. Namely, the syn molecule is nonplanar and the anti is planar, although there is a very small force constant for out-of-plane bending in the anti molecule. I will discuss this more below. The degree of bending is 13°, which is somewhat smaller than that found experimentally from X-ray structures of derivatives. If the same computations are carried out with the torsional interactions around the four C-C bonds from the bridgehead to the alkene carbons set equal to zero, the molecule goes flat! This is a computational test (proof?) of our hypothesis that the only effect causing the pyramidalization is the torsional effect around these four bonds. When a normal calculation is done with all of the torsional parameters present and the alkene is constrained to planarity but everything else is optimized, the flat structure is 1.8 kcal/mol less stable than the bent. Although this hardly proves our hypothesis, because everything is so highly parameterized that one cannot just cut out various interactions and know that they correspond to some physical reality, nevertheless, we believe these conformations are fully consistent with the idea that the pyramidalization is produced by this type of torsional interaction.

The unfavorable torsional interactions we are talking about are, in particular, the one involving the bridgehead C-H bond and the C-C bond from the alkene to the other bridgehead carbon. When the calculation is carried out without any constraints, the molecule goes nonplanar and the dihedral angle for this interaction changes from 22 to 44°. Much of the driving force for pyramidalization we feel is due to that particular interaction, since this distortion can be estimated to relieve about 2 kcal/mol of torsional strain.

The computations on _anti_-sesquinorbornene are a little more problematical. I have shown the perfectly planar species, but if one carries out the Allinger calculations using the small ring parameters, starting with different extents of pyramidalization, the dihedral angle or the nonplanarity around the alkene carbon stays essentially at the value where the calculations began. Begining at a planar geometry, symmetry prevents any pyramidalization. The molecule stays planar. Starting with the molecule with varying degrees of bending around the double bond, the geometry essentially stays there. To go from the flat structure to a 15° bend requires less than 0.2 kcal/mol according to the force-field calculations. This squishiness around the double bond is consistent with our torsional interaction ideas in the sense that upon bending the torsional interaction in one nobornene unit is improved but becomes worse in the other. We suggest that this force constant is low enough that it may be possible in certain derivatives to see the anti species significantly bent. There is not a strong preference in the anti compound for either planar or nonplanar geometry.

We investigated briefly what the consequences of pyramidalizations are on the shapes of π orbitals. It is tempting to say that because of pyramidalization there should be significant sp mixing and either orbital distortion in the Liotta phraseology [16], or "nonequivalent orbital extension" in the Fukui-Fujimoto sense [17]. This is a picture taken from one of Fukui's papers [17]. It is a contour plot for the π-electron density of the alkene

Non-Equivalent Orbital Extension

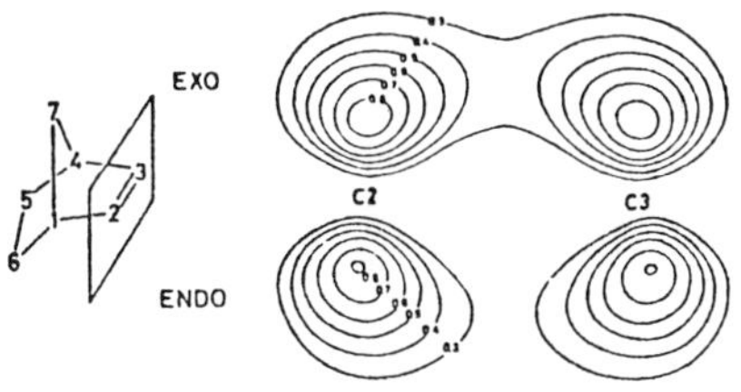

bond in norbornene. Fukui and Fujimoto suggested there was greater π density on the exo face than on the endo face because of sp mixing. Many people have done calculations on this. Rolf Gleiter, Paul Schleyer, and other groups, including our own, have all found that there is insignificant sp mixing even in a bent alkene. I will show you a computational experiment which emphasizes this.

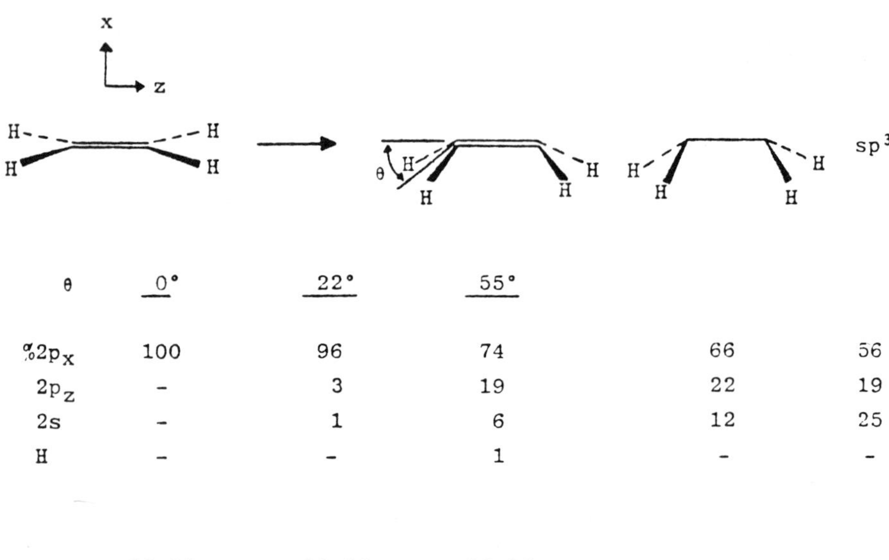

θ	0°	22°	55°		
%$2p_x$	100	96	74	66	56
$2p_z$	-	3	19	22	19
2s	-	1	6	12	25
H	-	-	1	-	-
ε_{HOMO}	-10.05	-10.05	-10.08		

This is a summary of 4-31G calculations, but similar calculations can be done with any basis set, and the results will be essentially identical. In ethylene, the π orbital is 100% $2p_x$, where the x axis is defined as perpendicular to the ethylene plane. Upon pyramidalization using a C_{2v} distortion, where the bisector between each C-H bond is moved out of the plane by 0°, 22°, and 50°, recomputation of the orbitals shows that the 22° bend causes only a 1% incorporation of "s character" into the HOMO! At 55°, only 6% s character has crept in. The main effect is p orbital tilting. The p orbitals follow the distortion in the sense that they remain more or less on a line which is equidistant from the three σ bonds to the ligands. For comparison, if a calculation on eclipsed ethane is carried out, but two syn hydrogens are clipped off, the HOMO has only 12% s character. We usually think of an sp^3 hybrid as possessing 25% s character. Of course this refers to a localized orbital, whereas our computation refers to the HOMO of this particular system. If a computation with localized orbitals were carried out on the last system, an orbital that had roughly 25% s character would be obtained. However, in the HOMO, the s character is resisted so that even for an essentially pure tetrahedral system, the HOMO has incorporated only 12% s character. This is about half of what it would have in a sp^2 hybrid. Even from very large distortions very little s character is mixed into the HOMO.

We do not believe there is significant rehybridization upon pyramidalization, but why do these molecules have greater π density on the exo face than on the endo face as indicated for example by Fukui's calculations on norbornene?

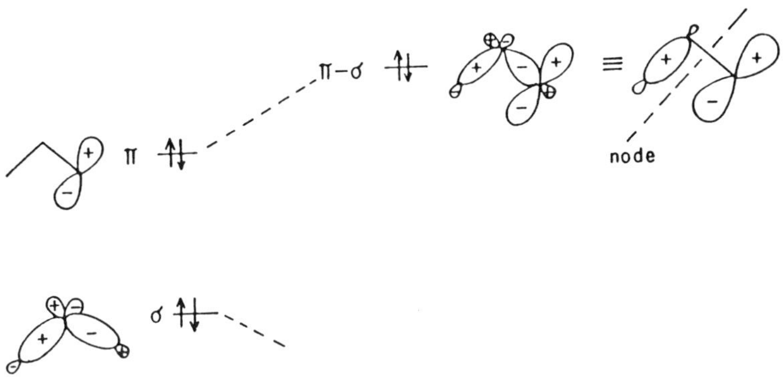

π−σ ⇅ ≡ node

π ⇅

σ ⇅

This is a side view of a π orbital and the highest occupied σ orbital of a norbornene skeleton which overlaps nicely with the π. These two are mixed significantly in the norbornene molecule, and the HOMO becomes a π orbital mixed in an antibonding fashion with this lower lying antisymmetric combination of σ orbitals. In terms of the orbitals involved the HOMO would look something like the picture in the center above. If we mixed together all of those orbitals and drew the resultant orbitals, it could look like the picture on the right. There is appreciable tilting of the p orbital away from the plane of the carbons. This is not primarily due to pyramidalization because the deviation from planarity is only about 5°. In reality there is less π electron density on the endo face than on the exo face, not only because of small s mixing, but because there is a node on the bottom side of this molecule due to π-allylic σ mixing. If these orbitals are plotted in terms of coefficients, then the top and bottom don't look much different, but if the total orbital electron density is plotted, as Fukui has done, then the bottom face has less electron density at a given location with respect to the alkene carbon than the top face. However, we do not believe that even this in itself is the origin of some stereoselectivities which I will discuss later.

I want to summarize some calculations we have done, and that others have done, in the following way.

1. Pyramidalization of these systems is caused by what we call asymmetric allylic torsional effects. When there is an unsymmetrical arrangement of allylic bonds with respect to an alkene or carbonyl plane, or to any other unsaturated linkage, there is a driving force for pyramidalization in order to achieve partial staggering with respect to the allylic bonds. The magnitude of bending is influenced by several things.

2. Calculations suggest that as the π bond becomes more electron rich the degree of pyramidalization increases. That is, if donor substituents are substituted on the alkene, then there is an appreciable increase in the electron density of the π system, particularly at the terminus remote from the donor. In effect the molecule becomes more carbanion-like [2]. The out-of-plane bending force constants are diminished thereby, and the degree of bending increases. On the other hand, an electron acceptor substituent on the alkene makes it more carbonium ion-like. The out-of-plane bending force constant increases and the molecule should have a lesser tendency to pyramidalize [2]. This is the reason that alkenes are pyramidalized more than similarly substituted carbonyls.

3. The pyramidalization should be significantly influenced by cis angle strain. Schleyer and Pople did computations on distorted ethylenes in which they constrained the HCC angles [18]. They found that eventually as the vicinal hydrogens are moved toward each other the molecule becomes nonplanar, even in the absence of any asymmetric torsional

interaction. For example, bicyclo[2.1.0]pentene with a double bond between the bridgehead carbons is nonplanar because of this kind of strain. So when cis substituents are moved toward each other the out-of-plane bending force constant is decreased [18].

In syn-sesquinorbornene, all of these effects are present. Four of the eclipsing or torsional types of interactions which cause the molecule to bend in an endo fashion are present. The double bonds are quite electron rich because the norbornene σ bonds are located nicely to overlap in a hyperconjugative fashion with the π bonds. The cis strain is present also. The C-C=C angles are on the order of 105°, so there is strain on both sides of the alkene which also decreases the out-of-plane bending force constant. Factors 1-3 listed above rationalize the bending found here, but these rules can also be used to at least qualitatively determine what kinds of molecules should be highly pyramidalized and in what direction.

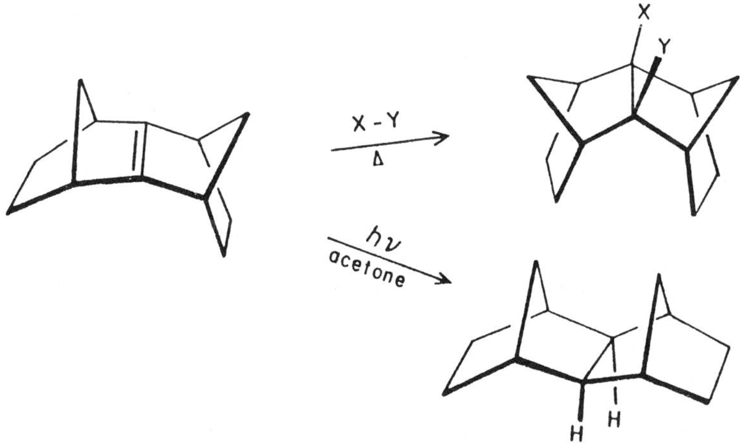

X-Y
Δ
X
Y
hν
acetone
H H

Now we will turn to the question of reactivity. Professors Bartlett and Paquette have studied thermal reactions of syn-sesquinorbornene and found that thermal reactions generally go stereoselectively with attack from the exo face of the molecule. On the other hand Professor Bartlett found the interesting result that the photosensitized reduction of syn-sesquinorbornene occurs from the endo face [19]. Now let me first talk about the thermal reactivity which is true not only for syn-sesquinorbornene but for norbornenes in general.

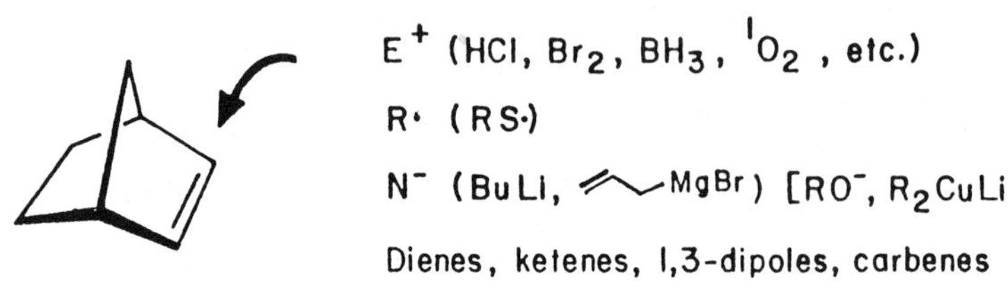

Norbornene is shown here with a summary of some reactions that go preferentially on the exo face. For example, electrophiles are known to add from the exo face, and this brings us unavoidably into the nonclassical norbornyl cation issue. Radicals and nucleophiles and concerted cycloaddends preferentially add to norbornene from the exo face. We feel that the exo reactivity of syn-sesquinorbornene is related to the same thing that occurs in norbornene itself [20]. In order to understand this in a more or less comprehensive fashion, I will have to go into a somewhat extended digression.

The next picture summarizes a considerable amount of computational effort on the transition structures for additions of various reagents to propene. We wanted to know in these studies what the arrangement of allylic bonds is with respect to forming bonds in addition reactions. Transition structures were obtained by 3-21G calculations using Pople's gradient program [20]. (A) is for hydride attack on the substituted carbon of propene. Note that the C-C-H angle is 126°, somewhat larger than that deduced by a number of people for nucleophilic addition reactions. The most significant feature is that the methyl group rotates so that it becomes staggered with respect to the pyramidalized carbon undergoing attack and with respect to this partially formed bond between carbon and the hydride. This transition structure is only slightly

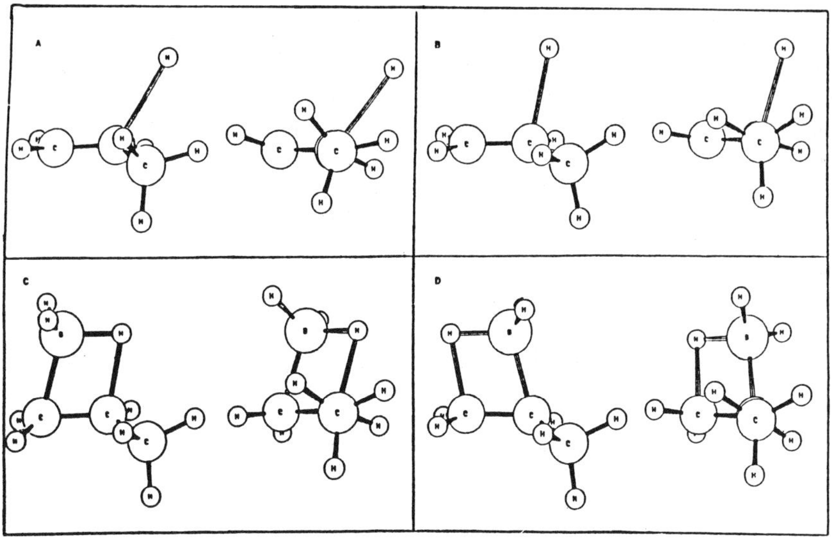

higher in energy than the reactant, and the forming bond between C and H is long and weak. Nevertheless, there is a large pyramidalization of the carbon undergoing attack and a significant driving force for the allylic bonds to become staggered with respect to the bonds to that carbon. The barrier to rotation of the methyl in the transition structure is 3.5 kcal/mol according to calculations, and this is of course the same order of magnitude one expects for the rotational barrier in the product of this reaction. So we deduce from this that torsional interaction involving the forming bonds and the allylic bonds as well as the bonds to carbon undergoing attack are very significant in the transition structure. Picture (B) shows the transition structure for a hydrogen atom attacking propene. The trajectory of attack is different and is nearly perpendicular. The methyl again rotates into a staggered arrangement. It is a different staggered arrangement, but it is staggered with respect to the bonds to the carbon. The last two are regioisomeric hydroboration transition structures. The first (C) is the preferred hydroboration transition structure with the methyl staggered with respect to that partially formed carbon-hydrogen bond. The second (D) is the transition structure for the formation of the minor product of hydroboration of propene. Again the methyl is staggered with respect to the partially formed boron-carbon bond. In nucleophilic, radical, and electrophilic reactions there is a significant driving force for allylic bonds to be staggered with respect to partially formed bonds. Felkin [26] deduced this from nucleophilic attack on carbonyls in 1968, without the benefit of *ab initio* calculations! In fact the barrier or the forces involved are of the same order of magnitude as of those in the fully formed products. For example, if one has a rigid system in which attack can occur in a staggered arrangement with respect to allylic bonds, it should be of the order of 3 kcal/mol or better than attack on a rigid system where it must occur eclipsed with respect to the allylic bonds.

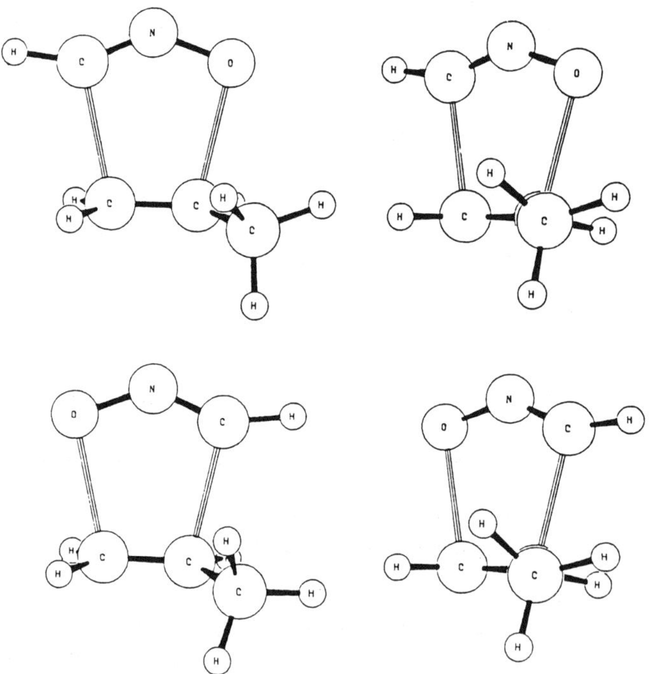

Above are shown model transition structures for fulminic acid attack on propene. They indicate that for concerted reactions the same sort of staggering occurs. Even though this C-O bond is very weak in the transition-state structure, the methyl rotates into a staggered arrangement not only with respect to that partially formed bond, but with respect to the partially pyramidalized carbon.

Let me continue my digression by mentioning some data amassed by Professor Huisgen on the cycloadditions of various reagents to bicycloalkenes [21]. He noted the fact that norbornene not only reacts stereoselectively from the exo face but it seems to be anomalously reactive as compared to other similarly strained hydrocarbons. These are the relative rates of reaction of mesitonitrile oxide and a tetrazine with these molecules [21].

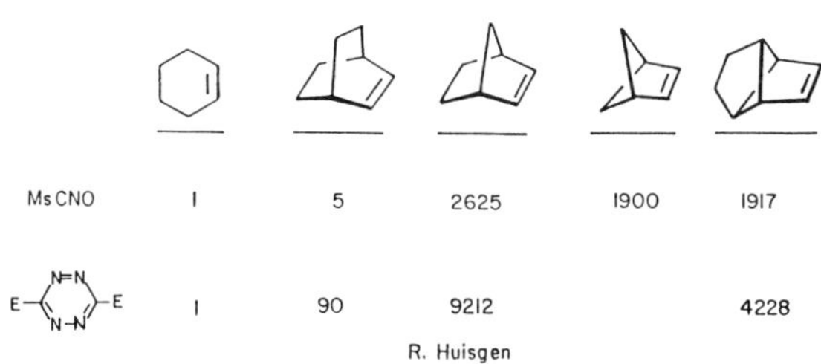

Note especially that norbornene is about 500 times more reactive than bicyclooctene and is slightly more reactive than several more highly strained molecules. Reagents such as other 1,3-dipoles and electron-deficient dienes also react with norbornene at an anomalously high rate. After correcting for strain release in the transition-state structure, Huisgen and Allinger decided there was still a mystery involved in the reactivity of norbornene [21]. There was still something like 2 to 4 kcal/mol lower activation energy than there should be. They attributed this to the quantity named factor "x". We would like to present our hypothesis for the origin of factor "x". Below are Newman projections looking down one bond from the alkene carbon to a bridgehead carbon of norbornene, bicyclooctene, and bicyclohexene. These extra balloons added on the exo and endo faces of norbornene and at one side of the bicyclooctene and bicyclohexene are placed at the same angles as the forming bonds are located in the transition structure for fulminic acid attack on propene. The striped lines represent the approach of either the carbon or the oxygen atom of fulminic acid to the alkene carbons. Exo attack on nornornene occurs in a nicely staggered arrangement. With respect to the C1-C7 bond, the torsional angle is about 68°.

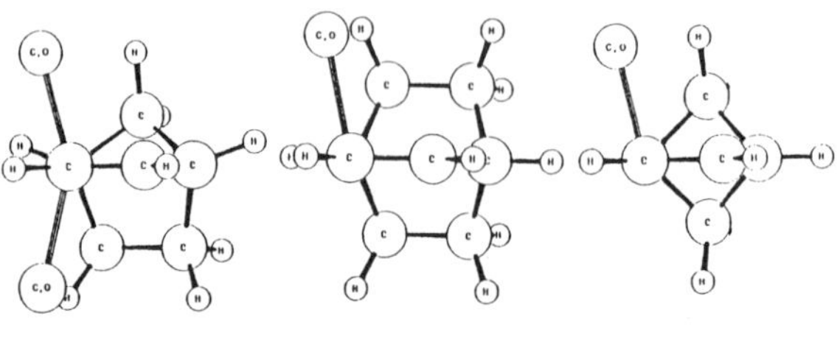

k_{rel} = (MsCNO) = 525 1 380

Endo attack is more nearly eclipsed. This angle is around 35°, and in bicyclooctene the arrangement is even worse than endo attack on norbornene. The dihedral angle is less than 30°. In bicyclohexene this dihedral angle is slightly less than the dihedral angle in the case of exo attack on norbornene. So these torsional interactions should make this a little less reactive than norbornene.

There is another postulate that has been suggested for factor "x" which has to do with the possibility that there is something special about the forming bond being anti-periplanar to the 1-6 and 4-5 strained σ bonds of norbornene. This is the effect that causes *exo*-norbornyl tosylates or other molecules of this type to solvolyze more rapidly than *endo* ones. Torsional effects would have some influence there, as shown by Schleyer [22], but there is definitely an influence of the strained σ bond that is not accounted for simply on the basis of torsional effects. The question is, are the cycloaddition effects accounted for solely on the basis of avoidance of torsional interactions here, or is there something magic about the *anti*-periplanar arrangement of the partially formed bond and the strained, and therefore good hyperconjugative, C-C bonds?

Jiri Mareda devised an experimental test to dissociate the torsional and hyperconjugation effects, and to determine which was more important. The idea was to compare norbornene with bicyclo[3.2.1]-oct-6-ene, because as shown below the torsional arrangements with respect to forming bonds are very similar in the two molecules.

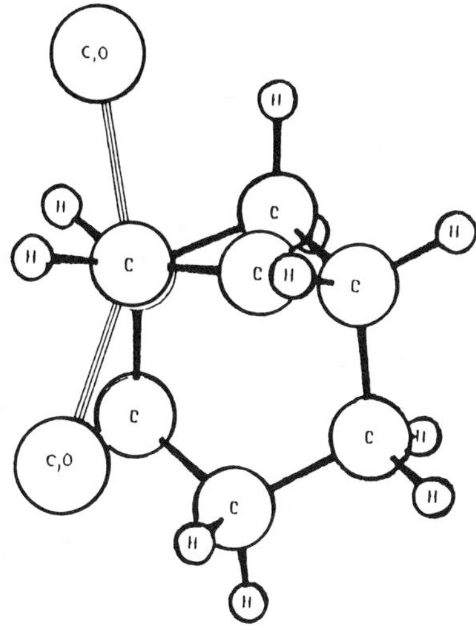

This is a MM2 structure of bicyclo[3.2.1]oct-6-ene. You can see that the torsional interactions are very similar to those in norbornene. The C8C1C7-(C,O) torsional angles are slightly large than the C7C1C2(C,O) torsional angles in norbornene and the HCC(C,O) angles are slightly smaller. Based on torsional interactions, we would expect these two molecules to have very similar reactivities toward cycloaddends. If hyperconjugative interactions were important, that is, if the 2-1 and 4-5 σ bonds entered into hyperconjugation with the partially formed anti-periplanar σ bond in the transition state, we would expect the [3.2.1] system to be less reactive. For example, solvolysis of exo-2-norbornyl tosylate is appreciably faster than that of exo-6-bicyclo[3.2.1]octyl. The latter solvolyzes at about the same rate as endo-norbornyl derivatives. This is proof that the 4-5 bond of the [3.2.1] system is less capable of the kind of hyperconjugative donation than the 1-6 σ bond in norbornyl. As far as strain release goes, norbornene should be more reactive. Using Huisgen's technique, we find that bicyclo[3.2.1]oct-6-ene relieves 3.5 kcal/mol less strain than is relieved upon addition to the double bond of norbornene. That is to say, strain effects alone should make the norbornene slightly more reactive.

Paul Mueller carried out an experimental study of this reaction. The reaction of mesitonitriloxide at 25^0 in carbon tetrachloride with norbornene is only 1.3 times faster than the reaction with bicyclo[3.2.1]oct-6-ene. We take this as proof that the staggered arangement of the bonds on the exo face of norbornene and on this molecule is the thing that makes them so reactive. At least for this reaction there is not much if any influence of the interaction of the anti-periplanar σ bond with the partially formed bond. I should caution slightly against going too far with this interpretation, because of course the two effects are geometrically inseparable. If a partially formed bond avoids eclipsing interactions with two allylic bonds, it must also be necessarily anti-periplanar to the third allylic bond. Geometrically there is no simple way to separate these two effects. It is similar in essence to the closed shell repulsion versus filled-vacant interactions in staggered ethane versus eclipsed. The two always must parallel each other geometrically, and so there is no geometric way to design a system that has one and not the other. The same sort of thing is true here. One would always expect some anti-periplanar hyperconjugation, if you like, to accompany the avoidance of eclipsing of partially formed bonds with respect to allylic bonds.

So let me conclude this section by saying the connection between the pyramidalization of alkenes and the stereoselectivity of similar molecules we believe is the following. Unsymmetrical alkenes are pyramidal to avoid torsional interactions with allylic bonds. In transition structures this effect becomes much larger. Attack tends to occur from the side which you might call the point of the pyramid not necessarily because of the pyramidalization, but because this happens to be the side that attack can occur and maintain staggering between the partially formed bond and the vicinal bonds.

Now let me say something briefly about the triplet photochemistry of syn-sesquinorbornene. In the ground state of the molecule we say that attack occurs at the top not because of hybridization, but because attack from the top avoids torsional problems with allylic bonds. Bartlett has found that in the triplet state, syn-sesquinorbornene must be bent up, because hydrogen is extracted from the bottom. The triplet state of ethylene has the perpendicular structure. At the 3-21G basis set level, the structure comes out as shown below.

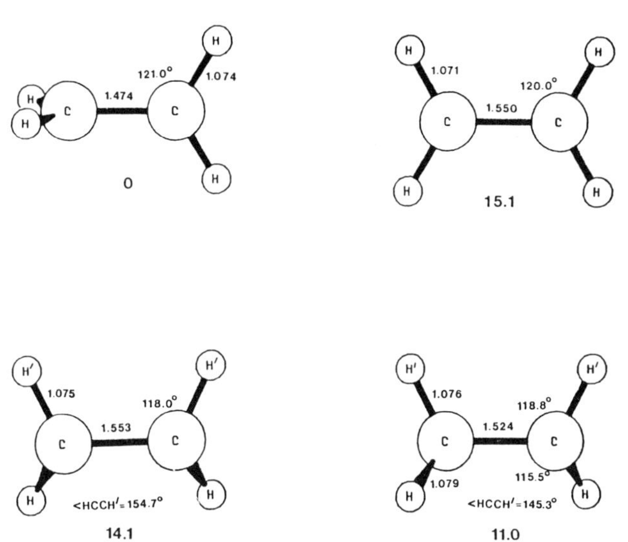

The perpendicular triplet is 15 kcal/mol more stable than an optimized planar ethylene triplet. There is a significant driving force to avoid overlap of the two p orbitals in triplet ethylene. If the molecular framework constrains the molecule so that it cannot rotate to the perpendicular structure, other distortions occur. We carried out calculations to model such systems by fixing the H-C-C-H dihedral angle of ethylene to zero and then optimizing all other geometrical parameters in the triplet state. There are two minima found with these constraints. There is a cis bent triplet with a dihedral angle of 155°. The trans bent species is a little lower in energy and has a dihedral angle of 145°. So if the alkene is constrained by incorporating it into a small ring, the triplet cannot go perpendicular. In a model cyclic small ring structure you would expect two isomeric triplets, one with a cis and one with a trans structure. In syn-sesquinorbornene, it is not really possible to rotate in a trans fashion to any significant degree. We would expect the molecule to be bent or folded in the cis fashion even more significantly than the ground state of syn-sesquinorbornene. Some simple steric considerations suggest that the only way the syn-sesquinorbornene triplet can bend is in an exo fashion.

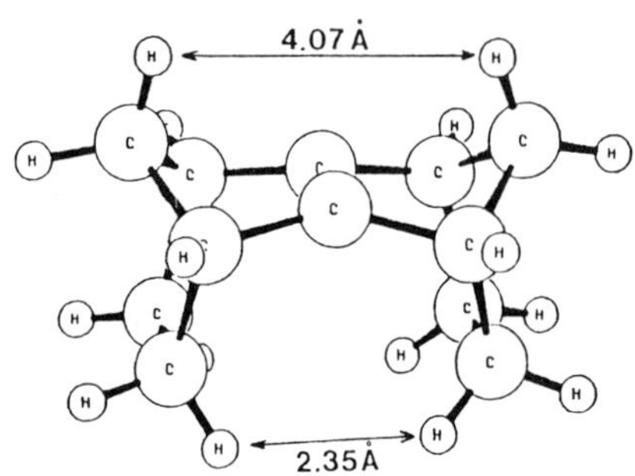

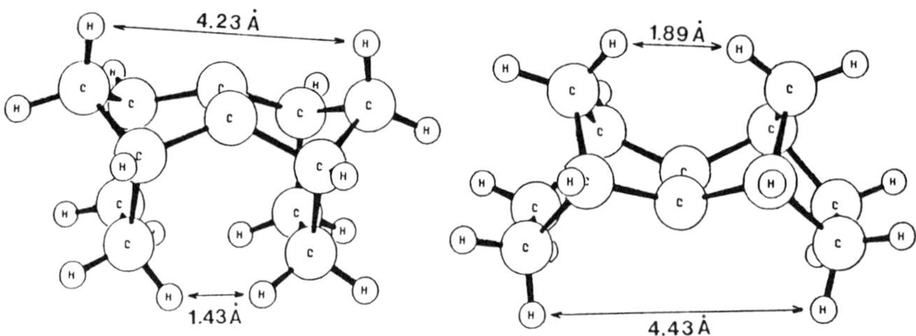

These are ORTEP drawings of syn-sesquinorbornene which are from the Watson, Bartlett structure of the phenylsulfonyl derivative [13]. We clipped off the $PhSO_2$ and substituted a hydrogen. We then bent the structure by 25^0 which is the amount of bending of triplet ethylene product from our calculation. Upon downward bending by 25^0, of course the endo-ethano hydrogens, which are separated only slightly more than van der Waals radii in the ground state, crash into each other extremely severely, so this endo bending is not possible. A 25^0 exo bent geometry can be produced without any severe problems. We feel that endo attack is a result of the driving force for the triplet to be more pyramidal than in the ground state and the geometrical constraints of syn-sesquinorbornene which allow it only to bend in an exo fashion. This model predicts that norbornene could bend either way and should exist as three isomeric species. There should be an exo-cis bent, an endo-cis bent, and a trans bent triplet of norbornene. In fact, triplet norbornene reacts in cycloadditions nonstereospecifically, as expected from this analysis [23].

Another way to look at it is in terms of products of reduction of the molecule. These are Allinger MM2 computations on the products obtained

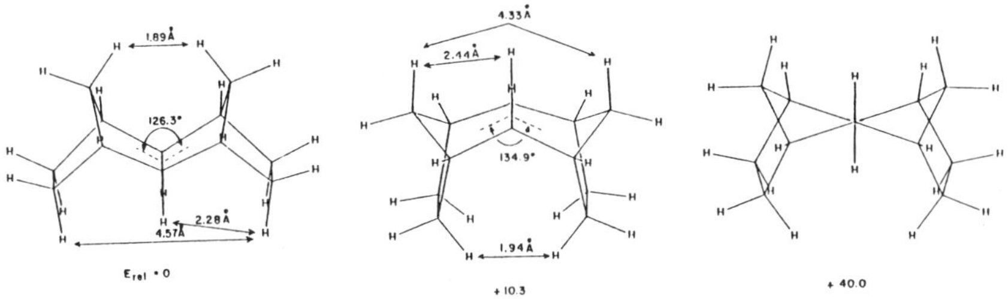

$E_{rel} = 0$ +10.3 +40.0

by reducing syn-sesquinorbornene in an endo fashion, exo fashion, or trans fashion. The most stable species is the endo bent species. It is calculated to be 10 kcal/mol more stable than the molecule obtained by reduction from the exo face, and 40 kcal/mol better than the trans-reduced species. We did these calculations primarily to find out how significantly one might be able to bend the alkene of sesquinorbornene in a trans fashion. The results indicate that it is not possible to bend much in the trans fashion. Bending in an exo fashion starting from the endo bent sesquinorbornene geometry is easier than bending in an endo fashion.

Finally, let me rush through somthing that is still in progress. We published a large, difficult paper on the subject already [1], but further studies in collaboration with Paul Mazzochi at Maryland have revealed some interesting things. Many groups have studied hexachlorocyclopentadiene cycloadditions to norbornadiene. These occur preferentially at the exo face, as everything else does. However, there were a number of reports of different stereochemistries when different 7-substituents were present in norbornadiene. Let me briefly summarize what we have found, and offer some hypotheses that perhaps will be worthy of our discussion We published a graph similar to the one below. This is a plot of the rate constants for attack on the three accessible faces of 7-substituted norbornadienes. The top line is drawn through the points which represent the second-order rate constant for exo-anti attack on 7-substituted norbornadiene. These are plotted against Grob's $\sigma_I{}^q$, an inductive substituent constant. It will become apparent

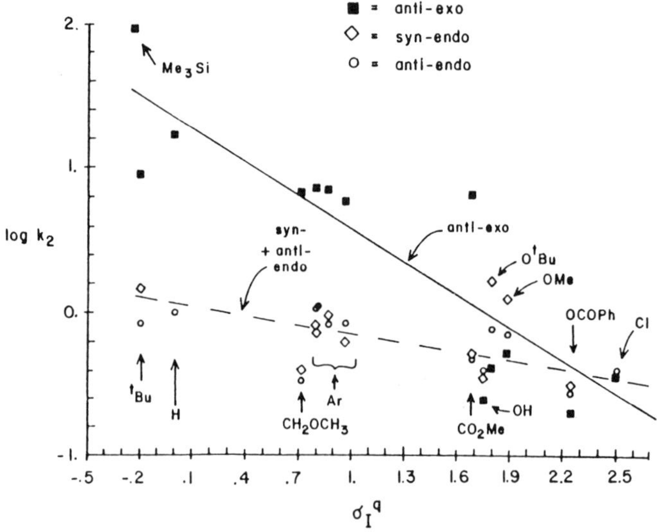

in a moment why we plotted rates versus this type of σ. The rate of anti-exo attack diminishes as the electronegativity of the 7-substituent increases. For endo attack we can draw a straight line indicating that there is a smaller effect of the 7-substituent electronegativity upon the rate of endo attack. Now in our published paper we attempted to show that every other explanation of this phenomenon was wrong, but we did not supply anything much better. We have done some further experiments which indicate in particular that some electropositive substituents like t-butyl and trimethylsilyl further enhance the rate of formation of the anti-exo adduct. Electron-withdrawers slow down exo attack, and endo attacks to a lesser extent, except for 7-t-butoxynorbornadiene, which is an oddball compared to the other 19 derivatives.

This compound gives anomalously fast syn-endo attack. We have studied 7-methoxy, and it is the same story. The syn-endo attack rate is anomalously large. Progression from electropositive substituents to electronegative substituents, produces a rapid decrease in rate of exo attack and a lesser decrease in the rate of endo attack. The only significantly anomalous compounds are 7-butoxy and 7-methoxy which give too much syn-endo attack. These are quite similar trends to those found in what at first sight would seem to be a very different reaction. Grob has published a number of papers on solvolysis of exo and endo-norbornyl tosylates in which the substituent "X" is varied [24]. Basically, the same kind of behavior is found as in norbornadienes. The rate of exo-tosylate solvolysis is linearly related to the electronegativity of "X". The rate of endo solvolysis is also decreased, but it is influenced to a smaller extent. For example, if X is t-butyl, the exo solvolysis is 2400 faster than the endo, but if X is bromo the endo compound actually solvolyses more rapidly than the exo [24]. Now our results are for a quite different reaction, but there is a startlingly similar trend. For hexachlorocyclopentadiene the rate of exo attack on 7-substituted norbornadienes is decreased to a large extent by electron-withdrawing substituents. The rate of attack on the endo side is affected less. Gerhard Klumpp has also found the same sort of thing for dichlorocarbene cycloadditions to 7-substituted norbornadienes [25]. That is, the rate of exo attack decreases rapidly as the electronegativity of "X" is increased. The rate of endo attack, and now there are actually three reactions, 1,2-anti-endo, 1,2-syn-endo, and the homo-1,4-cycloaddition is decreased in all three reactions by electron withdrawing substituents, but to a smaller extent than for exo attack.

Let me see if I can explain all of this, or at least present a hypothesis. Let us start first with the norbornyl derivatives and mention Grob's explanation for this system. If there were nothing stereospecific about the electronegativity effect, all of these reactions would be decreased to the same extent by electronegative "X". We would obtain parallel lines for rates of reaction plotted versus $\sigma_I{}^q$. However, there is some effect which differentially influences the exo versus the endo rates.

In the case of the norbornyl solvolyses it presumably is the fact that the <u>exo</u>-tosylate solvolysis is influenced by significant bridging involving the 1,6-sigma bond. If a strongly electron withdrawing group is placed at C(6), that bridging goes away. So there is both the overall inductive slowing down of the reaction because partial positive charge is developed in the transition state, and in addition the diminution of bridging occurs for the exo case. On the other hand the endo case is slowed down by the electron withdrawing effect, but the extent of bridging is not changed. Perhaps there can be bridging by the exo bridge, but that would be affected much less because the electron-withdrawing group is much further away.

What about the hexachlorocyclopentadiene case? Attack at any of these three sites, anti-endo, syn-endo, or anti-exo, should be influenced by the general electronegativity of the 7-substituent. As the substituent is more electron withdrawing, the π bonds become less electron rich, which shows up in the ionization potential, and should react more slowly with the electrophilic diene, hexachlorocyclopentadiene. We feel that the rate of decrease of exo attack is just a manifestation of this. It is more or less a pure inductive effect of the 7-substituent. However, there is a counteracting effect which causes endo attack to be diminished less than exo attack. Let me present the details of that hypothesis here for the first time. We have talked about the fact that norbornene and norbornadiene are attacked preferentially from the exo face because attack from the endo side experiences severe torsional interactions. One of the most prominent of these is the interaction between the partially formed bond and the 1,6-sigma bond of norbornene or norbornadiene. If an electron withdrawing group is placed at C(7) (or elsewhere), we predict that the electron density in the 1,6-bond is decreased, and the energy of the 1,6-sigma bond is lowered. This means that the torsional interactions involving the partially formed bond and the 1,6-sigma bond are diminished. The rate of reaction is diminished, but not as much as it is for exo attack, because closed shell-repulsion is now diminished. That is a hypothesis which perhaps can be tested by other experiments.

Finally, as many have suggested, the t-butoxy, and the methoxy group apparently have a special effect. Namely, interaction of the lone pairs on oxygen, or the negative region due to them, with the syn double bond causes the HOMO to be bigger here on the syn side of the alkene. This is a type of anchimeric assistance to the attack of any electrophile on the syn-endo site.

I have told you a little more than we actually know about the subject, especially here at the end! Nevertheless, I hope it will at least be food for discussion. I am looking forward very much to the rest of the talks.

Finally, let me acknowledge my debt to Professor Bartlett, whose undergraduate physical organic lectures at Harvard started me on the path I am traveling today. All of the research I reported was carried out by my fantastic group of co-workers, Nelson G. Rondan, Michael N. Paddon-Row, Pierluigi Caramella, Jiri Mareda, Frank Brown, and Paul Mueller. I am indebted to these men for their brillance and diligence. Our work in this area has been generously supported by the National Science Foundation and the National Institutes of Health.

Discussion

Paquette: With regard to the Schleyer hypothesis in norbornene, reference should really be made to those studies dealing with the 1-methyl derivative. If the theoretical conclusions are relevant to the transition states for electrophilic attack, reaction should occur preferably at the C-2 position. You do not seem to derive any contribution from the methyl-substituted bridgehead, preferentially, to relieve this torsional interaction. I think Schleyer has shown that in normal reactions there is little discrimination between C-2 and C-3. I was wondering if you had anything to say about that?

Houk: Yes. I think that the Schleyer philosophy was to compare the torsional interactions in the ground state to those in the transition state. That is, he focused on the relief of torsional strain going from the reactant to the transition state. He did not know at the time, and there really was not any way for him to know, (the first to suggest it was Felkin in 1968 who decided this would be important), that there are very significant torsional interactions between the forming bond and the allylic bonds in the transition state. These are roughly the same for C-H and C-CH_3, so that any difference between hydrogen and methyl at the bridgehead would have to be due to other steric interactions, which must not be significantly different for CH_3 and H. Knowing the experimental result, it is easy for me to say that it is consistent with the way we view things. I think that the only problem with the Schleyer hypothesis is that he considered the relief of torsional interactions exclusively, whereas we feel the torsional interactions in the transition state can be even more important than those in the isolated reactants or products.

Paquette: Let's make an extrapolation based on that. Suppose we take *cis*-sesquinorbornene, and we substitute the four bridgehead positions with methyl groups. What would be your prediction? Would there be an effect?

Houk: I don't think so. Not much. I don't think that torsional interactions would be much different from those in the parent system, and the pyramidalization would be about the same. I might have to look at the molecule carefully to see if methyl substitution does anything else, but just shooting from the hip, I don't think there would be any effect, since the CCCC torsional effect is very similar to the HCCC torsional effect.

Paquette: In the transition state, you think it would go in the same direction as hydrogen?

Houk: Yes.

Gleiter: We have published in *Tetrahedron Letters* some calculations on sesquinorbornene. We think that the electronic effects like hyperconjugation are as important as steric effects. The structure is due to a minimization of orbital interactions (σ versus π). If you put methyl groups on those positions of the σ frame with large contributions to the HOMO, the σ orbital is shifted toward lower binding energy and thus I would expect that the substituted sesquinorbornene will be a little bit more bent.

Houk: How much? A degree or two?

Gleiter: Two to three degrees.

Houk: I would go along with that; essentially no effect!

Gleiter: May I extend this subject. Our extended Huckel calculations on norbornene reproduce the bending very well, and we get a very nice potential for bending. Our analysis indicates that you do get a minimization of the hyperconjugative interaction between the σ frame and the π frame. You really see this if you plot this σ frame, that it is really moving when you do this bend.

Houk: Yes. I mentioned before that what you are saying and what I am emphasizing are inseparable geometrically. But I also feel, although I have no direct evidence, that the hyperconjugative effect, if I may call your hypothesis that, is not very important. Whenever you have relief of eclipsing interactions, which I keep talking about, you must simultaneously have a bond lined up perfectly to hyperconjugate. I feel that the energetic significance, the driving force for bending from hyperconjugation, is not as large as the relief of eclipsing effects. The only support I have for that is not a proof, it is only from molecular mechanics calculations. These are calculations in which there is nothing included about π interactions at all. But of course, force-field methods are highly parameterized and we don't really know what all of these different parameters correspond to physically. So I think we would agree in the sense that you need one bond perpendicular to the alkene plane for bending. I am focusing on the top (exo) two interactions causing bending to avoid repulsion, and you are saying that bending occurs to minimize hyperconjugation with the π orbital. Those two effects necessarily parallel each other, just as all three bonds of a methyl in ethane must simultaneously become staggered with respect to the other three in staggered ethane. My view is the "top" is energetically more significant than the bottom, while your view is the opposite.

Vogel: If I understood you correctly, one would also predict out-of-plane deformations for exocyclic diene or methylene groups grafted onto norbornane skeletons. In one of your recent papers [N. G. Rondan, M. N. Paddon-Row, P. Caramella, and K. N. Houk, J. Am. Chem. Soc. 103, 2436 (1981)] you calculate a nonplanar geometry for 2-methylenenorbornane where the terminal hydrogens are bent in the exo direction. To my knowledge there are no reported cases of 2-methylene- and 2,3-dimethylenenorbornanes where significant out-of-plane deformations have been measured by X-ray analysis for exocyclic olefins. From the data available, one can state that the double bond in norbornene deviates by about 10° from planarity, the substituents at C-2,3 being bent in the endo direction, whereas an endocyclic diene grafted onto norbornane systems does not deviate from planarity.

Houk: I have several comments on that. The torsional effect dictates the direction of pyramidalization, but all of these other factors determine how great the extent. For example, in the exocyclic diene you don't have any cis strain and the out-of-plane bending force constants are much larger than they are for the endocyclic alkene. So there is lesser tendency to pyramidalize in those systems. We predicted only 1-2° for methylenenorbornene. Furthermore, as pointed out by Perrin, an alkene likes to have one syn-eclipsed allylic bond, and so an exo-methylene bends little, while an endocyclic double bond bends more.

Vogel: Is that the result calculated by the MMPI2 technique?

Houk: No. MM2 gives an absolutely flat structure for exo-methylenes as in isodicyclopentadienes. Syn-sesquinorbornene comes out bent, while exocyclic dienes come out flat. Now there are several reasons for that. The endocyclic strain is absent, and the torsional interactions are different for endocyclic and exocyclic double bonds.

Vogel: When you say endocyclic strain do you refer to bond angle deformations?

Houk: Yes, cis bond angle deformation decreases the out-of-plane bending force constant. *Syn*-sesquinorbornene and norbornene have the proper torsional arrangement to drive the carbons nonplanar in an endo fashion. But the only reason this bending becomes so significant in *syn*-sesquinorbornene is that the CC=CC angles are constricted. The first

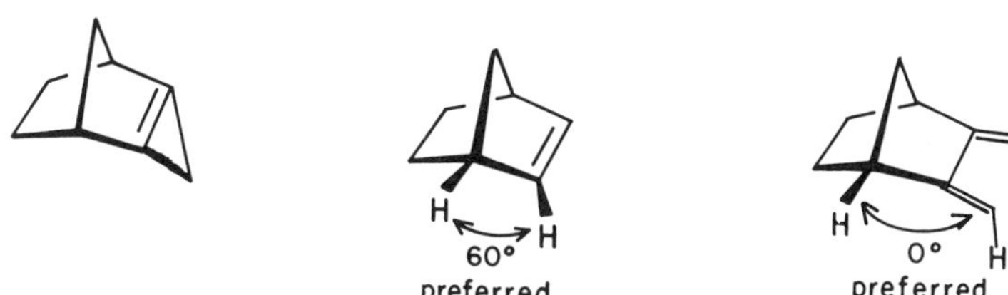

molecule above would be super bent. In your exocyclic dienes, the strain is low, and I think that means that the tendency for the carbon to pyramidalize is small. Secondly, because the main interaction reduces to the torsional interactions shown above, it is different for *exo*-methylenes, as pointed out by Charles Perrin. I think in the exocyclic diene case the smaller pyramidalization is alright. I think it should go *endo*, but there is not that very great a tendency to pyramidalize, as your experiments show.

Vogel: What do you expect to get by your calculations for the following molecule? Do you maintain the C_{2V} symmetry?

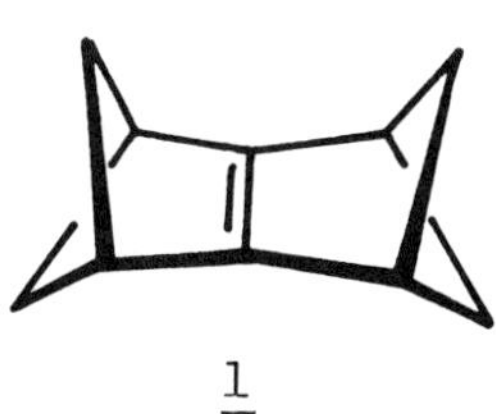

1

Houk: Well, I think that it will be flat except the out-of-plane bending force constants will be small. I think you have to have the asymmetric torsional interaction as in syn-sesquinorbornene and not present in anti-sesquinorbornene to get an equilibrium nonplanar structure. Here are curves which would characterize syn and anti-sesquinorbornene. This molecule (i) will be like anti-sesquinorbornene, flat but with an even smaller out-of-plane bending force constant.

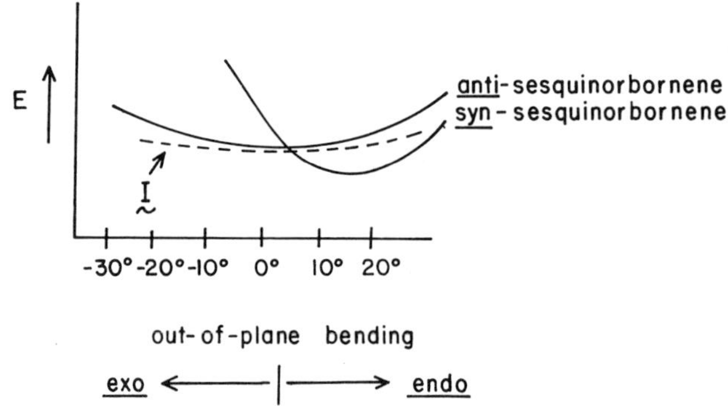

Vogel: In his book on applications of force field calculations, Ermer [O. Ermer, "Aspekte von Kraftfeldrechnungen", W. Bauer Verlag, Munich, 1981, pp. 313] calculates nonplanar double bonds for both the syn- and the anti-sesquinorbornenes and also for bicyclo[3.3.0]oct-1(5)-ene.

Houk: I have to be a little careful because eventually you do go nonplanar even without asymmetrical torsional interactions.

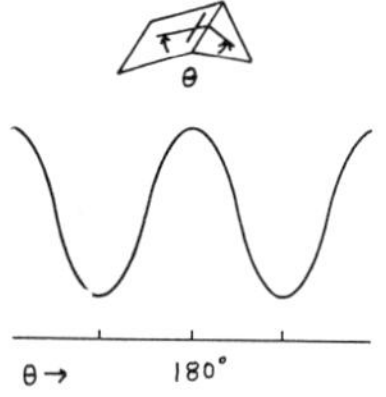

Bicyclo[2.1.0]pent-1(4)-ene is nonplanar. There are no asymmetric torsional interactions. I do not know exactly where the line falls between planarity and nonplanarity. If you squeeze angles in a cis fashion far enough, eventually you get a double minimum potential. Syn-sesquinorbornene is not a double minimum. I think the minimum is just shifted as shown above, but if you force enough cis-strain, it is the same as making the double bond very electron rich. Both of those effects eventually give nonplanarity even in the absence of torsional effects, but you always have a double minimum. The torsional effects always shift it away from planarity in one direction and maintain a single-minimum potential.

Vogel: If I understood correctly, your calculations gave only one energy minimum for these species. Ermer gets two. Mr. Birnbaum in our research group applied the MMPI-1 technique to syn- and anti-sesquinorbornene and oxasesquinorbornenes. He also obtained two energy minima for these systems. In some cases, extremely large out-of-plane deformations have been calculated. They appear unreasonable to us. For instance (see Ermer, p. 315) for the syn-sesquinorbornene: my feeling and that of Mr. Birnbaum is that the force-field calculations are not ready for the problem of the sesquinorbornene structure.

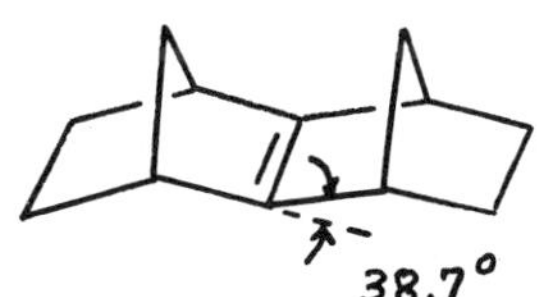

Houk: I agree there are problems.

Vogel: You can get whatever you wish!

Houk: Yes, you can, since there can be many arbitrary parameters. We did a syn-sesquinorbornene STO-3G calculation but only optimized the out-of-plane angle. It only bends 5° if you do not optimize anything else. There is only one minimum.

Vogel: I guess you have neglected configuration interaction!

Houk: I do not think you need configuration interaction to correctly predict these small distortions. This effect is much like the methyl tilt. All of these small deformations can be accurately predicted with single-configuration double-zeta calculations. They reproduce experimental results very nicely. Unfortunately, we have not done any double-zeta calculations on any of these systems, but when we do 4-31G or 3-21G calculations on *syn*-sesquinorbornene, I believe that they will give a result that is very similar to experiment.

Perrin: You have convinced me that norbornene and these molecules do distort, and I am willing to grant they do so because of torsional effects, and I am willing to grant that the transition state is on its way toward becoming eclipsed, but how about the energetics? How can you get the full factor "X" that Huisgen sees? How can you get these large endo-exo selectivities with what amounts to only a kcal/mol or so from a portion of the torsional barrier?

Houk: It is more than a kcal/mol because it really is not just the torsional interaction of the forming bond with the vicinal bonds. It is really the whole thing. The carbon undergoing attack is pyramidalized significantly. I think all the torsional interactions for a partially formed bond are as significant as for a fully formed bond.

Perrin: How much energy does the fully formed bond give you?

Houk: Three Kcal/mole. But for exo-endo you get much more than you need. The exo:endo ratio is probably 100 for these reagents with norbornene. You can get 6 kcal/mol if you have an ethane-like barrier, but you do not need that much. There is only a few kcal/mol preference of exo over endo attack.

Marchand: The problem of the Diels-Alder reaction with 7-substituted norbornadienes has been a concern of several groups for a great many years, and the preference for syn-endo selectivity when X is an electron donating group (like *t*-butoxy or methoxy) dates back to MacKenzie in England in 1964. Then, further Diels-Alder reactions with hexachlorocyclopentadiene were studied by Battiste and reported about 1976. I have to mention Houk and Mazzocchi in 1980, where you did calculations which indicated that for X = *t*-butoxy or methoxy, there was a polarization of the double bond such that the energy of the highest occupied molecular orbital was slightly higher for the syn than for the anti double bond. This would account for a slight preference of endo-syn addition versus anti-exo or anti-endo addition in this system. That is a kind of capsule of history. We became interested in this in a totally unrelated way. That is, if you react 7-substituted norbornadienes with iron pentacar-

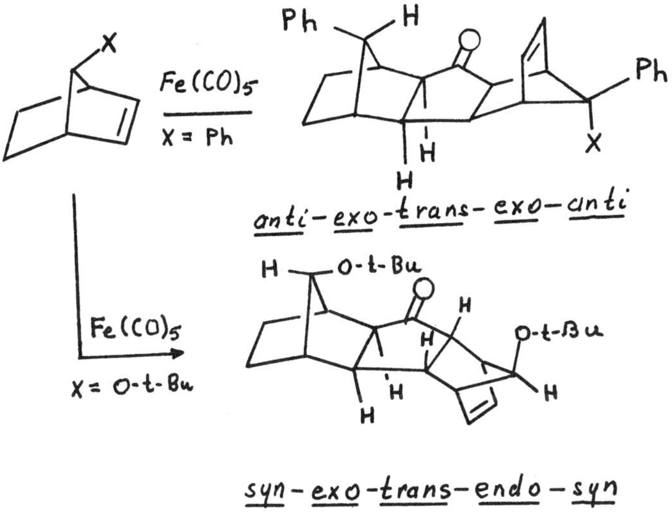

anti-exo-trans-exo-anti

syn-exo-trans-endo-syn

bonyl, eg, one finds similar selectivity. If X = phenyl, the predominant dimer lactone product (the only product of coupling to CO) has the phenyl group anti to the cyclopentanone ring and has an exo-trans-exo arrangement of the five-membered ring relative to the two norbornenes. So, this would be anti-exo-trans-exo-anti. If X = *t*-butoxy without going into all of the different products of the reaction, syn-exo-trans-endo-syn is the only dimer ketone formed. It is exactly the same selectivity, i.e. it is the syn double bond which reacts in both

7-*t*-butoxynorbornadiene moieties. One of these moieties reacts at the exo face, but the other most significantly reacts through its endo face. Now, in view of this result and in view of all of this history, back in 1980, Dr. Dennis Dougherty and I did the calculations and came to the same conclusions you published (which kind of slows down our ability to publish). At least the results reinforce one another. Now there is an ever so slight preference for addition on the syn double bond in 7-*t*-butoxynorbornadiene from the endo face. I think this is an extremely subtle effect. Incidentally, our iron carbonyl-promoted couplings of 7-substituted norbornadienes to carbon monoxide are stereospecific reactions; whether this subtle effect is sufficient to explain the stereospecificity, I don't know.

Recently, we have observed that 2-nitrobenzoquinone adds *non*stereospecifically to cyclopentadiene. It is *the* only case we know of this type

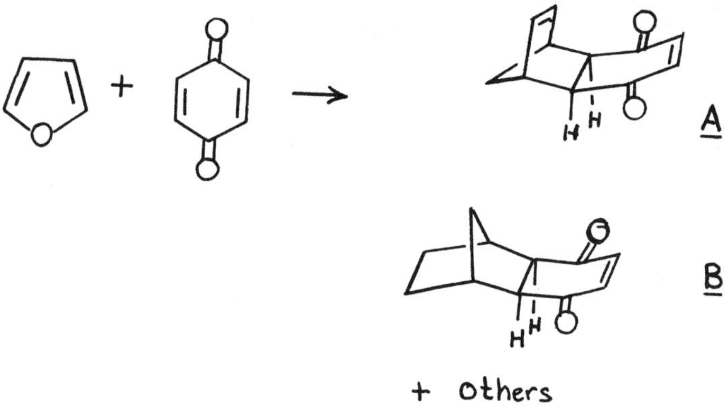

+ others

that is nonsterospecific. We get two 1:1 Diels-Alder adducts. Both result via additions to the nitro-substituted double bond in the dienophile, (see *A* and *B* above). Now, whether calculations are yet sufficiently sophisticated to account for the very subtle energy difference which gives rise to tremendous stereochemical consequences in terms of relative reactivities, I do not know. I do know what we had to spend on STO-3G calculations to optimize our geometry in 7-methoxynorbornadiene and in 7-*t*-butoxynorbornadiene. There may be a point of diminishing returns, because these are very expensive calculations and the energy differences are extremely small.

Herndon: Acrylonitrile and methylacrylate add nonstereospecifically to cyclopentadiene.

Marchand: There are a number of examples of nonstereospecific additions to cyclopentadiene. This impressed us because we have had experience with Diels-Alder reactions of cyclopentadienes with benzoquinones. We have never found one such case yet that was nonstereospecific until this. It is the only example which we've encountered of a nonstereospecific reaction of cyclopentadiene with benzoquinone, but there are many nonstereospecific Diels-Alder reactions known wherein cyclopentadiene reacts with acyclic dienophiles*.

*Note: A subsequent literature search has turned up the following information, quote from V. Mark, [J. Org.Chem. 39, 3181 (1974)]: "It is becoming increasingly more apparent from thoroughly investigated systems that when not in competition with hydrogen, unsaturated substituents tend to violate the rule of endo addition in kinetically controlled Diels-Alder reactions." Several useful references which substantiate this statement appear in Mark's paper.

References

[1] P.H. Mazzochi, B. Stahly, J. Dodd, N.G. Rondan, L.N. Domelsmith, M.D. Rozeboom, P. Caramella and K.N. Houk, J. Am. Chem. Soc. 102, 6582 (1980).

[2] N.G. Rondan, M.N. Paddon-Row, P. Caramella, and K.N. Houk, J. Am. Chem. Soc. 103, 2436 (1981).

[3] G. Wipff and K. Morokuma, Chem. Phys. Letters 74, 400 (1980); Tetrahedron Lett. 21, 4446 (1980).

[4] M.D. Newton, J.M. Schulman, and M.M. Manus, J. Am. Chem. Soc. 96, 17 (1974).

[5] N.L. Allinger, J. Am. Chem. Soc. 99, 8127 (1977). Quantum Chemistry Program Exchange, No. 395.

[6] O. Ermer, Tetrahedron 30, 3103 (1974).

[7] U. Burkert, Angew. Chemie. Int. Ed. Engl. 20, 572 (1981).

[8] E.E. Burnell and P. Diehl, Can. J. Chem. 50,

3566 (1972); J.W. Emsley, and J.C. Lindon, Mol. Phys. 29, 531 (1975); K.C. Cole and D. F.R. Gilson, J. Mol. Structure 82, 71 (1982).

[9] Reviewed in R.S. Mulliken and W.C. Ermler, "Polyatomic Molecules", Academic Press, New York, N.Y., 1981, pp. 182-184.

[10] J.P. Lowe, J. Am. Chem. Soc. 92, 3799 (1970).

[11] G.A. Jeffrey, J.R. Ruble, R.K. McMullan, D.J. Defrels, J. Binkley, and J.A. Pople, Acta Cryst. B36, 2292 (1980).

[12] G.A. Jeffrey, H. Huber, and K.N. Houk, unpublished results.

[13] W.H. Watson, J. Galloy, P.D. Bartlett, and A.A.M. Roof, J. Am. Chem. Soc. 103, 2022 (1981).

[14] L.A. Paquette, R.V.C. Carr, P. Charumilind, and J.F. Blount,J. Org. Chem. 45, 4922 (1980).

[15] A.A. Pinkerton, D. Schwarzenbach, J.H. Stibbard, P.-A. Carrupt and P. Vogel, J. Am. Chem. Soc. 103, 2095 (1981).

[16] C.L. Liotta, Tetrahedron Lett., 519, 523 (1975).

[17] S. Inagaki, H. Fujimoto, and K. Fukui, J. Am. Chem. Soc. 98, 4057 (1976).

[18] H.-U. Wagner, G. Szeimies, J. Chandrasekhar, P.v.R. Schleyer, J.A. Pople, and J.S. Binkley, J. Am. Chem. Soc. 100, 1210 (1978).

[19] P.D. Bartlett, A.A.M. Roof, and W.J. Winter, J. Am. Chem. Soc. 103, 6520 (1981).

[20] J.S. Binkley, R.A. Whiteside, R. Krishnan, R. Seeger, D.J. DeFrees, H.B. Schlegel, S. Topiol, L.R. Kahn, and J.A. Pople, GAUSSIAN 80, Department of Chemistry, Carnegie-Mellon University, Pittsburgh, PA 15213.

[21] R. Huisgen, P.H.J. Ooms, M. Mingin, and N.L. Allinger, J. Am. Chem. Soc. 102, 3951 (1980).

[22] N.G. Rondan, M.N. Paddon-Row, P. Caramella, J. Mareda, and P. H. Mueller, submitted for publication.

[23] P.J. Kropp, J. Am. Chem. Soc. 91. 5783 (1969); 95, 4611 (1973) and references therein.

[24] C.A. Grob and A. Waldner, Tetrahedron Lett. 4429, 4433 (1980).

[25] G.W. Klumpp and P.M. Kwantes, Tetrahedron Lett., 813 (1981).

[26] H. Felkin and C. Lion, JCS Chem. Comm. 60 (1968).

2. LONG-RANGE STEREOELECTRONIC CONTROL BY NORBORNYL FRAMEWORKS OF CYCLOADDITION AND ELECTROPHILIC PROCESSES

Leo A. Paquette

Evans Chemical Laboratories, The Ohio State University, Columbus, Ohio 43210

The question concerning whether an isolated bond or 1,3-dienyl unit can, <u>under conditions of purely electronic perturbation</u>, exhibit a measurable propensity for π-facial stereoselectivity is a fascinating one. At the outset of this study, we were intrigued by the fact that no purposeful examination of this phenomenon had yet been reported. Certainly, it had been recognized for a long time that norbornene and its derivatives exhibit a strong predilection for exo attack. However, our inability to dissect steric and electronic contributions to this stereoselectivity [1,2] has caused several hypotheses to be advanced and controversy to persist.

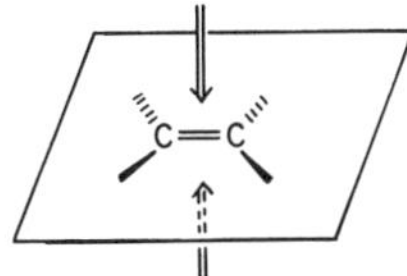

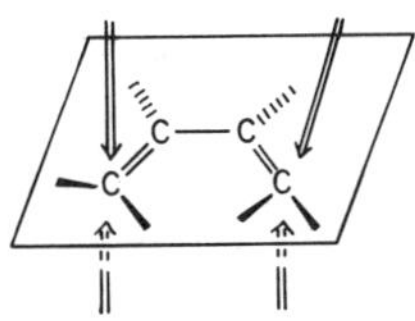

The last two decades have brought forth such concepts as differential steric shielding [3], preferential torsional strain relief [4], and more favorable secondary interaction between reactants and framework bonds flanking the π system [5]. Only very recently has the possible existence of electronic factors within norbornyl systems been given serious theoretical consideration. Fukui deduced on the basis of molecular orbital theory that unsymmetrical π-orbital extension exists as a consequence of prevailing interaction with the anti C(7)-H bond [6]. In independent studies, Houk [7], Wipff [8], and Gleiter [9] have proposed that σ-π interaction induces significant pyramidalization of the π bond, the resulting tilt giving rise to favored exo stereoselection.

We held the opinion that the evidently strong orbital interactions existent in bridged bicyclic systems should have, at least to a first approximation, a recognizable impact at more remote sites which do not suffer from comparable steric bias. To test this concept, we first turned our attention to isodicyclopentadiene (1), its dehydro congener 2, and tricyclo[5.2.2.0^{2,6}]undeca-2,5,8-triene (3). This choice of substrates was predicated on several factors, not the least of which were the earlier

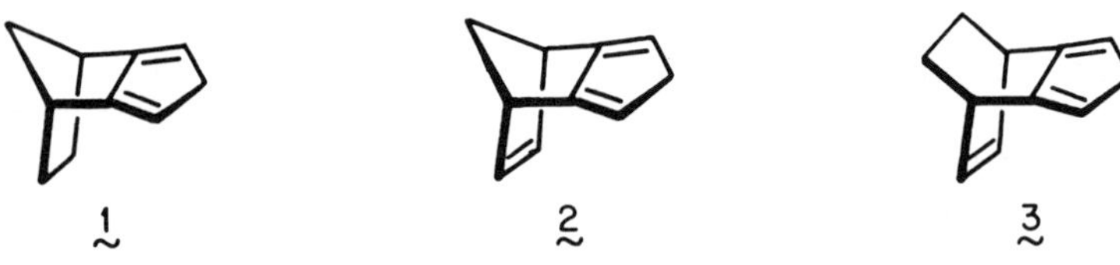

reports by Alder [10], Kobuke [11], Feast [12], and their coworkers who demonstrated the ability of 1, 2, and perfluorinated 3 to serve as 4π components in Diels-Alder reactions. Since varying stereoselectivities were claimed by these research groups, it was obvious that detailed examination of their products and many others by modern spectroscopic techniques was warranted.

When 1 was allowed to react with a host of dienophiles including those shown in Scheme I, below-plane dienophile capture occurred exclusively in all but two examples [13]. Structural assignments to the adducts were made on the basis of their high-field ^{1}H NMR spectra, X-ray crystal structure analysis in selected examples, and chemical intercorrelation where feasible. As an example of the latter, the endo dienophile stereoselection which occurs during the formation of 4 was established in a five-step reaction sequence. Following diimide reduction which occurs by clean exo delivery of hydrogen to the central double bond, the ester function was saponified and converted to the thallium carboxylate 9.

SCHEME I

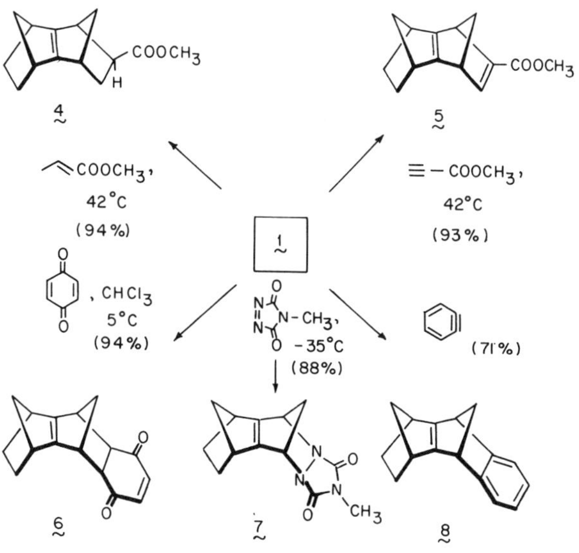

Hunsdiecker degradation of 9 with elemental bromine followed by reductive debromination afforded 10 which was independently prepared by catalytic hydrogenation of the well-known diene 11 [13].

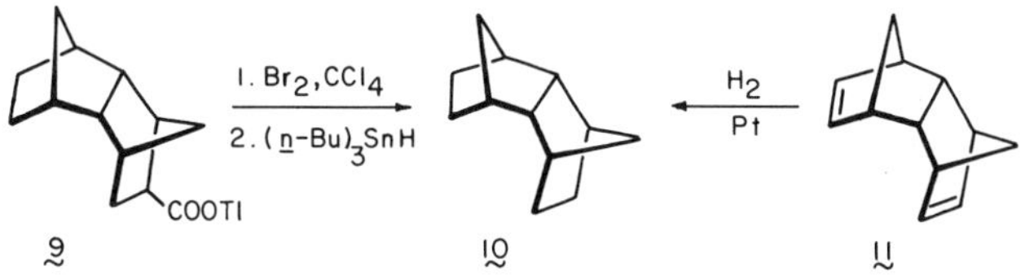

By making recourse to phenyl vinyl sulfone as dienophile, reaction with 1 leads stereospecifically to 12, the reductive desulfonylation of which delivered syn-sesquinorbornene (13) for the first time [13,14]. This fascinating hindered olefin is now recognized to possess an appreciably folded double bond, such that it reacts readily with various reagents from its exo surface. This deviation from planarity was originally deduced from an X-ray crystal structure of 12 which was found to exhibit a dihedral angle of 162-164° between the planes of the two rings sharing the double bond [15]. This topological phenomenon is characteristic of all syn-sesquinorbornenes [16-18]. One interesting molecule to arise from studies involving 12 is the p-nitrobenzoate 14

151.6 ppm

SO_2Ph $\xrightarrow[\text{Na}_2\text{HPO}_4,\ \text{CH}_3\text{OH}]{\text{Na(Hg)}_x}$

H ODNB

12 13 14

which qualifies as the most highly reactive saturated tertiary derivative presently known [19].

We have also noted that introduction of incremental degrees of unsaturation into syn-sesquinorbornenes leads to progressively deshielded C-13 NMR shifts of the internal olefinic carbon atoms. In syn-sesquinorbornadiene (15), a 6.1 ppm downfield shift is seen relative to 13 [14]. The significant increase in strain which materializes in the trienyl diester 16 is adequate to shift its internal trigonal carbons to a value below that of the prior record holder, bicyclohexene 17 [20]. The C-13 shift of the unsubstituted triene, which remains unknown to this time, will be of particular interest. The varied energetic situation in this series also reveals itself in chemical reactivity differences. For example, whereas 13 does not react with triplet oxygen, 15 is responsive to this reagent and 16 reacts exothermically in air to give

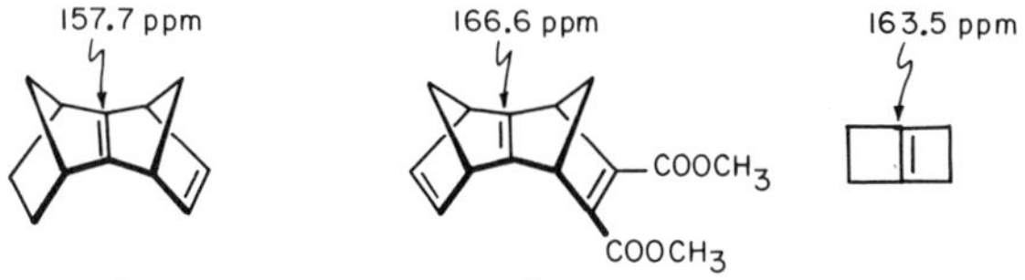

the exo epoxide 18. The ready conversion of 18 to cage compound 19 under conditions of triplet sensitization substantiates the stereochemical assignment [14].

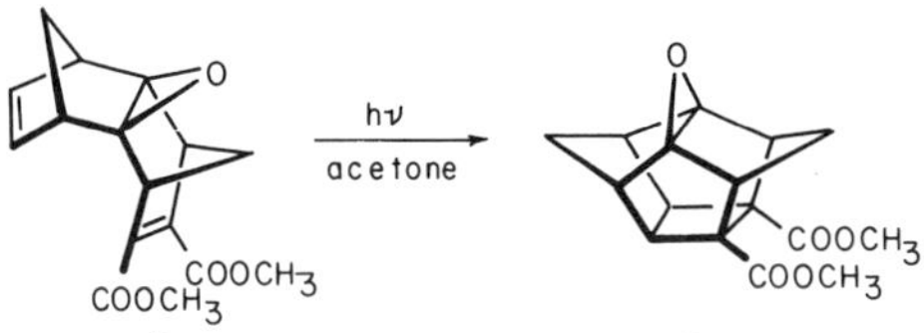

The two dienophiles which do not add to 1 with clean endo stereoselectivity are maleic anhydride [13,15] and singlet oxygen [21]. The minor anhydride adduct which possesses the anti,endo structure 20 has been shown to be endowed with an interplanar angle of 177.9° which is entirely normal for an isolated double bond [15]. Other anti-sesquinorbornenes likewise have essentially planar features about their internal olefinic centers [17,18]. Diepoxides 21 and 22 are produced in an approximate ratio of 2:1 upon capture by 1 of $^1O_2$, followed by thermal rearrangement of the endo peroxides so produced. While it remains unclear why maleic anhydride deviates from the norm, the loss of stereochemical control during singlet oxygenation is attributed to energetic factors arising

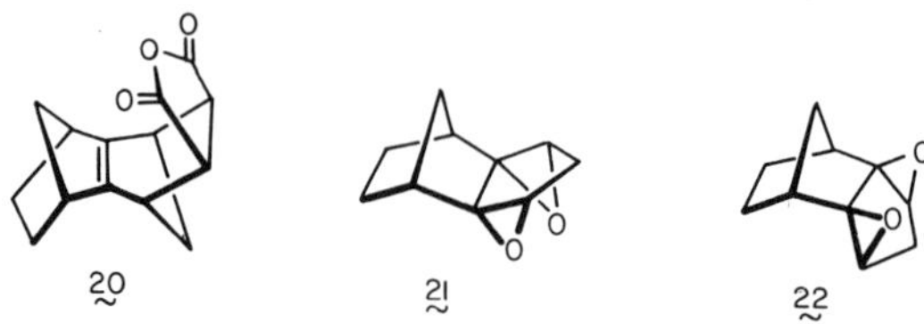

from the ionization potential of $^1O_2$ which differs considerably from those of normal dienophiles [21].

Methyl acrylate, methyl propiolate, maleic anhydride, o-benzoquinone, N-methyltriazolinedione, and benzyne cycloadd to 2 with formation of endo products in each instance [13,22]. In the last example, only the 2:1 adduct 25 was obtained without regard for the relative amounts of benzyne employed. Evidently, the central double bond of initial adduct 23 is adequately reactive to capture a second benzyne molecule by a [2+2] mechanism more rapidly than [4+2] addition to 2.

2 23 24

Urazole 25 could be selectively reduced to dihydro derivative 26 with diimide. Two fundamental stereochemical questions were now resolved simultaneously by hydrolysis-oxidation of 26 to the unsaturated azo compound 27 and its photocyclization to 28 [22]. The excited-state intramolecular $[\pi^2 s + \pi^2 s]$ cycloaddition of 27 represents only the second example of azo involvement in such a reaction [23].

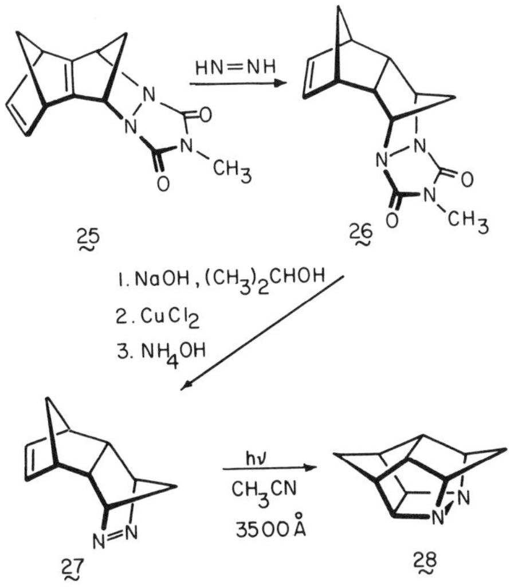

Reaction of 3 with methyl propiolate resulted in the formation of 29a (21%) and 30a (79%). A somewhat comparable product distribution was observed with dimethyl acetylenedicarboxylate: 29b (14%); 30b (86%) [13]. The obvious crossover in stereoselection is not restricted to α,β-unsaturated ester dienophiles, as evidenced by the response of 3 to benzyne (31:32 = 19:81). As before, those reactions are considered to proceed under normal

Diels-Alder frontier orbital guidelines where the dienophile LUMO and diene HOMO orbital energies comprise the dominant interaction [24]. In this context, perturbation theory clearly denotes that the earlier the transition state, the more successful will be the application of molecular orbital analysis [25]. For this reason and because early transition states are thought to resemble most closely a two-plane orientation complex of

29 30

a, R = H; b, R = $COOCH_3$

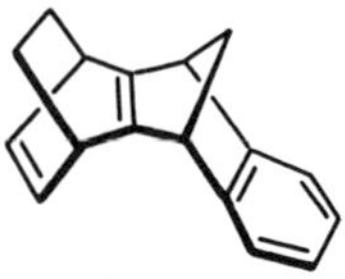

31  32 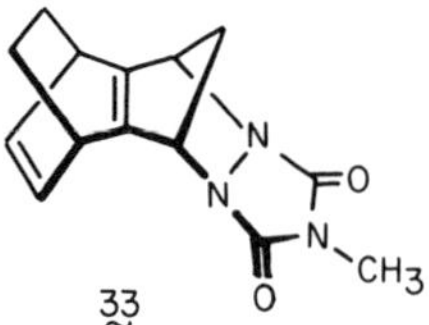 33

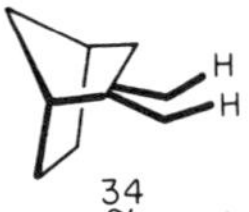

34 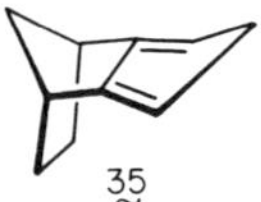 35

the reactant pair, the low-temperature cycloaddition of 3 to N-methyltriazolinedione was examined. The lone adduct produced under these conditions was confirmed to be 33 by X-ray crystal structure analysis. This urazole also suffers endo pyramidalization in its ground state, the downward tilt giving rise to a 168.2° angle under the central flap [22].

These findings were viewed to be demonstrable evidence that important electronic perturbations which bear directly on product development were being strongly manifested in 1 and 2, but less so in 3. The stereoselectivity differences cannot arise from steric factors because C(1) and C(4) of the cyclopentadiene units in these substrates are simply too remote from either bridge. In any event, if such a working hypothesis were adopted, the behavior of 1 and 3 would have to be implausibly regarded as contrasteric.

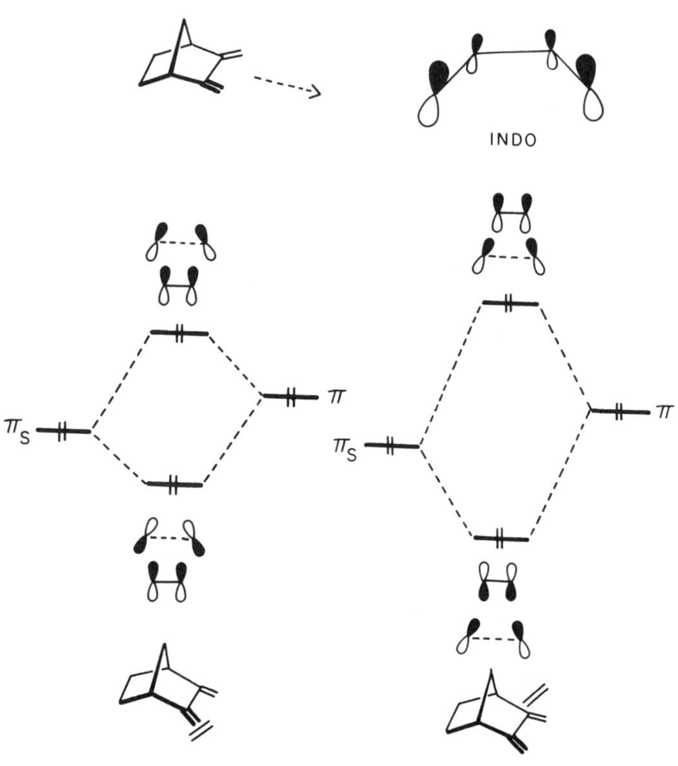

Figure 1. Interaction between butadiene π(s) and dienophile HOMO for approach from the ethano face (left; favored) and the methano face (right; disfavored).

To gain suitable insight into this question, we in collaboration with Professor Gleiter engaged in more sophisticated molecular orbital calcultions as supported by extensive photoelectron spectroscopic experiments. It was quickly determined that the HOMO of the diene fragment does not interact significantly with the bicyclic σ frameworks because of the great disparity in their energies. The stereocontrol must therefore be dictated by secondary factors. Quite evident here was the relatively strong interaction of the neighboring σ orbital framework with the subjacent π_S orbital in 1 and 2; the effect was very weak in 3. The resulting admixing effects a disrotatory tilt of the π orbitals at C(1) and C(4) of the butadiene moiety such that electron density is enhanced syn to the methano bridge (see Figure 1). Antibonding interaction between the π_S orbital of the exocyclic butadiene system and the HOMO of the dienophile is consequently smaller for attack from below [13].

The preceding analysis constituted a working hypothesis that served to guide us toward further relevant experimentation. Before proceeding to these highly revealing studies, we first consider two ancillary investigations which were designed to dismiss two other possible contributory influences.

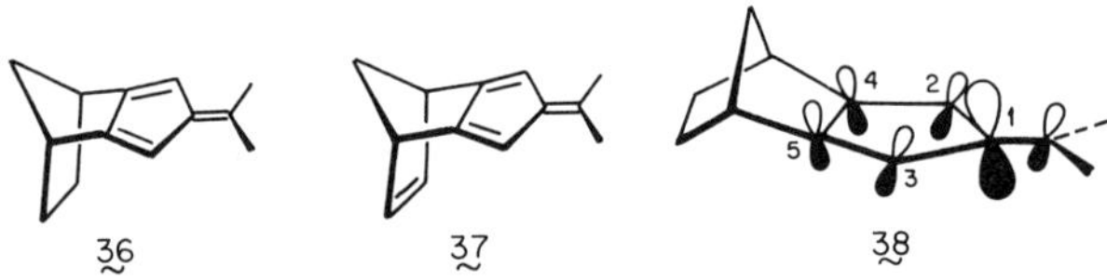

In an extension of his π orbital distortion arguments for norbornene, Houk also gave consideration to 2-methylidenenorbornane (34) and observed through calculation that pyramidalization of its exocyclic double bond occurs to bend the terminal hydrogens in the exo direction [7]. The obvious relationship of 34 to 1 (see 35) was cited as a possible reason for preferential approach of the dienophile from below the plane.

To resolve this question, we proceeded to examine the Diels-Alder behavior of the norbornyl-fused dimethylfulvene systems 36 and 37. This pair of compounds was selected principally for two reasons. First, X-ray crystallographic [26] and electron diffraction studies [27] of 6,6-dimethylfulvene and related molecules [28] distinctly reveal the five-membered ring to be planar. As a direct consequence of this planarity, π-orbital overlap of the cross-conjugated polyolefin is maximized and thought to provide 0.386 eV of stabilization energy per carbon-carbon bond [29]. Since any pyramidalization of the five-ring π bonds in 36 and 37 would result in diminution of this resonance energy, these compounds are unlikely to be otherwise distorted. Second, detailed molecular orbital calculations of 36 and 37 reveal the existence not only of strong σ-π interaction in subjacent orbitals, but also of π-lobe tilting reminiscent of that present in 1 and 2 (see 38).

These structural and electronic features remain conducive to fully stereocontrolled dienophile capture from below the diene plane [30]. The formation of adducts 39a, 39b, 41, and 42 is exemplary. Stereochemical assignments to the various products were arrived at on the basis of their ready air oxidation to epoxides (cf 39 → 40), C-13 NMR correlations, and chemical interconversions in certain cases. Since pyramidalization as in 35 is very likely absent, there exists no reason to invoke double bond distortions of this type as a possible source of stereochemical control. On the other hand, INDO calculations clearly reveal that π-σ coupling remains highly pronounced. Accordingly, antibonding four-electron, four-center interaction continues to be substantially reduced during endo attack.

Past experience has shown that Diels-Alder reactivity differences can be affected in a major way by the distance separating C(1) and C(4) in the diene [31]. This factor is well known to underlie the preeminent position of cyclopentadiene as a highly reactive 4π partner [32]. Does heightened reactivity also bring enhanced stereoselectivity? To answer this question, the π-facial stereoselective behavior of 43 and 44 was investigated. The presence of the methyl groups follows from the

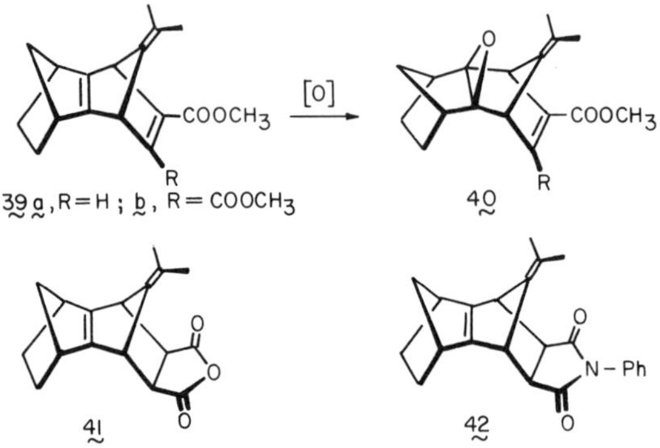

39a, R=H ; b, R=COOCH$_3$ [O] 40

41 42

obvious requirement that stereochemical markers be incorporated into the 4π component.

43 44

Diels-Alder addition of 43 to N-phenylmaleimide, p-benzoquinone, phenyl vinyl sulfone, and dimethyl acetylenedicarboxylate (DMAD) gave exclusively products resulting from below-plane dienophile capture as in 45-48, respectively [33]. In contrast, N-methyltriazolinedione (MTAD) added to 43 with overwhelmingly preferred above-plane stereoselection. Where the lesser reactive system <u>44</u> was concerned, below-plane stereoselectivity was noted to fall off when DMAD was involved (94% endo). However, reaction of <u>44</u> with MTAD led to a 55% endo:45% exo product distribution. Therefore, we

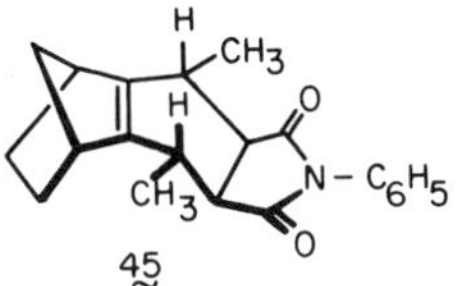

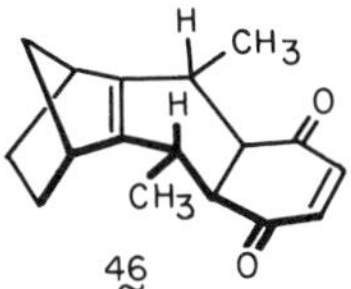

observe for the first time that the situation can become more complex if either the diene carries an additional norbornene double bond or the dienophile has a second occupied MO with a π system perpendicular to the first one as in DMAD or MTAD. With dienophiles of this type, possible coupling between additional occupied MOs and the symmetrical π combination of a diene unit must be taken into account. Such influences could well give rise to a dynamically induced σ-π coupling that modifies the orientation of the π lobes. In cases of this type, a detailed investigation of those interactions prevailing in the transition state is necessary since the simple static model fails. Studies along these lines are currently under active investigation in Heidelberg.

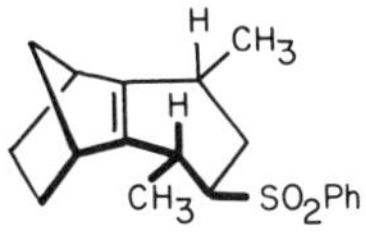

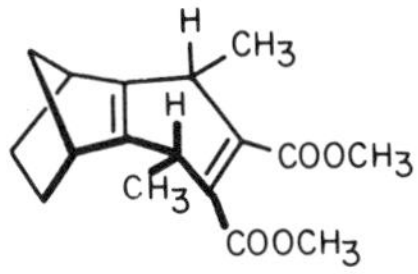

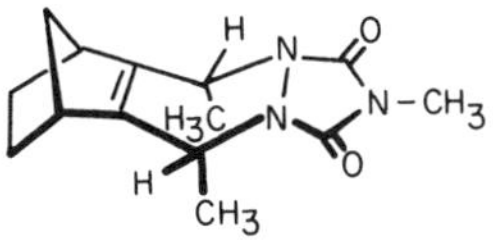

The preceding findings bear importantly on an additional relevant question. Vogel has argued that cycloaddition stereoselectivity in these systems may be governed by the relative stabilities of the isomeric adducts [34]. This phenomenon is not generally encountered in Diels-Alder chemistry (adherence to the Alder rule, etc) and is considered to be of little consequence during cycloadditions to the norbornyl- and norbornenyl-fused diene systems presented here. Thus, 43 and 44 lack that methylene carbon which would generate a cyclopentadiene ring and consequently do not ultimately produce sesquinorbornene-type adducts. For this reason, there should exist little thermodynamic difference between 45-49 and their isomers. Also, it is unlikely that crossovers *syn-/anti*-sesquinorbornene stabilities should occur with small structural changes in the diene. Yet, stereoselectivity differences of this type have been uncovered (see below).

As the direct result of Heilbronner's investigation of the photoelectron (PE) spectra of 50 and 51 over a decade ago [35], it is now obvious that substituents placed at C(5) of a cyclopentadiene ring exert a strong influence on the orbital energies of the π electrons. For example, the second PE band of 50 which is assigned to ionizaton from the π_S level is appreciably destabilized due to

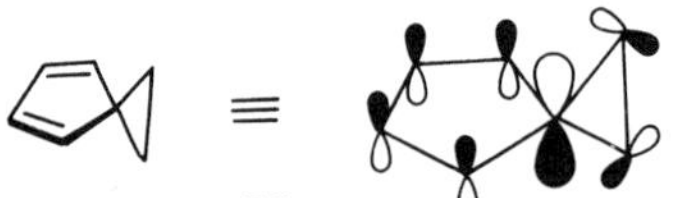

≡

50

51

conjugative interaction with the symmetric Walsh orbital of the three-membered ring. Of course, 51 lacks this particular type of σ-π interaction. These differences were expected to persist in the ground state structures of 52-54. When the long-range interactions of the norbornyl and norbornenyl fragments are superimposed upon the symmetrically substituted cyclopentadiene networks in this manner, the two faces of each of the diene rings become nonequivalent and differentiatable. We had to contend with the possibility that nonbonded steric hindrance to product formation might now modulate π-facial stereochemistry. For this reason, diene 55 [36], was studied in order to examine the complementary state of affairs depicted in transition states A and B.

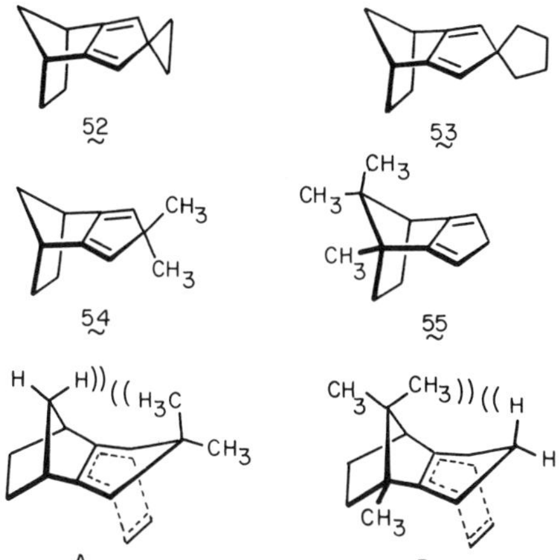

The spirocyclopropane derivative 52, like 1 and 2, exhibits a strong predilection for below-plane attack with a wide range of dienophiles to give adducts, such as 56 and 57 , having syn-sesquinorbornene geometry [18]. In contrast, spirocyclopentane 53 enters into [4+2] cycloaddition totally by top-face bonding to generate anti-sesquinorbornene derivatives (eg, 58 and 59) except when DMAD is involved [18]. X-Ray crystal structure analyses of selected adducts revealed that torsion angles on the order of 17° with tilting in the endo direction persisted in 56 and related compounds. On the other hand, the dihedral angle associated with 59 was within experimental error of 180°.

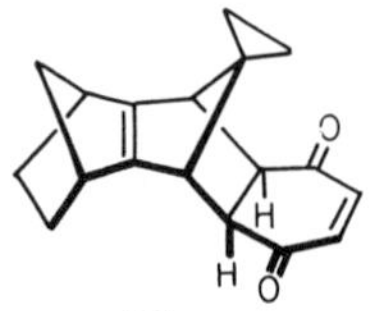

56 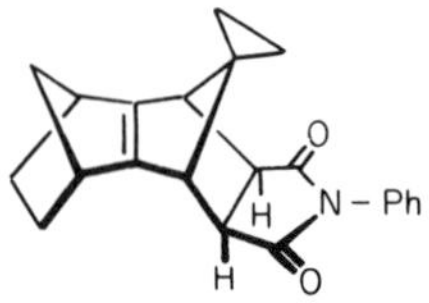 57 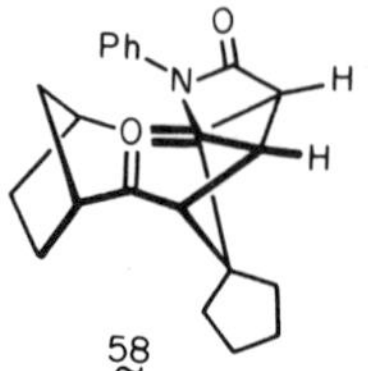 58

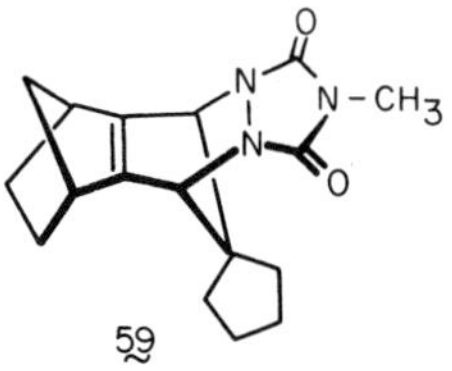

59 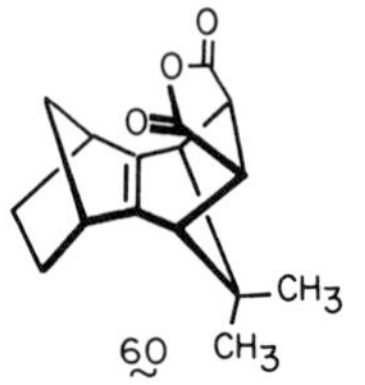 60 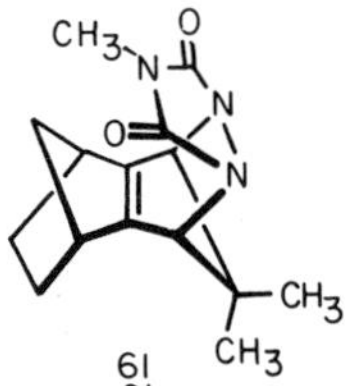 61

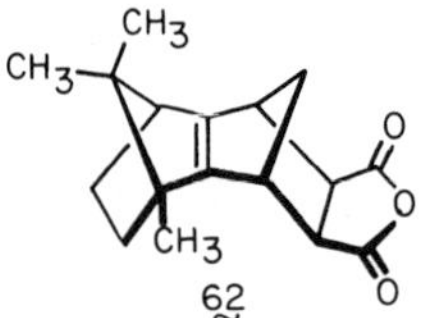

62 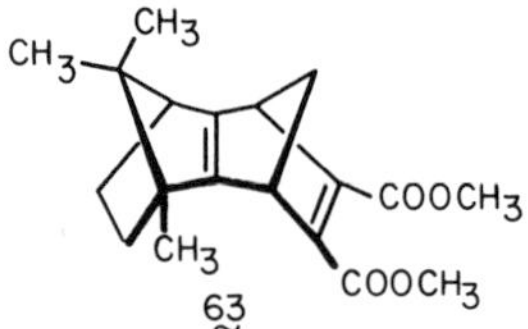 63

With the placement of geminal methyl substituents on the cyclopentadiene ring as in 54 came the opportunity to examine noncyclic alkyl influences. When reaction was carried out with such dienophiles as maleic anhydride, MTAD, and the like, a single adduct of the anti-sesquinorbornene type (eg, 60 and 61) was isolated in each instance [17]. Accordingly, the response of 54 is comparable to that of 53 and opposite to that of 1, 2, and 52. The interesting question relating to the π-facial stereoselective behavior of 55 was quickly answered. When condensed with various dienophiles, efficient cycloaddition occurred to produce only syn-sesquinorbornenes such as 62 and 63 [17].

Clearly, the π-facial stereoselectivity shown by isodicyclopentadienes in Diels-Alder reactions can be modified by substitution of the tetrahedral carbon of the cyclopentadiene ring. These striking chemical phenomena could have their origin in steric hindrance changes which selectively restrict dienophile approach from above or below the diene plane and variations in timing of the respective transition states with a corresponding crossover from reactant-like to product-like characteristics, in addition to modifications in the energies and shapes of the frontier and subjacent MOs as a consequence of through-bond and through-space orbital interactions. In order to acquire background kinetic information, we have measured the rates of endo addition of DMAD to many of our substrates (Figure 2). When the rate constants for 1, monomethyl derivative 64, and 54 are compared, a reactivity order of 950:300:1 is seen at 25° . Thus, the presence of an _endo_-methyl substituent on the cyclopentadiene ring causes an approximate threefold dropoff in cycloaddition rate. On the other hand, placement of a pair of methyl groups at this position causes an additional 300-fold dropoff in reactivity. Whereas the rate retardation observed for 64 might be attributable to added primary steric encumbrance, this argument must be inordinately stretched for 54 unless secondary nonbonded steric interactions such as pictured in A gain importance in the transition state. However, our measurements indicate the kinetic behavior of 55,

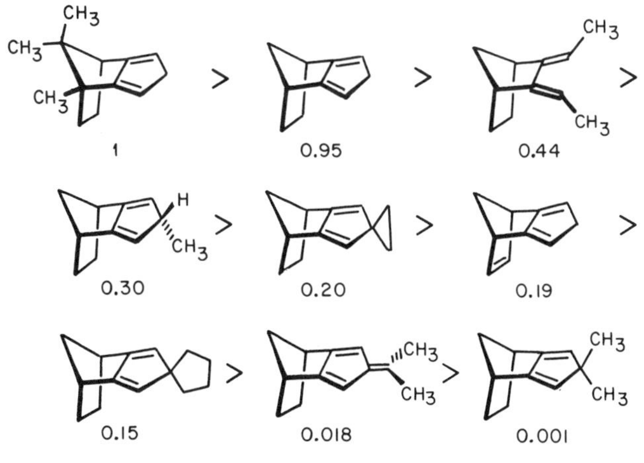

CH_3 CH_3 CH_3 > > CH_3 CH_3 >

1 0.95 0.44

H CH_3 > > >

0.30 0.20 0.19

> CH_3 CH_3 > CH_3 CH_3

0.15 0.018 0.001

Figure 2. Relative rates of DMAD cycloaddition from the below-plane direction.

whose reactivity is probably attenuated to a comparable extent as indicated in B, to be slightly faster than that of isocyclopentadiene itself. As a consequence, we do not view the secondary influences delineated in A and B to be the primary factor behind the low reactivity of 55. This conclusion is supported by the X-ray data on 38, which show the inside apical methyl group to be well beyond the van der Waals range of the opposed apical hydrogen.

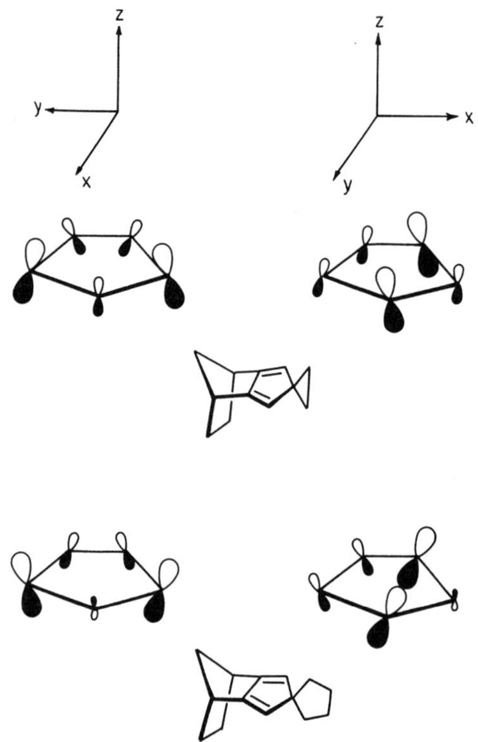

Figure 3. Schematic representation of the π_s orbital in 52 (top) and 53 (bottom).

Following recording of the PE spectra of 52-55 and detailed INDO computations, it was recognized that the terminal π lobes of 53 and 54 in their individual π_s MOs are disrotatorily rotated away from the methano bridge, unlike the situation prevailing in 1, 52, and 55 (Figure 3). The fate suffered by alkyl-substituted isodicyclopentadienes of the first type (including 64, see below) is avoided by 52 only because σ-π interaction is of greatest importance in the π_s + Walsh linear combination (13a') of this hydrocarbon. On the basis of Figure 3, antibonding interaction between the diene system and an attacking dienophile should be minimized if the electron-deficient reagent enters below-plane in the case of 52 and above-plane in the case of 53.

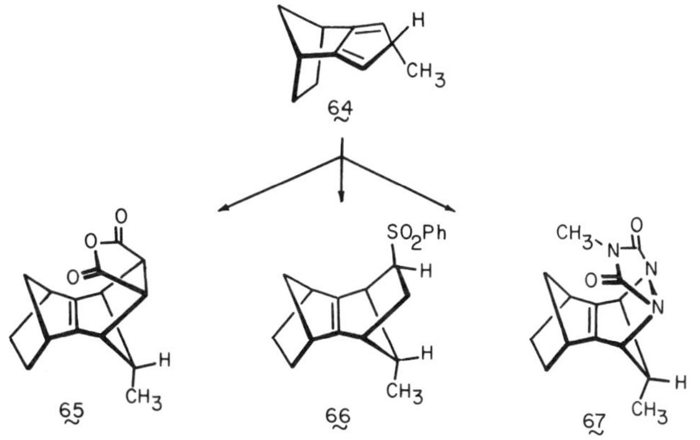

64

65 66 67

It was briefly mentioned above that 64 shares an orbital construct in common with 54. If this is so and secondary orbital effects are indeed important in controlling π-facial stereoselectivity, then 64 should prefer to undergo exo bonding under Diels-Alder conditions. Adducts 65 and 66, which are produced in isomerically pure condition within our limits of detection, show that this trend is followed. With MTAD, a 93:7 mixture of 67 and its endo isomer was isolated [17].

It now becomes relevant to inquire whether the prediscussed differing stereochemical responses can be attributed to a single cause. In actuality, there are at least five phenomena that could becloud the overall influence of electronic factors:

(a) The possibility that the cycloaddition reactions in question are reversible. In actuality, all attempts to detect reversibility and to observe equilibration of norbornyl-fused dienes with their cycloadducts have failed except in two clearly delineated examples [30,34]. Thus, while reversibility cannot be discounted, it is required that product stereochemistry not be affected by the existence of an equilibrium as already noted for 41 and 68.

COOMe
C
C
COOMe
O
COOMe
COOMe
68

(b) The likelihood that factors, present in certain dienes and not others, may give rise to different stereochemical results. However, as the kinetic studies summarized in Figure 2 reveal, 1, 52, and 53 exhibit comparable reactivity toward DMAD. Thus, the crossover in π-facial stereoselectivity observed for 52 and 54 does not arise from gross kinetic imbalances.

(c) The possible intervention of π-complexes. If weak π complexes are indeed formed initially, as is likely in certain circumstances, they are considered to lie directly on the reaction coordinate leading to adduct formation.

(d) Changes in transition state timing may not be a sensitive probe of subjacent orbital effects. Although the concept of variable transition states has seen extensive application in explaining the properties of organic reactions, it is entirely possible that selectivity could sometimes be an insensitive probe of transition-state structure, although its general importance to this question cannot be underestimated. In short, no guarantee exists as to the actual relationship of these effects in the many reactions examined.

(e) A parallelism exists between kinetic and thermodynamic control as the result of fortuitous adherence to the Bell-Evans-Polanyi principle. Under these conditions, cycloaddition stereoselectivity would be governed by the relative stabilities of the isomeric adducts. Since these factors cannot plausibly be operative with 43 and 44, and because it is unlikely that a crossover in stability should occur as one proceeds down the series 52 → 53 → 54, we do not consider stereoselectivity to be connected with adduct stability.

Although all factors point to π-lobe deformation as the controlling element of stereoselection for isodicyclopentadienes, it becomes important to emphasize that we have used a static argument in analyzing the shape of the frontier orbitals in the unperturbed diene or triene systems. The frontier orbitals of the attacking dienophile are certainly capable of modifying this assumed time-independent π-σ interaction. Consequently, a complete answer demands an extended dynamical approach where coupling between all components is investigated as a function of changing internuclear distance. One such analysis, the cost of which is very high, is presently in progress in Heidelberg. Two situations should be distinguishable. In the first, no pronounced modification of the diene's MO amplitudes will be seen (static picture remains valid). In the second, strong perturbation is engendered in direct opposition to the ground-state interaction pattern. Dienophiles possessing an additional double bond (eg, DMAD) or nonbonded electron pairs (eg, MTAD) may belong to the latter category.

If isodicyclopentadienes do indeed have this penchant for π-orbital tilting the phenomenon may well be detectable by spectroscopic means. Although reliance must be placed upon long-range orbital interactions, our expectations ran high because of recently uncovered large orbital interactions through four or more bonds [37]. It should be noted that the probes which we shall use 69, (R_1 and R_2) have been substantively removed from the immediate vicinity of the norbornyl framework and arranged symmetrically about the fused model plane.

R_1 R_2

69

Expectedly, the carbon atoms within the three-membered ring of spirocyclopropane 52 exhibit different C-13 chemical shifts. The upfield position

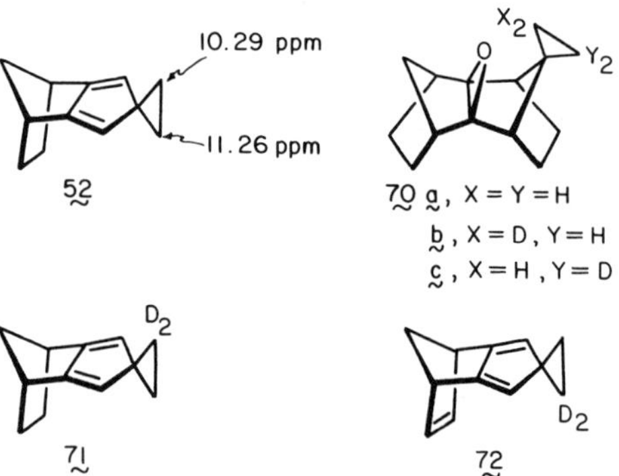

of the above-plane carbon atom in this hydrocarbon was ascertained by two complementary series of experiments [38]. In the first, 52 was transformed via its phenyl vinyl sulfone adduct into 70a where a clear-cut distinction between proximal and distal cyclopropane protons was realized by Eu(fod) shift studies (ΔEu values [39]: X_2 = 4.07, Y_2 = 1.75). With these observations as our basis for stereochemical assignment, the d_2 derivatives 71 and 72 were prepared and similarly transformed into 70b and 70c. With knowledge of the location of the isotopic labels in 71 and 72, the specific C-13 shifts of interest in 52 became clearly apparent.

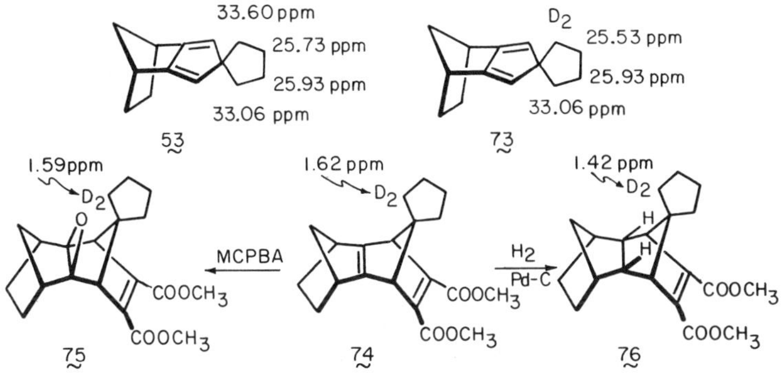

33.60 ppm
25.73 ppm
25.93 ppm
33.06 ppm
53

D_2
25.53 ppm
25.93 ppm
33.06 ppm
73

1.59 ppm
D_2
O
$COOCH_3$
$COOCH_3$
75

MCPBA

1.62 ppm
D_2
$COOCH_3$
$COOCH_3$
74

H_2
Pd-C

1.42 ppm
D_2
H
H
$COOCH_3$
$COOCH_3$
76

The four methylene carbons within the spirocyclic five-membered ring of 53 also emerge with distinctively different chemical shifts. While individual assignments had to await completion of the ensuing isotopic studies, it was again quite apparent that electronic influences within the norbornyl system extend to remarkably long distances. Following the preparation of 73, the lowest field C-13 signal of the set of four was not seen and a small geminal deuterium isotope effect was observed as expected [40]. In the case of 73, knowledge of the locus of isotopic substitution was gained by preparing 74 and examining ^{2}H chemical shift alterations as its central double bond was chemically altered in two different ways. The relevant data accompany the structural formulas for 74-76. Without doubt, the carbon shielding effects operating in 53 are reversed relative to those in 52!

This contrasting behavior appears to be general. Thus, the isomeric products prepared from the anion of 1 and epichlorohydrin exhibit C-13 cyclopropyl shifts entirely comparable to those of 71 and 72.

1. $NaNH_2$, NH_3 2. epichlorohydrin (CH_2Cl), THF → 77 (72%) [H, CH_2OH; 27.28 ppm; 16.55 ppm] + 78 (28%) [15.39 ppm; 28.01 ppm; H, CH_2OH]

It may be important to recognize that the C-13 spectral data parallel exactly the different stereoselectivities of dienophile capture exhibited by 52 and 53. *gem*-Dimethyl derivative 54 behaves comparably to 53, but no technique has yet been found to distinguish between its two alkyl substituents (C-13 shifts: 23.55 and 22.92 ppm). [4+2] cycloaddition is seen to occur preferentially from the less shielded face of these dienes. In our view, this stereoselectivity arises in order to profit from lower secondary antibonding influences along this pathway.

Any complications which might be introduced by adduct formation are seen not to be present during the capture by anion 79 of suitable electrophilic reagents. For steric reasons, bonding was expected to be directed to the cyclopentadienide center most remote from the bicyclic units to deliver C_s-symmetric products. In these examples, the causative factors which underlie the extent of endo/exo stereoselection must clearly be electronic in nature, since the reaction site is quite distant from the methano and ethano bridges which distinguish the π planes. Consequently, 79 and related carbanions offer the potential for providing additional independent experimental evidence for the fact that norbornyl ring systems can indeed exert long-range electronic perturbational effects.

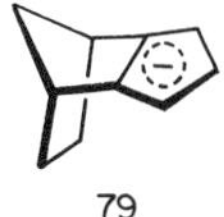

79

As matters turned out, sequential treatment of 1 with n-butyllithium and methyliodide afforded 64 as the only monomethylated product [41]. When geminal dialkylation was effected with three dideuterated 1,2-disubstituted ethanes, product distributions were obtained which indicated some loss of stereocontrol. However, the predominant product in each instance showed that anion 80 continues to favor C-C bond formation from the endo surface as in 79. Because the partitioning ratio is not large in these examples, a major question remains about the dropoff in stereoselectivity during cyclopropane ring formation, especially since this behavior persists upon comparable reaction with epichlorohydrin (77:78 = 72:28%).

Condensation of 1 with $Cl(CH_2)_3CD_2I$ led cleanly to 73. This exceptionally high stereoselectivity was also encountered with 1,4-dibromo-2-isopropylidene-butane which gave 81 as the only identifiable product [41].

Quenching of 79 with D_2O in tetrahydrofuran at -78° results in the regiospecific and π-facial stereoselective formation of 82 as denoted by the appearance of a lone singlet at 3.02 ppm in the proton-decoupled ^2H NMR spectrum of the product.

Following repetition of this procedure to give 83, a third exchange was carried out and this d_1 anion was protonated so as to give 84 which exhibits a

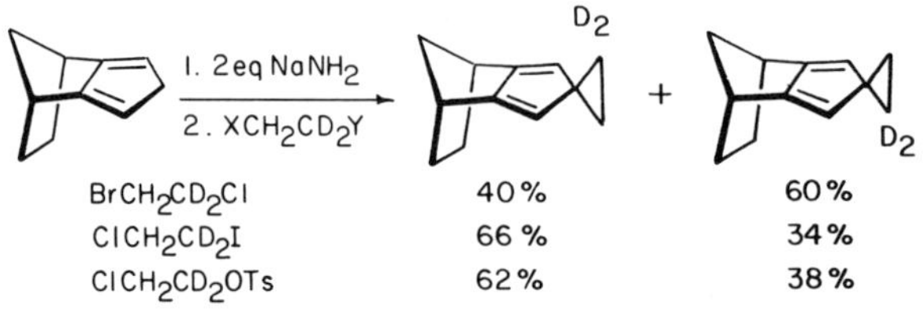
1. 2eq NaNH2
2. XCH2CD2Y
D2
+
D2
BrCH2CD2Cl 40% 60%
ClCH2CD2I 66% 34%
ClCH2CD2OTs 62% 38%

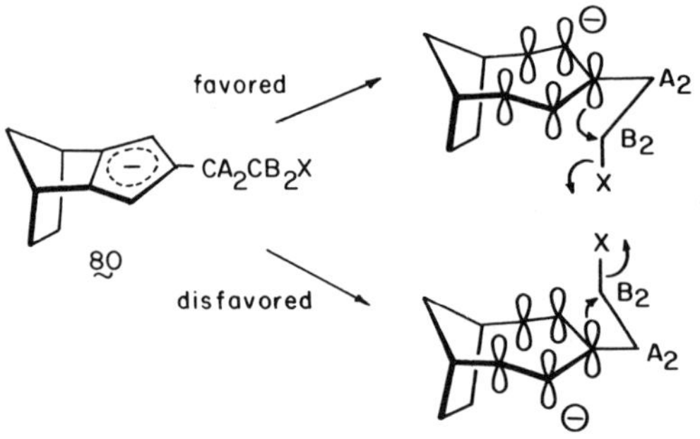
favored
disfavored
CA2CB2X
80
A2
B2
X
X
B2
A2

singlet at 3.14 ppm. Confirmation of these stereochemical assignments was achieved by means of independent cycloaddition of phenyl vinyl sulfone to 82 and 84, reductive desulfonylation of 85a and 85b, and peracid oxidation of the two d_1 *syn*-sesquinorbornenes. The well-recognized anisotropy contributions of the oxirane oxygen [42] in 87a and 87b causes the inner D atom to be deshielded by 0.49 ppm and the outer shielded by 0.46 ppm.

While C-13 shieldings are dominated chiefly by local paramagnetic electron currents [43], local

81

diamagnetic effects make the more important contribution to ^2H chemical shifts [44]. These differences may account for the spectral parameters of 82-84.

3.14 ppm D H 1. n-BuLi 2. D_2O D 3.02 ppm 82 1. n-BuLi 2. H_2O D 3.02 ppm 83 3.14 ppm D H 84

X Y SO_2Ph 85 1.39 ppm X 1.05 ppm Y 86 1.88 ppm X 0.59 ppm O Y 87

a, X = H, Y = D ; b, X = D, Y = H

The stereochemical preferences observed in the alkylation reactions of 79 are reminiscent of the cycloaddition behavior of isodicyclopentadienes. INDO calculations proved uniformly consistent in predicting the observed stereoselectivity, with π orbital tilting occuring as shown in Figure 4. As a result, bonding interaction between the π donor (D) of the anionic system and the acceptor (A) or the attacking electrophile is more efficient if alkylation occurs from the endo surface. The loss of stereocontrol experienced by 80 is attributed to the probable product-like nature of the ensuing transition states which reflect the rather unusual ground-state electronic properties of the spirocyclopropane products [41]. The esoteric question of whether ion-pairing effects have any impact on this chemistry may prove resolvable by lithium NMR experiments.

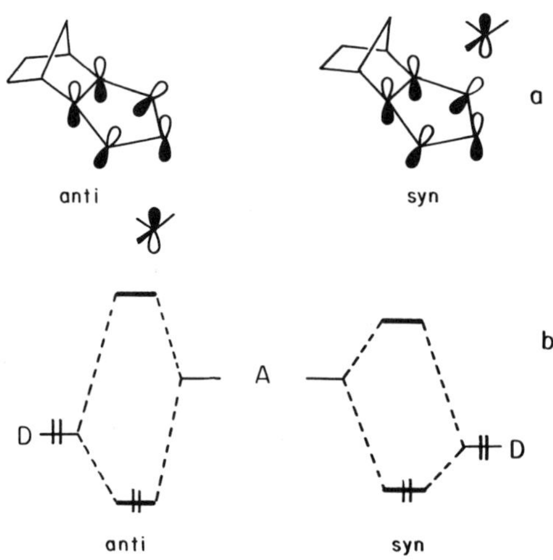

Figure 4. (a) Schematic representation of a transition state for S_N1 reaction between an electrophile and anion 79; (b) MO interaction diagram between the π orbital of a donor system and the empty orbital of a carbenium ion.

So encouraged were we by these remarkable reactions that we next sought to delineate the approximate limits of this phenomenon by preparing the previously unknown hydrocarbons 88 and 89. The anthracene subunits necessitate that [4+2]π bonding involve only the central aromatic ring. These reaction centers are more distal to the fused bicyclic rings than any heretofore examined. Also, an intervening benzenoid ring is present to attenuate long-range electronic effects. In the several cycloaddition examples which were examined, approximate 1:1 ratios of top- and bottom-face bonding were observed [45]. The adduct pairs 90/91 and 92/93 are exemplary. Without doubt, remote stereoelectronic effects were inoperative; perhaps it was too much to ask of these systems.

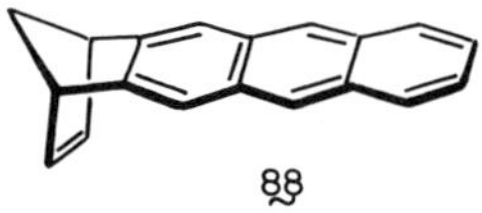

88 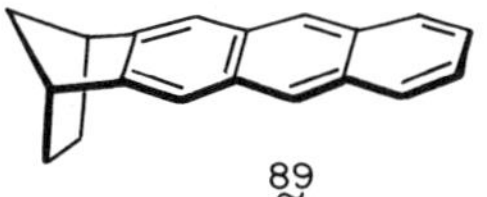 89

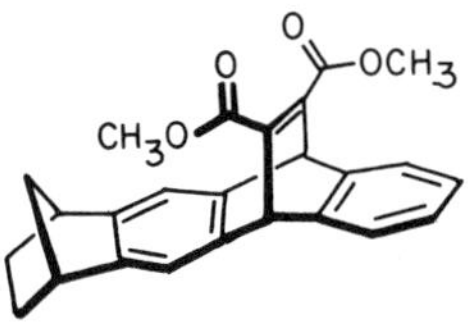

90 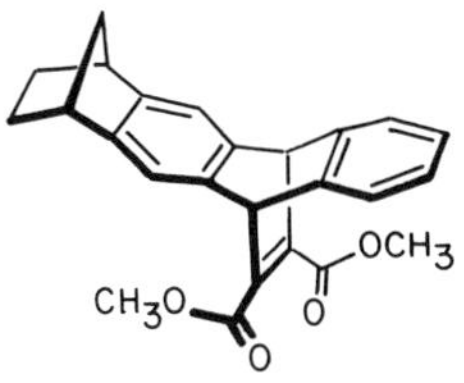 91

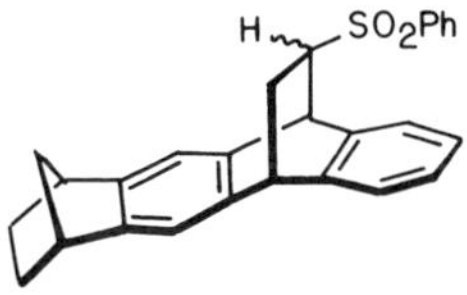

92 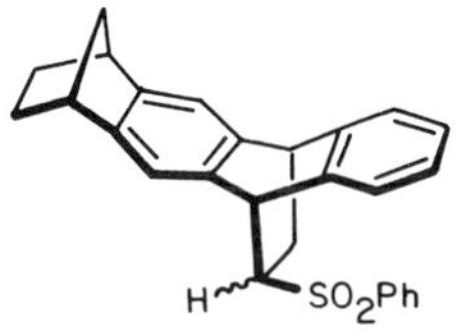 93

We close by reiterating that our subjacent π-orbital distortion arguments concisely explain the varied π-facial stereoselectivities exhibited by structurally varied isodicyclopentadienes. Nonetheless, additional assessment of this concept is most certainly warranted. For this reason, we are currently investigating possible crossovers in stereoelectronic control which might operate in 94 as X is varied from an electronegative to an electropositive group. Additionally, since the tilted cyclopentadiene π orbitals are held within reasonably fixed spatial limits, reagents having frontier π orbitals held at distances greater than those in olefinic reagents (see 95) might well cycloadd the different π-facial stereoselectivities. These and other studies are currently being pursued.

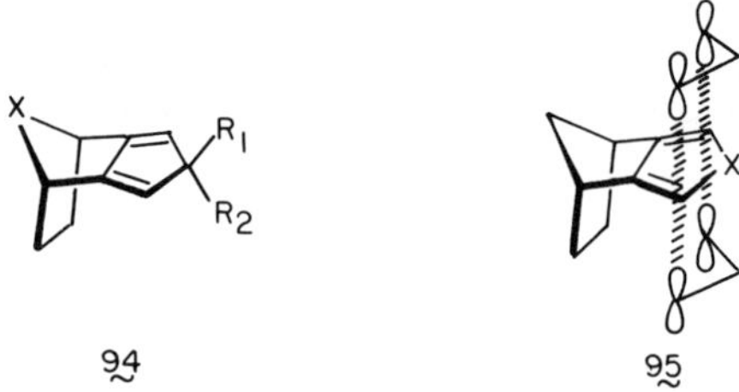

Acknowledgment. This research program has been greatly stimulated by the infectious enthusiasm of my students whose names are cited in the references and the cooperative interest of Rolf Gleiter. We are particularly grateful to the National Cancer Institute for their financial support.

References

[1] H.C. Brown, J.H. Kawakami, and S. Ikegami, J. Am. Chem. Soc. 89, 1525 (1967); H.C. Brown and K.-T. Liu, J. Am. Chem. Soc. 89,3898, 3900 (1967).

[2] T.T. Tidwell and T.G. Taylor, J. Org. Chem. 33, 2615 (1968).

[3] H.C. Brown, Chem. Brit. 2, 199 (1966); H.C. Brown, W.J. Hammer, J.H. Kawakami, I. Rothberg and D.L.V. Jagt, J. Am. Chem. Soc. 89, 6381 (1967).

[4] P.v.R. Schleyer, J. Am. Chem. Soc. 89, 701 (1967).
[5] K.N. Houk in "Reactive Intermediates," M. Jones and R.A. Moss (eds); Wiley-Interscience: New York, 1978, Vol. I, pp 326-327.
[6] S. Inagaki, H. Fujimoto, and K. Fukui, J. Am. Chem. Soc. 98, 4056 (1976).
[7] P.H. Mazzocchi, B. Stahly, J. Dodd, N.G. Rondan, L.N. Domelsmith, M.D. Rozeboom, P. Caramella, and K.N. Houk, J. Am. Chem. Soc. 102, 6482 (1980); N.G. Rondan, M.N. Paddon-Row, P. Caramella, and K.N. Houk, J. Am. Chem. Soc. 103, 2436 (1981); P. Caramella, N.G. Rondan, M.G. Paddon-Row, and K.N. Houk, J. Am. Chem. Soc. 103, 2438 (1981).
[8] G. Wipff and K. Morokuma, Tetrahedron Lett. 4445 (1980).
[9] R. Gleiter and J. Spanget-Larsen, Tetrahedron Lett. 927 (1982).
[10] K. Alder, F.H. Flock, and P. Janssen, Chem. Ber. 89, 2689 (1956).
[11] T. Sugimoto, Y. Kobuke, and J. Furukawa, J. Org. Chem. 41, 1457 (1976).
[12] W.J. Feast, W.K.R. Musgrave, and W.F. Preston, J. Chem. Soc. Perkin Trans. I, 1830 (1972); W.J. Feast, R.R. Hughes, and W.D.R. Musgrave, J. Chem. Soc. Perkin Trans. I, 152 (1977).
[13] L.A. Paquette, R.V.C. Carr, M.C. Bohm, and R. Gleiter, J. Am. Chem. Soc. 102, 1186 (1980); M.C. Bohm, R.V.C. Carr, R. Gleiter, and L.A. Paquette, J. Am. Chem. Soc. 102, 7218 (1980).
[14] L.A. Paquette and R.V.C. Carr, J. Am. Chem. Soc. 102, 7553 (1980).
[15] W.H. Watson, J. Galloy, P.D. Bartlett, and A. A.M. Roof, J. Am. Chem. Soc. 103, 2022 (1981).
[16] A.A. Pinkerton, D. Schwarzenbach, J.H.A. Stibbard, P.-A. Carrupt, and P. Vogel, J. Am. Chem. Soc. 103, 2095 (1981).
[17] L.A. Paquette, P.C. Hayes, P. Charumilind, M.C. Bohm, R. Geliter, and J.F. Blount, J. Am. Chem. Soc. 105, 3148 (1983).
[18] L.A. Paquette, P. Charumilind, M.C. Bohm, R. Gleiter, L.S. Bass, and J. Clardy, J. Am. Chem. Soc. 105, 3136 (1983).
[19] L.A. Paquette, K. Ohkata, and R.V.C. Carr, J. Am. Chem. Soc. 102, 3303 (1980).
[20] J. Casanova, J. Biagin, and F.D. Cottrell, J. Am. Chem. Soc. 100, 2264 (1978).
[21] L.A. Paquette, R.V.C. Carr, E. Arnold, and J.

Clardy, J. Org. Chem. 45, 4907 (1980).
[22] L.A. Paquette, R.V.C. Carr, P. Charumilind, and J.F. Blount, J. Org. Chem. 45, 4922 (1980).
[23] W. Borning and S. Hunig, Angew. Chem., Int. Ed. Engl. 16, 777 (1977).
[24] R. Sustmann, Tetrahedron Lett., 2717, 2721 (1971); K.N. Houk, Acc. Chem. Res. 8, 361 (1975).
[25] R. Sustmann and H. Trill, Angew. Chem., Int. Ed. Engl. 11, 838 (1972); R. Sustmann and R. Schubert, Angew. Chem., Int. Ed. Engl. 11, 840 (1972); R. Sustmann, Pure Appl. Chem. 40, 569 (1974).
[26] N. Norman and B. Post, Acta Cryst. 14, 503 (1961).
[27] J.F. Chiang and S.H. Bauer, J. Am. Chem. Soc. 92, 261 (1970).
[28] For selected examples, see: H. Shimanouchi, Y. Sasada, T. Ashida, M. Kakudo, I. Murata, and Y. Kitahara, Acta Cryst. B25, 1890 (1969); H. Burzlaff, K. Hartke and R. Salamon, Chem. Ber. 103, 156 (1970); R. Bohme and H. Burzlaff, Chem. Ber. 107, 832 (1974); H.L. Ammon, Acta Cryst. B30, 1731 (1974); L. Fallon, H.L. Ammon, R. West, and V.N.M. Rao, Acta Cryst. B30, 2407 (1974); H. L. Ammon and G. L. Wheeler, J. Am. Chem. Soc. 97, 2326 (1975). 97, 2326 (1975).
[29] D.H. Lo and M.A. Whitehead, Tetrahedron 25, 2615 (1969).
[30] L.A. Paquette, T.M. Kravetz, M.C. Bohm, and R. Gleiter, J. Org. Chem. 48, 1250 (1983).
[31] R. Sustmann, M. Bohm, and J. Sauer, Chem. Ber. 112, 883 (1979); L. Schwager and P. Vogel, Helv. Chim. Acta 63, 1176 (1980).
[32] J. Sauer and R. Sustmann, Angew Chem., Int. Ed. Engl. 19, 779 (1980).
[33] L.A. Paquette, A.G. Schaefer, J.F. Blount, M.C. Bohm, and R. Gleiter, J. Am. Chem. Soc. 105, 3642 (1983).
[34] J.-P. Hagenbuch, P. Vogel, A.A. Pinkerton, and D. Schwarzenbach, Helv. Chim. Acta 64, 1818 (1981).
[35] R. Gleiter, E. Heilbronner, and A. de Meijere, Helv. Chim. Acta 54, 1029 (1971).
[36] A.W. Burgstahler, D.L. Boger, and N.C. Naik, Tetrahedron 32, 309 (1976).
[37] H.-D. Martin and R. Schwesinger, Chem. Ber.

107, 3143 (1974); P. Pasman, J.W. Verhoeven, and Th.J. de Boer, Tetrahedron Lett., 207 (1977); R. Bartetzko, R. Gleiter, J.L. Muthard and L.A. Paquette, J. Am. Chem. Soc. 100, 5589 (1978); M.N. Paddon-Row and R. Hartcher, J. Am Am. Chem. Soc. 102, 671 (1980); M.N. Paddon-Row, H.K. Patney, R.S. Brown, and K.N. Houk J. Am. Chem. Soc. 103, 5575 (1981).

[38] L.A. Paquette and P. Charumilind, J. Am. Chem. Soc. 104, 3749 (1982).

[39] P.V. Demarco, T.K. Elzey, R.B. Lewis, and E. Wenkert, J. Am. Chem. Soc. 92, 5734 (1970).

[40] Y. Inamoto, K. Aigami, Y. Fujikura, M. Ohsugi, N. Takaishi, K. Tsuchihashi, and H. Ideda, Chem. Lett., 25 (1978).

[41] L.A. Paquette, P. Charumilind, T.M. Kravetz, M.C. Bohm, and R. Gleiter, J. Am. Chem. Soc. 105, 3126 (1983).

[42] K. Tori, K. Kitahonoki, Y. Takano, H. Tanida, and T. Tsuji, Tetrahedron Lett., 559 (1964); K. Tori, K. Aono, K. Kitahonoki,R. Muneyuki, Y. Takano, H. Tanida, and T. Tsuji, Tetrahedron Lett., 2921 (1966); L.A. Paquette, W.E. Fristad, C.A. Schuman, M.A. Beno, and G.G. Christoph, J. Am. Chem. Soc. 101, 4645 (1979). J. Chem. Soc. 101, 4645 (1979).

[43] F.W. Wehrli and T. Wirthlin, "Interpretation of Carbon-13 NMR Spectra," Heyden: London, 1980: Chapter 2.

[44] L.M. Jackman and S. Sternheld, "Applications of Nuclear Magnetic Resonance Spectroscopy in Organic Chemistry," Second ed., Pergamon Press: New York, 1969; Chapter 2-2.

[45] P.C. Hayes and L.A. Paquette, J. Org. Chem. 48, 1257 (1983).

3. STRUCTURAL AND GEOMETRIC FACTORS IN THE REACTIVITY OF Π SYSTEMS

Paul D. Bartlett, Andrew J. Blakeney, Gerald L. Combs, Jean Galloy, Antonius A. M. Roof, Ravi Subramanyam, William H. Watson, William J. Winter, and Chengjiu Wu

Department of Chemistry, Texas Christian University, Fort Worth, Texas

My talk will be divided into two parts: first, some of our experiences with the sesquinobornene system, to which a very good introduction has just been given; second, some things we got interested in along the way concerned with sigmatropic hydrogen shifts in the isodicyclopentadiene system.

The origin of our interest in the problem [1] was not totally different from that of Leo Paquette [2] except that we did not have any special designs on synthetic methods. We were intrigued with the fact that so very much had been done in the study of singlet oxygen using compounds such as biadamantylidene, in which a double bond joins two bicyclic ring systems with flanking bridgehead hydrogens to prevent ene reactions. Obviously, another way of doing this is the way that appears in sesquinorbornene, where the double bond is in common between two bicyclic ring systems. It is still flanked by bridgehead hydrogens so we have no trouble with the ene reaction. We were really thinking quite a lot about singlet oxygen when we started out, and it so happened that the first synthesis that we used led only to _anti_- sesquinorbornene while the first one that Leo used led only to syn. This has been influential in the emphasis that we have given; however, since we were interested in the special properties of these systems, an obvious way to get at it was to compare the behavior of syn- and anti-sesquinorbornenes toward the things in which we were especially interested.

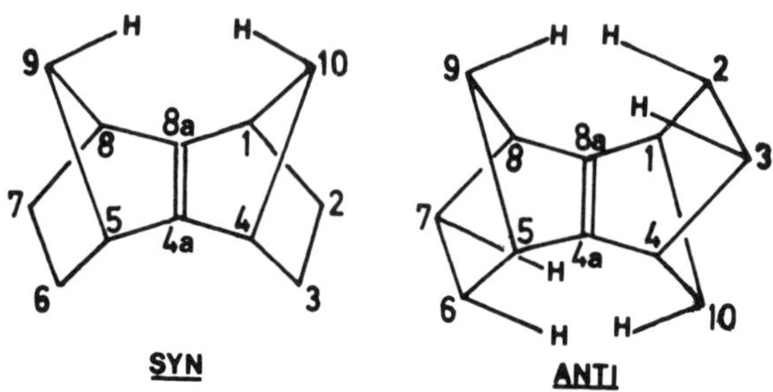

Figure 1 is a formal diagram which we built of a section through anti-sesquinorbornene. We are looking down on the top of the double bond and the ethylene bridge is on the left and the methylene bridge on the right. The circles are van der Waals radii drawn around the known locations of the

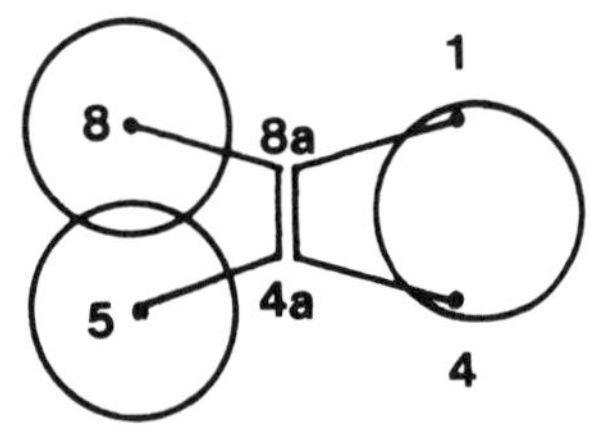

Figure 1. A view toward the C=C bond of anti-sesquinorbornene. Circles represent 1.2 A van der Waals radii of blocking hydrogen atoms.

hydrogen atoms. If you look at the diagram from the point of view of classical stereochemistry some hindrance is to be anticipated in approach of any reagent to the double bond if indeed the reaction is a concerted one. The reagent would have to ram its way in between the ethylene bridge and the methylene bridge and would encounter some quite serious van der Waals-type hindrance. So the first thing we tried after getting _anti_-sesquinorbornene was to make it react with singlet oxygen under the same conditions that worked well for biadamantylidene. The result was essentially no reaction with thermally generated singlet oxygen from triphenylphosphite ozonide. If we used photosensitization for a long period, we got some photooxygenation of the double bond to the epoxide, but we never saw any dioxetanes formed on this double bond.

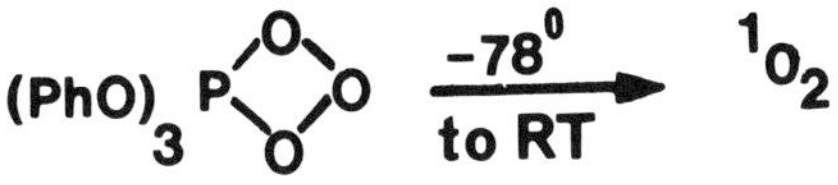

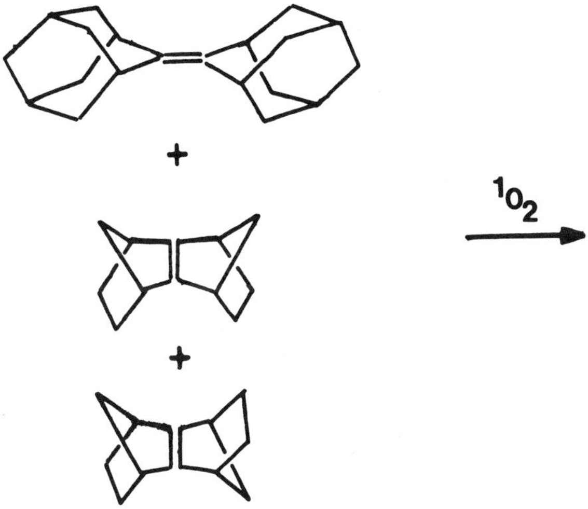

$(PhO)_3P(O_3)$ $\xrightarrow[\text{to RT}]{-78^0}$ $^1O_2$

\+

\+

$\xrightarrow{^1O_2}$

In comparison with biadamantylidene under identical conditions (similar amounts in the same reaction of biadamantylidene and _syn_- and _anti_-sesquinorbornenes), the biadamantylidene captures the singlet oxygen efficiently while the _syn_- and _anti_-sesquinorbornenes are recovered. So there's a big difference between these olefins. At first we thought this was the result of the concerted reaction of singlet oxygen being rendered impossible by the hydrogen blocking access to the double bond. This idea didn't last very long in that simple form because it turns out that one can epoxidize syn- and anti-sesquinorbornenes with a peracid just as easily as any tetrasubstituted double bond. It shows normal reactivity. One can also get a good

Cl, CO_3H

m.p. 41–44°

$CH_3COCOCH_3$

O_2 + hν (benzene)

yield of the epoxide from anti-sesquinorbornene by the photoepoxidation using an α-diketone, in this case biacetyl, oxygen, and light, carried out in benzene. I had always thought that epoxidation with a peracid must be a concerted reaction. It goes stereospecifically and it could go very well from the hydrogen bonded form of the peracid simply delivering an oxene oxygen in a simple process. However, that would certainly be a concerted reaction, so there's something wrong with the simple idea that the screen of hydrogens over the double bond would stop any concerted attack.

As you will see, we have not used molecular orbital calculations in our work, but our emphasis has been to find out what actually happens in the systems that we're interested in. So here was a reaction, probably concerted, which went very well.

There's another concerted reaction which Leo Paquette has carried out with syn-sesquinorbornene and diimide. The reaction goes very cleanly and gives a nice hydrogenation on the exo face, where the number of interfering hydrogens is less. It

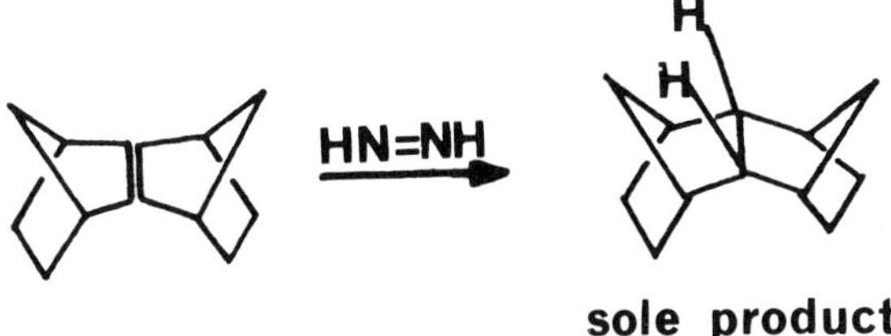

gives a well-characterized single product of hydrogenation, and I don't know anyone who thinks that diimide reacts any other way than concertedly with the carbon-carbon double bond. So we had to face the fact that here a kind of system which seems very hindered to one reactant is perfectly normal toward a number of others. This poses a problem of visualization as to how a concerted reaction can be taking place.

The most unexpected thing we found with _syn_-sesquinorbornene, in collaboration with Bill Watson and his group, was the bend in the double bond whereby the plane of the atoms on one side is at an angle from the plane of the atoms on the other [3]. We have two derivatives of _syn_-sesquinorbornene that are bent at 162-164° while the _anti_-sesquinorbornene system is, within the uncertainty, perfectly planar. This doesn't tell us anything about the stiffness of the system, but it does require that we think of reasons why it should be bent.

We have heard more sophisticated reasons this morning than the one that we have been working on. I think maybe I would tentatively advance the proposition that plain old stereochemistry isn't dead. We have a view, which has proved to be a very successful working hypothesis, which regards the filled π orbitals of the double bond as a unit having a space demand. Of course a great deal of stereochemistry is based on conflicting space demands at different parts of the molecule, and I will present the partial hypothesis to the effect that the filled π orbital on the double bond in the middle of *syn*-sesquinorbornene acts like a bulky group. It is of some interest that the bulk of this group is specifically related to the exact location of the thing on which it is pushing.

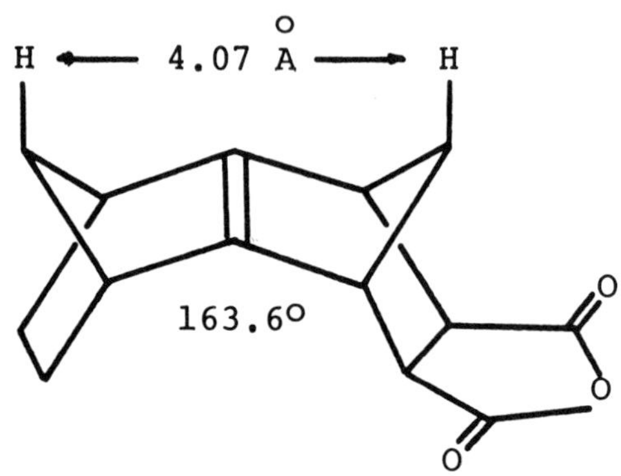

Dihedral angles:

syn-exo-sulfone 162.0°

anti-endo-anhydride 177.9°

Now another unexpected thing came soon after this. The major result from irradiating *syn*-sesquinorbornene in acetone is the addition of an H and an acetonyl radical on the upper or exo face, giving 55% of the product [4]. The second most abundant product is from the addition of two hydrogens on the lower (endo) face where nothing has ever been seen to add thermally, that is, a 25% endo-endo hydrogenation. There is also a trace of exo-exo hydrogenation and of endo acetone addition, but these are very small compared to the others. It's quite clear that in this reaction the addition of hydrogen goes in the endo, the nonpreferred way where the hindrance is known to be very large. The acetone addition goes the other way. We also found that we could initiate this acetone addition in the dark with a free radical initiator such as di-*t*-butyl diperoxyoxalate and get the exo acetone adduct with no hydrogenation and no endo adduct. This adds a little to our view of the mechanism.

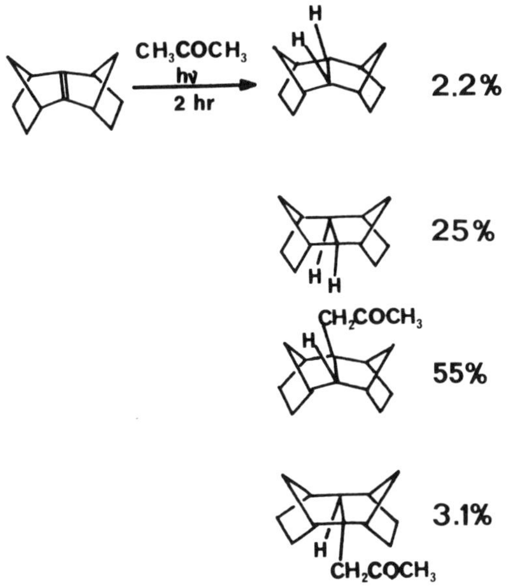

$$\xrightarrow[2\ hr]{CH_3COCH_3,\ h\nu}$$

2.2%

25%

CH_2COCH_3

55%

CH_2COCH_3

3.1%

If it is a thermal reaction, it goes exo like other

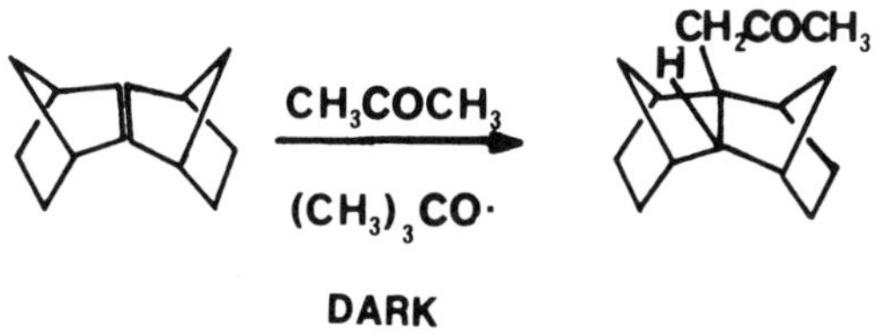

$$\xrightarrow[(CH_3)_3CO\cdot\ \text{DARK}]{CH_3COCH_3}$$

CH_2COCH_3

thermal reactions. Only when the photochemical process is involved does it go endo.

Scheme 1 shows our version of the mechanism which involves as a first step the delivery of excitation energy from the acetone to the _syn_-sesquinorbornene, producing an _excited syn_-sesquinorbornene, evidently a triplet. If you

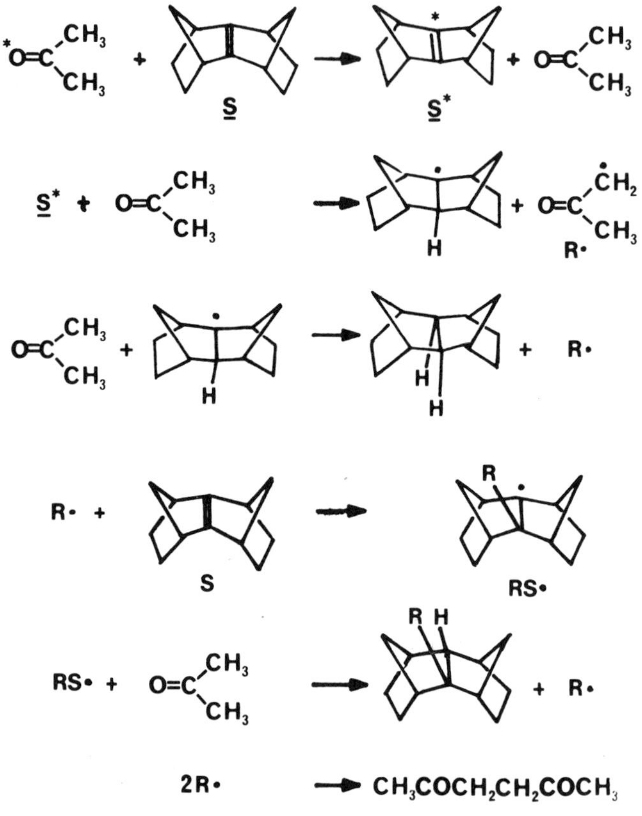

$$\overset{*}{O}{=}C(CH_3)_2 + \underline{S} \longrightarrow \underline{S}^* + O{=}C(CH_3)_2$$

$$\underline{S}^* + O{=}C(CH_3)_2 \longrightarrow \text{(radical, H)} + O{=}C(\dot{C}H_2)CH_3 \; (R\cdot)$$

$$O{=}C(CH_3)_2 + \text{(radical, H)} \longrightarrow \text{(H, H)} + R\cdot$$

$$R\cdot + S \longrightarrow RS\cdot$$

$$RS\cdot + O{=}C(CH_3)_2 \longrightarrow \text{(R, H)} + R\cdot$$

$$2R\cdot \longrightarrow CH_3COCH_2CH_2COCH_3$$

Scheme 1.

simply take the working hypothesis that excited sesquinorbornene has a bend in the opposite direction from that in the ground state, then we can account for everything that happened. In the next step excited *syn*-sesquinorbornene reacts with acetone and adds a hydrogen to what is now the exposed face since it is bent the other way. It adds it to the bottom side to produce the mono-hydrogen free radical, which reacts with another acetone to give the *endo*-dihydrogen adduct. That leaves two acetonyl radicals from the hydrogen capture and they react with the abundant ground state *syn*-sesquinorbornene to initiate the chain reaction which is responsible for the exo-exo additon of acetone to the double bond.

Everybody wants to know what is the reason for the bend in the *syn*-sesquinorbornene. I will expound probably the most simple-minded idea that will be heard at this conference. Essentially, I will treat the π orbital as a kind of space-filling pillow or pair of pillows which produces the anticipated result. Here is a view not from the top but

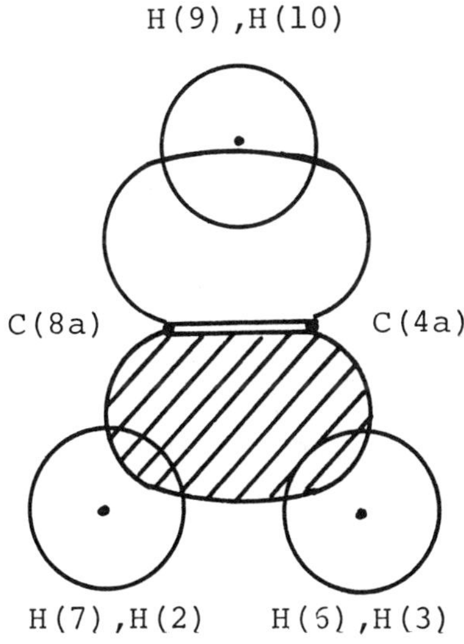

Bonding π orbital

from the side looking through the first norbornene ring across the two ring junction carbons and then through the second norbornene. What we have is the two endo hydrogens of the ethylene bridge (one pair in front of the double bond and one behind it) leaving everything else out, and the endo hydrogens of the methylene bridges similarly.

When this molecule is in the ground state and the π orbital is filled, it occupies continuous space above the two unsaturated carbons. From the fact that the methylene hydrogen is a little bit closer to the double-bond axis than the ethylene hydrogens, we propose that the competition in space between the methylene hydrogen and the filled π orbital is greater than that between the ethylene hydrogens and the filled π orbital. Therefore, in

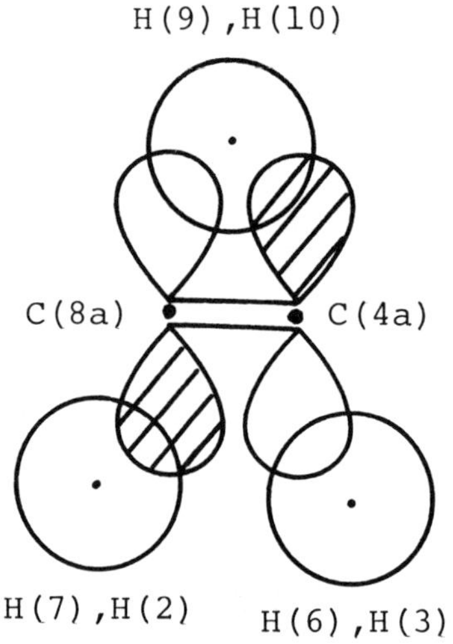

Antibonding π orbital

the ground state, there is more of a push from the space-filling π orbital against the methylene bridge than against the ethylene bridge. The situation reverses in the excited state when an electron of the double bond is promoted to the antibonding orbital with a node down the middle. This node comes where the methylene hydrogen has the greatest amount of space occupancy and ethylene hydrogen the least amount. No matter what the relative magnitudes of the push on these hydrogens, it will be relatively less in the excited state on the methylene hydrogens than it was in the ground state. Hence, whatever the balance position of the molecule with respect to the bend of the double bond in the ground state, it is going to be shifted in the excited state in the direction of the ethylene bridges being pushed farther apart. The methylene hydrogens collapse closer together. This can be seen from the front side looking along the two unsaturated carbons. You don't see the carbons but you are looking from one carbon back toward the

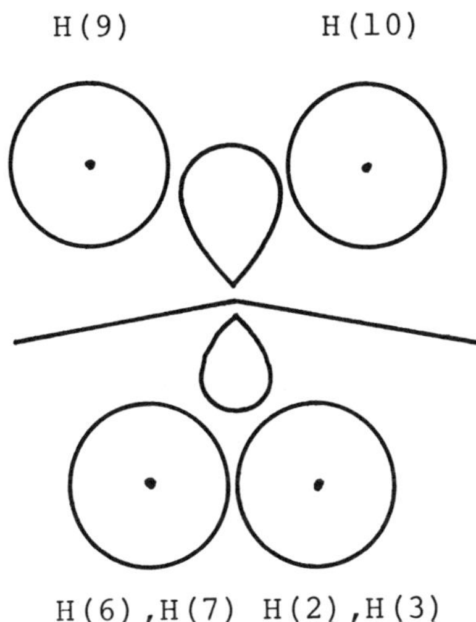

Ground state

next. Here is the proposed orbital redistribution in the ground state with the corresponding bend. The methylene bridge hydrogens are moved farther apart and the ethylene bridges are moved as close together as they comfortably can be. The excited state has a node right where the methylene hydrogens are and they come together more easily than the ethylene Hs that are pushed apart. There is a corresponding bend in the double bond for the excited state.

That is the simple hypothesis as to why we've got (a) a bend that we know we have in the ground state from X-ray; and (b) the bend that we propose to explain the excited-state reactions. Now if

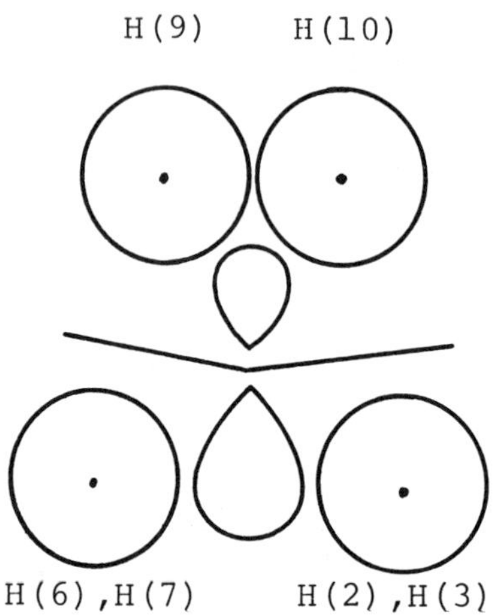

Excited state

this is true and if a relatively small perturbation like that can produce a geometrically quite appreciable effect in the molecule, it follows that anti-sesquinorbornene is in a balance of forces which are identical at top and bottom. The planarity of anti-sesquinorbornene is not so much because of the energy minimum in the planar double bond as because of a balance of distorting forces coming from the competition between the π orbital and the space occupied by the hydrogen. There could be quite a difference in the chemistry of syn and anti and we have done some experiments with direct competition between syn and anti-sesquinorbornenes with respect to their reactivity toward a series of reactants.

In Table 1 we have equal concentrations of syn and anti competing for an amount of the reactant that is only enough for one of them. The first case is the thermally initiated reaction with acetone, a chain reaction. The next one is phenylazide. It is known to be a concerted reactant for the double bond and there was an interesting little series of approximations on that. The first *anti*-sesquinorbornene derivative was formed from maleic anhydride along with *syn*-*exo*-sesquinorbornene anhydride. The *anti*-*endo*-sesquinorbornene anhydride would not react with phenylazide at all and

Table 1. Competitive Reaction Rates with *syn*- and *anti*-sesquinorbornenes.

Reagent	Product	k(syn)/(anti)
Acetone (+ free radical initiator)	Acetone adduct	4.0 - 4.6
$C_6H_5N_3$	Phenyltriazoline	3.4
CO_3H, Cl	Epoxide	2.2
HN=NH	H_2-adduct	1.2
$Br_2(CCl_4$, dark)	Dibromide	<0.05
HCl in ether	HCl adduct	<0.05

since that was the first phenylazide reaction that we tried it gave us a temporary feeling that it proved something about a concerted reaction being screened in this compound. As soon as we had

unsubstituted *syn*- and *anti*-sesquinorbornenes they both reacted beautifully with phenylazide and therefore we couldn't hold to the screening idea.

We also used m-chloroperbenzoic acid and diimide. These first four reactants give reaction rate ratios of syn- to anti-sesquinorbornene that are all less than five. The diimide, which is presumably the smallest of the reactants, gives about an even balance between rate of attack on syn- and anti-sesquinorbornenes. This is not a big enough difference between the syn- and anti-sesquinorbornenes to justify the kind of first-approximation blocking diagram that I showed you or justify extrapolating from that diagram to rates of reaction. What we conclude is that with the bend in syn-sesquinorbornene the hydrogen overhead is of no importance at all as a hindrance. In anti-sesquinorbornene the restoring force to bring it down to planarity is nothing like that in an open unhindered olefin. The molecule could be flapping, with a very low restoring force, back and forth all the time. If it spends half its time flapped down and half its time flapped up, anti-sesquinobornene is going to offer perfectly good access to attacking reagents.

Now when we look at certain other reactants that presumably go through an ionic mechanism, we get a different story. Bromine in carbon tetrachloride, HCl or acetic acid react with a mixture of syn- and anti-sesquinorbornenes in such a way that the anti all reacts and less syn product than we can detect is formed. We say this is at least a ratio of 20:1 in favor of greater reactivity of the anti, which of course is reversed from what we had before. How do we correlate this with the ionic nature of this reaction? To be sure, there is something special about the norbonyl cation, which could be doubled or squared in the sesqui system. However, we went through a bit of that and the answer turned out to be something else, namely, that this is not just a kinetic difference. Actually in the addition of these reagents and also in the addition of methanol or water under acid catalysis to syn- and anti-sesquinorbornenes the reaction with anti-sesquinorbornene proceeds to completion, and that with syn-sesquinorbornene goes to an unfavorable equilibrium. We can prove this by preparing the adducts in a roundabout way. We have the methyl ether from adding methanol to syn-sesquinorbornene. Paquette,

Ohkata and Carr [5] have measured the high rate of solvolysis of the p-nitrobenzoate made from the *exo*-alcohol of *syn*-sesquinorbornene. All such derivatives revert to *syn*-sesquinorbornene under acidic conditions.

Table 2. Ethylene Bridge Interactions

	Syn-exo anhyd.	*Syn-exo* sulfone	*Syn*-ϕN_2 adduct
H(2en)-H(7en)	2.35	2.32	1.76,1.81
	Anti-endo- anhyd.		
H(6en)-H(10en)	3.35		

Table 2 shows that on the underface of an exo adduct to *syn*-sesquinorbornene the hydrogens of the ethylene bridges have come very close together, as close as the van der Waals radii will allow. In the anti adduct we have about an Angstrom to spare. This critical distance can make an important difference in the stability of the adduct to *syn*- and the adduct to *anti*-sesquinorbornene.

We started out doing this as part of a study of singlet oxygen and it is worthwhile to look back. We saw no evidence in our first experiments that singlet oxygen did anything either to *syn*- or to *anti*-sesquinorbornene. One thing that distinguishes the mechanism of singlet oxygen addition is that it is a concerted [2+2] cycloaddition reaction. Unlike the other concerted addition reagents that we had, singlet oxygen is forbidden to add with the active atoms all in the same plane. It's got to have an antarafaciality about it [6]. It can either come on [2s+2a] or it can go through a perepoxide in which again the two oxygens are spread across the space above the double bond, perpendicular to the double-bond axis. Although singlet oxygen is no bigger a reagent than diimide, it appears to be a much more bulky reagent when it comes to adding to a double bond. It can't add unless it spreads out in the way that is most bulky and most interfering between the bridges. That raises a question. Is the rate of reaction of singlet oxygen with sesquinorbornene really zero or is it just very small?

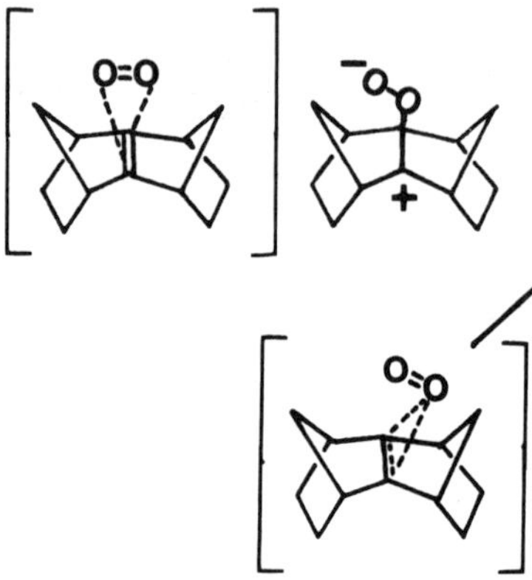

I mentioned that we did get some epoxide when we used vigorous photochemical generation of singlet oxygen. We also found some of the diketone which

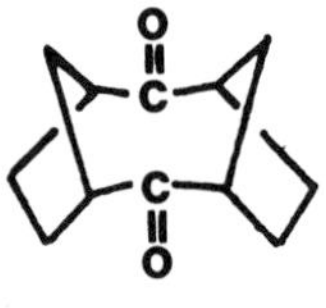

is a product that could come from a short-lived unstable dioxetane, and we were thrown off at first because the dioxetane of biadamantylidene is so exceedingly stable [7]. The dioxetane of sesquinorbornene is so unstable that you never see any of it. Was it there and is that the reason why this diketone was formed? Well, there's a test for this [8]. 9,10-Dibromoanthracene can pick up energy from an excited triplet ketone. When tetramethyldioxetane is decomposed thermally in the presence of 9,10-dibromoanthracene, luminescence occurs which can only be connected with the presence of an excited triplet ketone. When syn-sesquinorbornene reacts with singlet oxygen generated

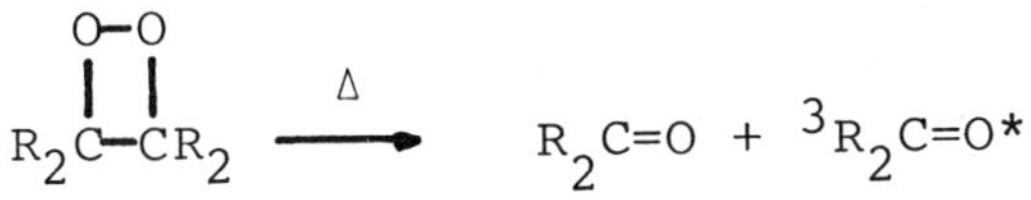

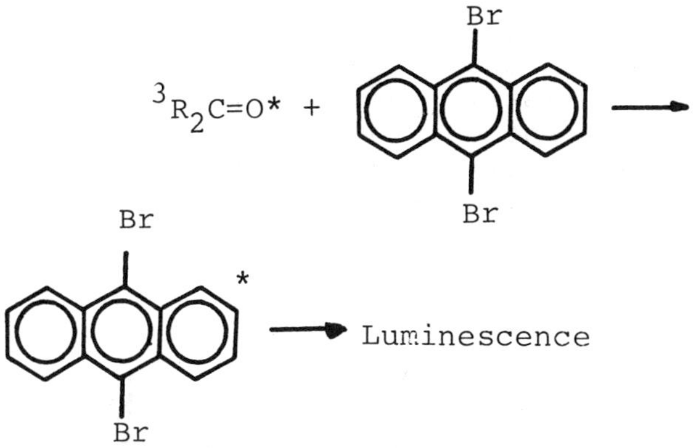

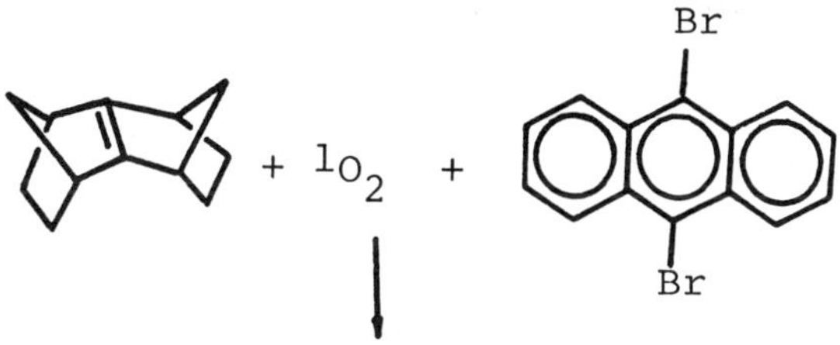

$$R_2C(-O-O-)CR_2 \xrightarrow{\Delta} R_2C{=}O + {}^3R_2C{=}O^*$$

$${}^3R_2C{=}O^* + \text{9,10-dibromoanthracene} \longrightarrow \text{9,10-dibromoanthracene}^* \longrightarrow \text{Luminescence}$$

$$\text{syn-sesquinorbornene} + {}^1O_2 + \text{9,10-dibromoanthracene}$$

Bright blue luminescence

80% recovered olefin

10% 9

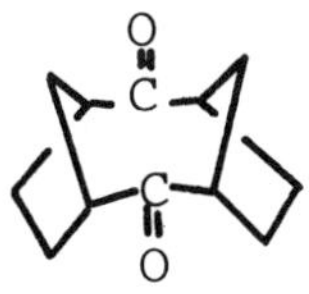

thermally from triphenylphosphite ozonide in the presence of 9,10-dibromoanthracene a bright blue luminescence is produced. One can isolate as much as 10% of the diketone and about 80% of sesquinorbornene is recovered. Not very much of the singlet oxygen survived undeactivated to react with the olefin, but what did react left its signature in the formation of an excited ketone that could be registered by this bright luminescence.

Because of the importance of the isodicyclopentadiene in the preparation and study of sesquinorbornene we became quite interested, as Leo Paquette did, in the unique features of isodicyclopentadiene chemistry. Of the three isomers of isodiclopentadiene only one (1) can be prepared and isolated under normal conditions. This has been known since the early work of Alder and co-workers [9].

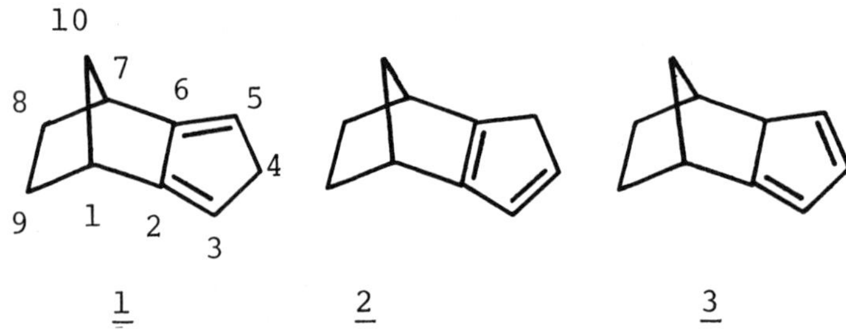

In order to make this diene one starts out with a partly hydrogenated dicyclopentadiene and hydroxylates at the CH_2 group. The subsequent elimination might give isomer 3 first, but in fact it doesn't. If formed at all, 3 shifts to 1 in what must be a very easy reaction. It belongs to the long interesting category of 1,5-sigmatropic shifts in which the shift of hydrogen from the saturated to an adjacent unsaturated carbon is concerted with relocation of the diene system [10]. To get from 3 to 1 requires two such shifts in succession.

This particular part of the project started when we thought we would get some interesting compounds if we could add either vinylene carbonate (4) or dichlorovinylene carbonate (5) to isodicyclopentadiene. These reagents have been known before [11] and they are very weak dienophiles compared

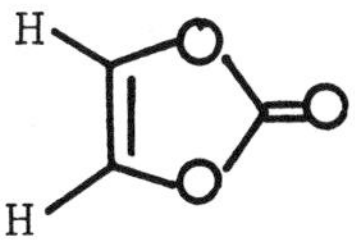

4 5

to maleic anhydride. There is no reaction on heating either of these compounds up to 100° with isodicyclopentadiene. But if we go to 180° with

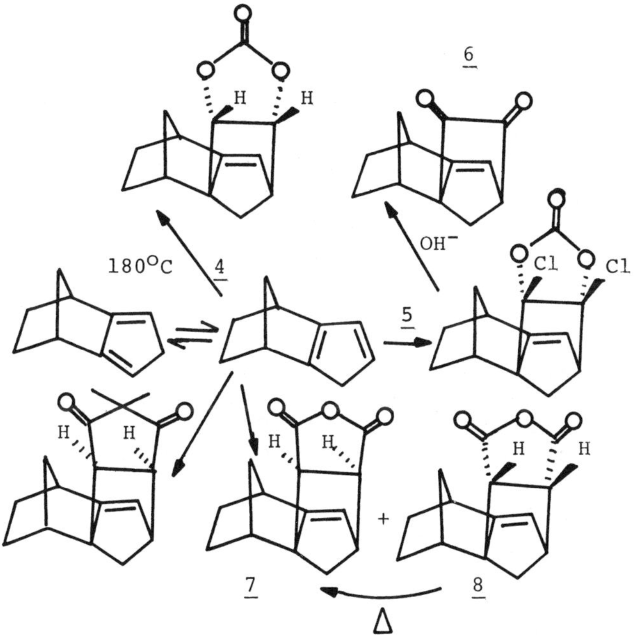

180°C 4 6 OH⁻ 5 Cl Cl H H 7 8 Δ

isodicyclopentadiene we get some products with vinylene carbonate and dichlorovinylene carbonate which look very much as if a thermal equilibrium had been established at that temperature between 1 and 2, and these dienophiles selectively reacted with isomer 2. We never saw any adduct of the vinylene carbonates with isomer 1 of isodicyclopentadiene but only with 2 and in conformity with Alder's rule. The dichlorovinylene carbonate adduct can be converted by strong base into (6) which has a bridge of two carbonyls. Maleic anhydride, which adds very well to ordinary isodicyclopentadiene (1) at lower temperatures, at 180° gives only adducts of isomer 2, with none of the normal adduct at all. Another interesting feature is that whereas vinylene carbonate goes strictly according to Alder's rule, maleic anhydride goes both Alder and anti-Alder and the Alder's rule adduct is unstable relative to the anti-Alder's rule adduct. If the adduct is held at 180° for a while the kinetic (Alder) product goes over into the thermodynamic (anti-Alder) product. One can rationalize that very nicely on the grounds of the nature of the dienophile. The maleic anhydride has a carbonyl group that, as models show, gets in the way of the bridge in the Alder's rule product and not in the anti-Alder's rule product. Although the orbital reasons for Alder's rule existing are still there and play a part in the original reaction, it turns out that that's only a kinetically favored product. The thermodynamically favored product is the one that gets rid of this interference.

There is a dienophile which won't react with skewed structure number 2 at all, and that is dimethyl acetylenedicarboxylate which sticks to its role as a normal endo-face dienophile. At 180 or at any other temperature the DMAD will single out the stable structure 1 of isodicyclopentadiene and add to that on the under face and pay no attention to the rearranged one.

Now if we take that diketone 6, which was made by way of the dichlorovinylene carbonate, and photolyze it at 0° in the presence of maleic anhydride, we get the same mixture of adducts of maleic anhydride to the rearranged diene system, 7 and 8, the Alder and the anti-Alder, that we get at 180° from thermal equilibration. In no other way have we seen any evidence of enough of the rearranged structure to do anything at that low temperature. So we have actually photochemically generated this diene at 0° under conditions where it will survive to react with maleic anhydride. If we don't put the maleic anhydride in there and warm to room temperature, 2 is converted to 1.

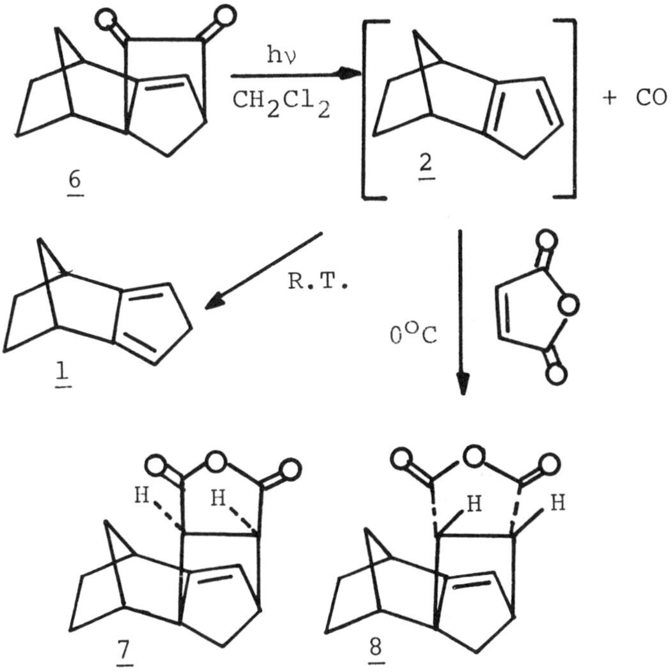

Now you may have noticed that all the additions to this rearranged isodicyclopentadiene are on the exo face. We have never seen any addition on the endo face of 2 and, as I said, with dimethyl acetylenedicarboxylate we don't even get any addition on the rearranged diene at all. Why do dienophiles, many of which like to add on the endo face of the symmetrical isomer 1, insist on adding on the exo face of isomer 2? We have postulated [4] that if there is a node in the highest occupied molecular orbital at the ring junction in sesquinorbornene, it influences the relative interaction of that filled orbital with the single methylene hydrogen or the two ethylene bridge hydrogens in such a direction that the node permits more coexistence between this single central hydrogen and the orbital than it does with the two hydrogens of the ethylene bridge. In the symmetrical structure 1 the HOMO has in fact such a node because it is up-up down-down for the p orbitals constituting these two double bonds. In the rearranged structure where one double bond is shared by the rings, there's no node in either the lowest or the highest occupied moleculr orbital. Therefore, isomer 2 is the one that pushes hard on the one hydrogen and isomer 1 is

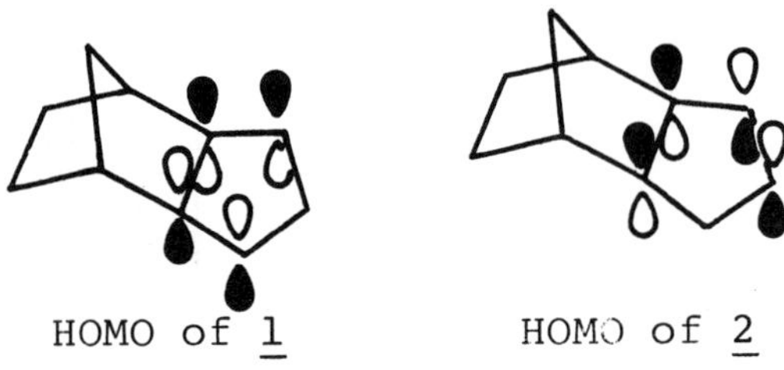

the one that pushes hard on the two hydrogens. That bends isomer 1 up and bends isomer 2 down, which is pretty much what we see in the Diels-Alder reaction.

We got very interested in the possible sigmatropic shifts and Figure 2 is the diagram of what could be happening in a molecule of this sort undergoing 1,5-sigmatropic shifts of hydrogen. As the Paquette group has done, we have labeled isodicyclopentadiene with deuterium and we can tell the deuterium on the α face from the deuterium on the β face and from the deuterium at the identical positions 3 and 5. We generally get about 80% on the endo (α) side of carbon 4 and maybe 17% on the exo (β) side of carbon 4 and about 3% distributed betweeen carbons 3 and 5. We took samples made from isodicyclopentadienyl lithium and methanol-d in tetrahydrofuran at temperatures between -78° and 0°. We followed such samples with time to see how those isotopes redistributed themselves at 100° to the eventual 1:1:1:1 equilibrium. The deuterium at 4α can move over to C-3 or the hydrogen at 4β can move over to C-3, yielding in each case the unstable diene structure 2, deuterated at C-3 or C-4. They must go to the symmetrical structure 1 much more quickly than they form because we never find any 2 at equilibrium. By moving back with a shift of H or D they can give redistribution. One of these return shifts will go to deuterium at the

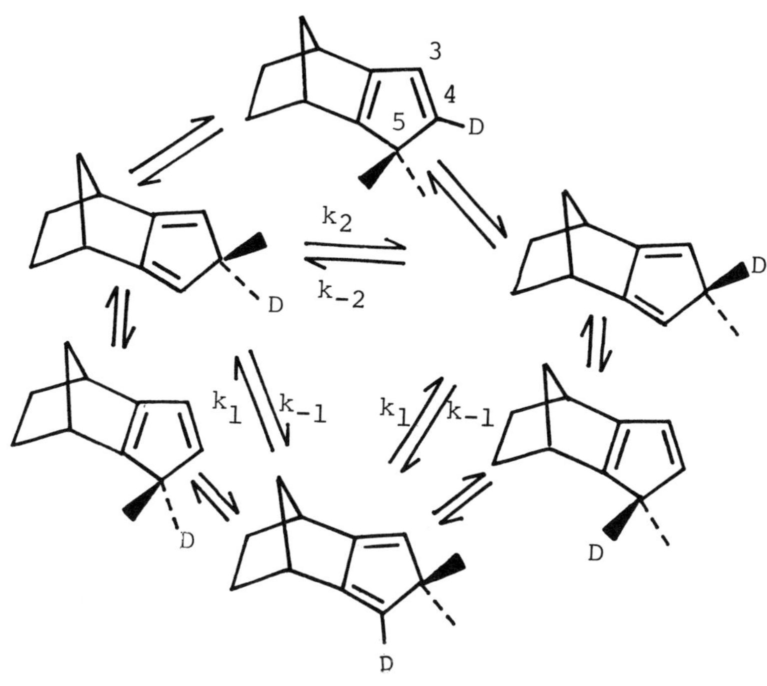

Figure 2. Diagram of possible 1,5-sigmatropic shifts.

3-position of the symmetrical diene structure and the other, by a hydrogen jumping back from C-3 to C-4 on the lower face, gives the isomer in which the deuterium is on the upper face, at 4β. Figure 3 is a plot of the course of the redistribution. The upper curve is the D-4α, with deuterium on the endo face. The next is the D-4β (deuterium on the exo face) and the third is the (D3 + D5). The squares, triangles, and circles represent the experimental determinations with time. The solid lines represent the results of an incremental computer calculation in which the differential equations governing this process were broken down into very small time increments. We took 170 points for the incremental calculation as a practical substitute for

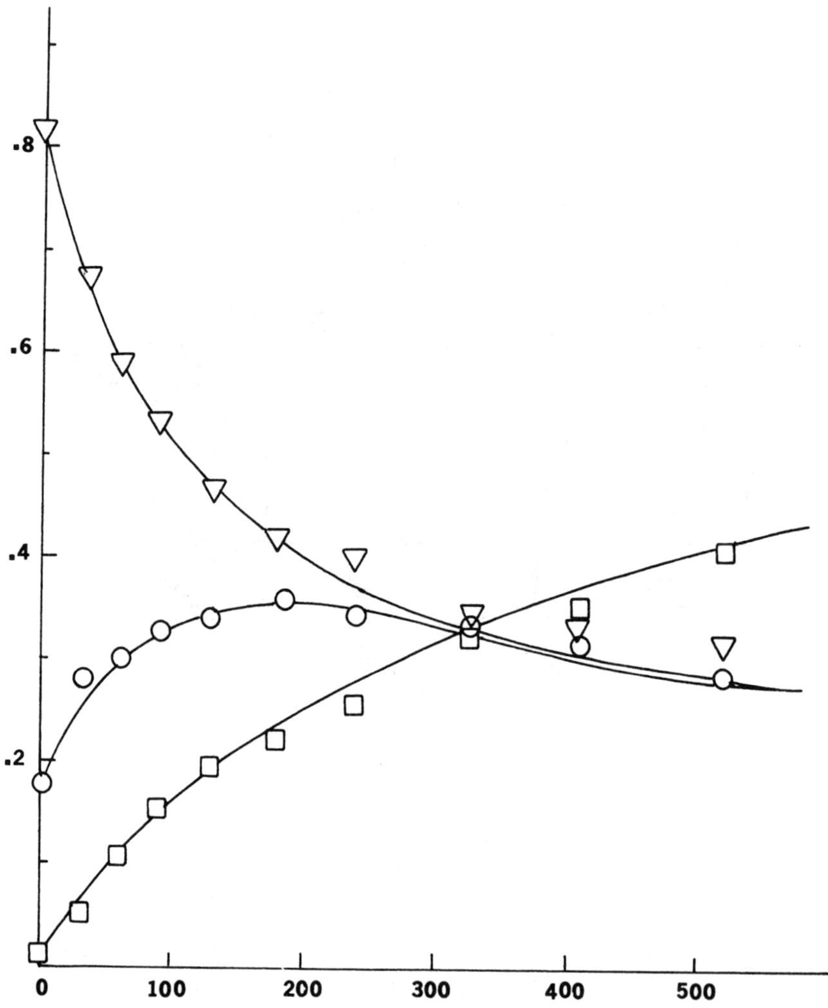

Figure 3. Redistribution of deuterium in 1-d.

the unavailable integrated equation. The results show that by successive approximations we have been able to calculate the course of the reaction with two rate constants. One rate constant is for the shift of deuterium between C-4 and C-3 or C-5 and the other rate constant is for the interconversion of D-4α and D-4β. The second rate constant is greater, as it should be, because D-4α can go to D-4β and back without any deuterium moving. But deuterium cannot go from 4 to 3 or 4 to 5 without deuterium moving. We expect therefore that the interconversion of D-4α and D-4β will be faster than the interconversion of D-3 and D-4 because of an isotope effect. Indeed the factor was 7.5 according to these measurements. We could have told from just the experimental results that the interconversion of D-4α and D-4β was faster than the others, because 4β passes through a maximum.

Since a return from 2 to 1 has to be so much faster than the conversion of 1 to 2 (the equilibrium being immeasurably one sided) the 1 → 2 rearrangement is the trigger reaction in each sigmatropic isomerization between deuterated isomers of 1. The simplest possible kinetic scheme is to assume that all such H shifts from 1 to 2 proceed with a common rate constant k_1, while all the corresponding D shifts have rate constant k_2. The role of the rapid return shifts from the 2 to the 1 system is to partition the sequels to the rate-determining 1 → 2 shift into return to starting material and completion of the isomerization (Figure 2). It would be understandable if there were more diversity than this in the elementary rate consstants (for example, faster shifts on one face of the π-system than on the other), but since the facts we have are compatible with this simplest assumption of two rate constants only, we cannot detect other inequalities. We do anticipate that H or D shifts to C-2 or C-6 will be slower than k_1 and k_2 respectively, since the five-membered ring in 3 is forced to be nonplanar and hence has extra strain.

We did the deuteration of the lithium isodicyclopentadienide under a number of conditions and it was eventually found that if instead of using methanol-d_1 we used D_2O as a deuterating agent, and did it in ether over the temperature range from -78° to 0°, we got increasing amounts of a new isomer that we had never seen before. It turns out to be deuterated 3, the third of those isomeric diene structures. It has deuterium in a very specific place, namely, on the β(exo) face at the ring junction on carbon 6. This is as much as

62% of the deuteration product when we do it at 0° with D_2O in ether as a solvent. This structure can

1) n-BuLi/THF/0°
2) ROD/T°

1 1-d 3-d

ROD	T°C	1-d	3-d
D_2O	0	51	49
	-78	75	22
CH_3OD	0	79	21
	-78	>98	<2
t-BuOD	0	>98	<2
	-78	>98	<2
D_2O/Et_2O	0	38	62

be made with or without deuterium on it. When made with H_2O in ether, with just enough maleic anhydride then added to react with the unconverted or with the returned symmetrical isodicyclopentadiene 1, the remaining diene does not immediately isomerize to the other structure. That diene can actually be separated by low-temperature distillation from the maleic anhydride product and the distillate can be made in turn to react with maleic anhydride to give a product for which we have an X-ray crystal structure (10). This structure shows incidentally that whereas the preferred point of attack of deuterium at C-4 of the anion was on the α(endo) face, the deuterium in this new isomer is on the β(exo) face of C-6.

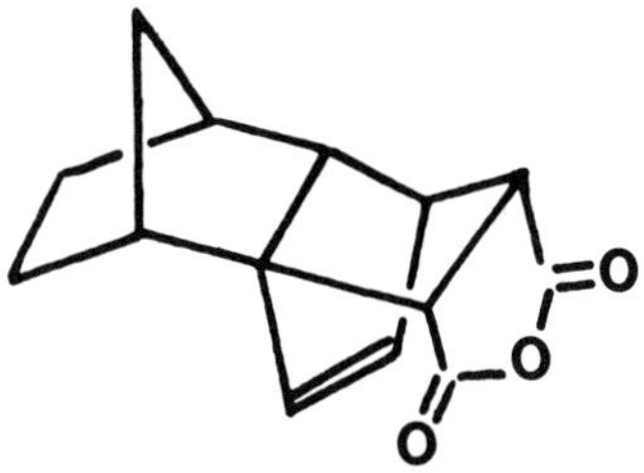

So by one way or another we have obtained all three isomers of isodicyclopentadiene. They turn out to be quite different from one another in their properties. In the equilibrium 1↔2↔3, the amounts of 2 and 3 are immeasurably small. However, whereas the reactivities in sigmatropic isomerization fall in the order 2>3>1, the reactivities in the Diels-Alder reaction with maleic anhydride are in the order 2>1>3, while toward dimethyl acetylenedicarboxylate the stablest isomer 1 remains the most reactive.

Acknowledgments. Support for the research reviewed here was provided by The Robert A. Welch Foundation and the National Science Foundation.

Discussion

Adam: You report that photo-hydrogenation with acetone proceeds mainly from the endo side, postulating a triplet being involved. Have you explored this type of photoreactivity further, by trying to carry out photocycloadditions?

Bartlett: We have not found anything else that will do that yet.

Adam: What does the sesquinorbornene do by itself on irradiation? I guess at 254 nm it does not absorb, but how about its photo-reactivity at 185 nm?

Bartlett: It certainly does not dimerize. There is not room for a bond across there. If it doesn't react with something reasonably mobile from outside, it just goes back to the ground state.

Paquette: I want to pick up on the last point you made. What are your thoughts regarding the stability of isomer 3? Why do you feel that it is so stable? That is a remarkable observation.

Bartlett: These things of course are made from the anion. I think it is very consistent with the other simple orbital ideas that the hydrogen comes on the underface first.

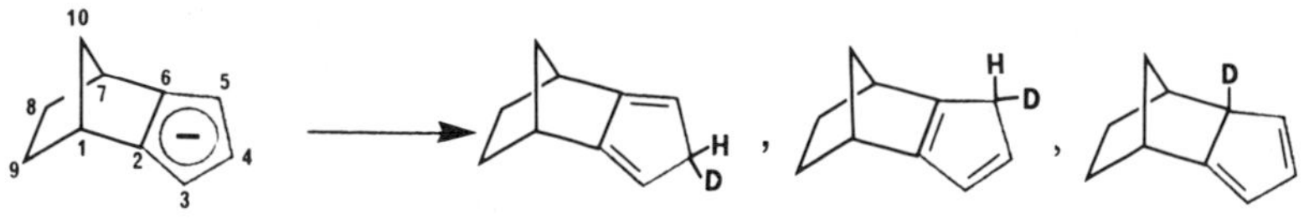

First of all it should come on C-4 because when it comes there it produces what is know to be the most stable diene system (1). If it had to come on C-3 it would produce system 2 which we know is so unstable relative to 1 that it goes right over to it. Likewise if it were coming at C-6 it would produce 3. There is a built-in preference for hydrogen attacking at position C-4. Now I haven't really checked the dimensions, but if you consider that this hydrogen on C-10 is stretched over the middle of the diene ring, a hydrogen or deuterium entering at this point (C-4) can either come in β (exo) and be eclipsed with H(10), or it can come in (endo) and be staggered between the two hydrogens of the ethylene bridge. That is at least one way of rationalizing why it prefers the α face. It does not prefer it all that much, but by a factor of 4-5. There is about 1.5 kcal greater activation energy for formation of the deuterated 3 than of D-4α, based on product compositions over a range of temperatures. So this fits with the idea that the energy barrier is lower at 4α. It also fits with the idea that both faces are available and small differences in the circumstances can shift the product composition. Now suppose the deuteron is going to come on at C-6; it's now even closer to the bridge hydrogens than it was at C-4. If it comes on at this point on the α face it is going to be directly eclipsed by H(8) endo. If it comes on the β face it is staggered with the single cis hydrogen (H-10) of the methylene bridge. The same general line of thought would suggest that it is reasonable that α attack is associated with approach at C-4 and β attack with approach at C-6.

Foote: What is the isomer distribution in the 1,2-dimethylcyclopentadienes? I have forgotten. Does anyone remember? There is a predominant one and I think it is the one in which the methyl surrounds the double bond.

Unknown: Yes, it is the 1,2 isomer.

Greene: Was this 60% figure that you mentioned the isolated yield after you removed the other isomers?

Bartlett: The relative amounts of D-4α and D-6β could be determined by NMR immediately after the reaction. The 62:38 refers to the immediate product of the reaction. When we isolate the compounds we do not get all 52%, but we get a lot more than we thought we would.

Houk: Does the change in preferred site of protonation with different solvents and protonating sources suggest the site of protonation as well as stereochemistry could be determined by ion pairing?

Bartlett: Yes. We have not made any observations that tell us anything about ion pairing, and we don't even know for sure whether the proton comes on the face where the lithium is or the opposite face. I would think it very likely that it would come on the face where the lithium is located since lithium ion is a very good thing to take hold of the oxygen of the deuterating agent. I suspect that is what happens. The sensitivity to solvent and to the change between methanol as deuterating agent and D_2O as a deuterating agent is unexpected, but it tells us that there are some quite critical phenomena here.

Greene: Is the structure of the dibromo adduct from the bromination of the *anti*-sesquinorbornene known?

Bartlett : We think that all the adducts to these double bonds are cis because it would put quite a strain in the sesquinorbornene ring system to put them on trans. We are having second thoughts about that as we have run into an unknown structure that may possibly turn out to be trans. I would say that this is an unanswered question.

Greene: You mentioned that syn was slow. Does it nevertheless react with bromine? I think that in the competition experiments you indicated that in bromination of the sesquinorbornenes the anti was faster than the syn by at least 20 to 1. Did you react pure syn with bromine?

Bartlett: In the competition experiments we do not get any adduct of syn; however, you can make the syn adduct, but it doesn't last very long.

References

[1] P.D. Bartlett, A.J. Blakeney, M. Kimura and W. H. Watson, J. Am. Chem. Soc. 102, 1383 (1980).
[2] L.A. Paquette, R.V.C. Carr, M.C. Bohm, and R. Gleiter, J. Am. Chem. Soc. 102, 1186 (1980).
[3] W.H. Watson, J. Galloy, P.D. Bartlett and A.A. M. Roof, J. Am. Chem. Soc. 103, 1186 (1981).
[4] P.D. Bartlett, A.A.M. Roof and W.J. Winter, J. Am. Chem. Soc. 103, 6520 (1981).
[5] L.A. Paquette, K. Ohkata and R.V.C. Carr, J. Am. Chem. Soc. 102, 3303 (1980).
[6] R.B. Woodward and R. Hoffmann, "The Conservation of Orbital Symmetry", Verlag Chemie, Weinheim (1970).
[7] J.H. Wieringa, J. Strating, H. Wynberg and W. Adam, Tetrahedron Lett. 169 (1972).
[8] T. Wilson, D.E. Golan, M.S. Harris and A.L. Baumstark, J. Am. Chem. Soc.98, 1086 (1976).
[9] K. Alder, F.H. Flock and P. Janssen, Chem. Ber. 89, 2689 (1956).
[10] Ref. 6, pp. 114-140.
[11] C.H. Depuy and E.F. Zaweski, J. Am. Chem. Soc. 81, 4920 (1959); C.H. Depuy and C.E. Lyons, J. Am. Chem. Soc. 82, 631 (1960).

4. REGIO- AND STEREOSELECTIVITY IN DIELS-ALDER REACTIONS. THEORETICAL CONSIDERATIONS

Rolf Gleiter and Michael C. Böhm

Institute für Organische Chemie der Universität
D-6900 Heidelberg (West Germany)

A large number of mechanistic investigations on Diels-Alder reactions [1] indicate that the new bonds are formed at the same time, ie, that we are dealing with a one-step process [2]. This is indicated in Scheme 1 using the cycloaddition reaction between cyclopentadiene (1) and cis- and trans-1,2-dichloroethylene (2 and 3) as an example [3]:

Scheme 1

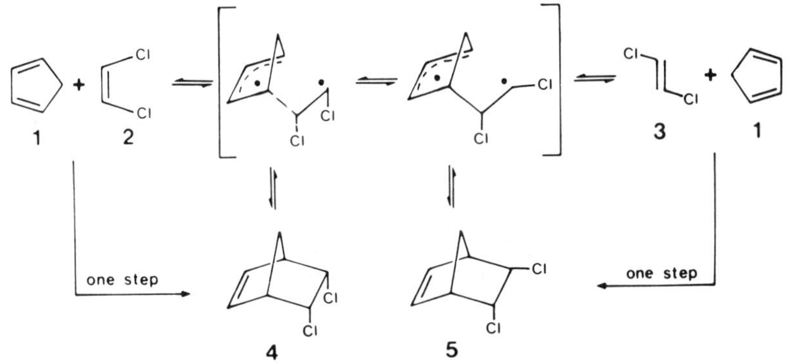

(1)

In the case of the reaction shown in Scheme 1 one can only detect (within the limits of error (< 0.5%)) those products which are expected in the case of a one-step reaction. This observation can be explained by assuming synchronous bond formation. In agreement with this explanation are theoretical investigations by Woodward and Hoffmann

[4] using a molecular orbital (MO) model. Correlating the MOs of the adducts (*cis*-butadiene and ethylene) and the product (cyclohexene) of a model Diels-Alder reaction they concluded that a concerted process should be thermally favored. The same answer is provided if we consider the frontier orbitals [5] of the diene and the dienophile as is done in Figure 1.

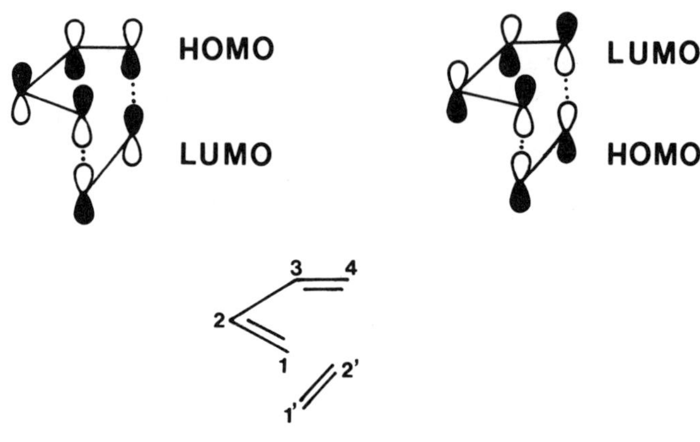

Figure 1. HOMO of *s-cis*-butadiene and LUMO of ethylene (left) and HOMO of ethylene and LUMO of *s-cis*-butadiene (right).

In this figure we show the highest occupied molecular orbitals (HOMO) and the lowest unoccupied molecular orbitals (LUMO) of butadiene and ethylene. We find for both frontier orbital pairs (HOMO(diene)-LUMO(dienophile) and HOMO(dienophile)-LUMO(diene)) a bonding interaction, ie, an *in phase relation* between the atomic orbital (AO) amplitudes at those centers where the new bonds are formed.

This *in-phase relation* just described is responsible for the low activation energy of a synchronous process. We will call this effect (*in-phase relation*) a *first-order* orbital interaction. This effect determines the *stereospecificity* of a concerted cycloaddition reaction.

Regio- and Stereoselectivity in Diels-Alder Reactions

A still growing number of investigations in this field of cycloaddition reactions reveal a preferred formation of one regioisomer [2,4,5]. In analogy to the first-order orbital interactions we will call those electronic effects which determine the regio- and stereoselectivity second-order orbital interactions. In nonpolar cycloaddition reactions we can discriminate between secondary orbital effects, substituent effects, and polar group effects. Before we discuss these effects in a more general frame we will present examples for each of the three categories.

a. Secondary Orbital Effects

The classic example for the preferred formation of the thermodynamically less stable endo product is the reaction between cyclopentadiene (1) and maleic anhydride (6) [6].

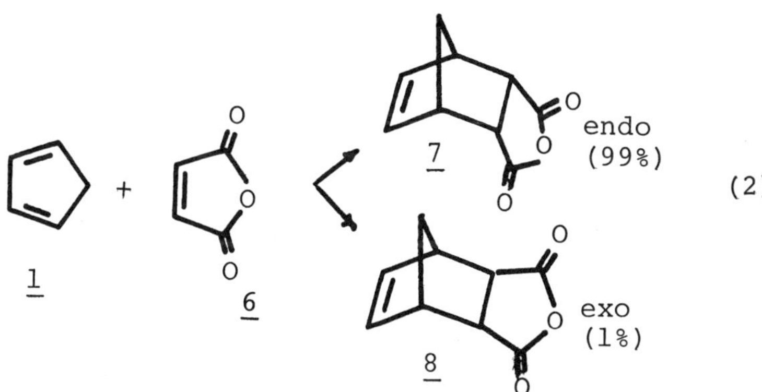

(2)

b. Substituent Effects

1. Formation of Ortho and Para Products

Dienes substituted in the 1 position yield with electron-deficient dienophiles preferably the ortho product [2]. The reaction of 2-substituted butadienes with dienophiles yields preferably the para-substitution product (15) as indicated in eq. (4).

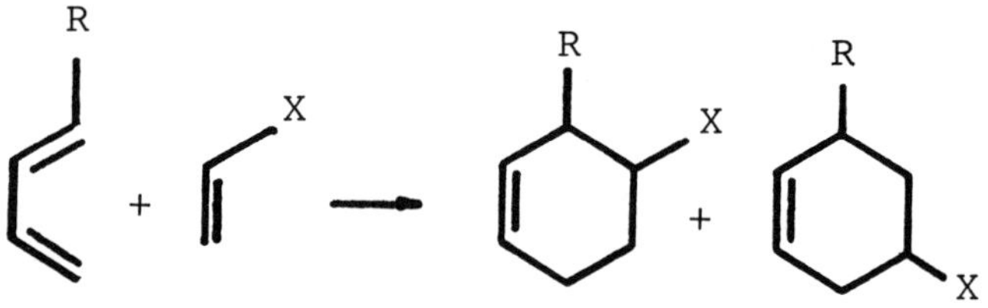

R X R X R + + (3) X

9 10 11 12

R = OCH_3, CH_3, C_6H_5, COOR

X = CN, COOR

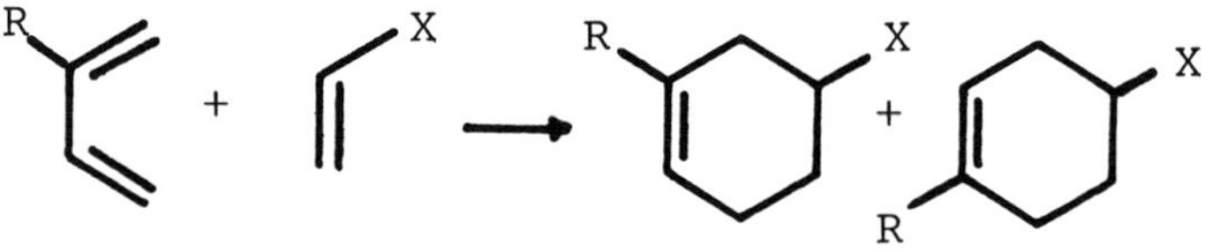

R X R X X + + (4) R

13 10 14 15

R = OCH_3, CH_3, C_6H_5, COOR

X = CN, COOR

2. Syn-Anti Addition on Mono- or Disubstituted Cyclic Dienes.

In eqs. (5) and (6) we summarize two cycloaddition reactions in which the dienophile approaches either from the sterically more hindered side, eq. (5) [7], or in which there seems not much steric hindrance at all but nevertheless the dienophile attacks exclusively from one side [8].

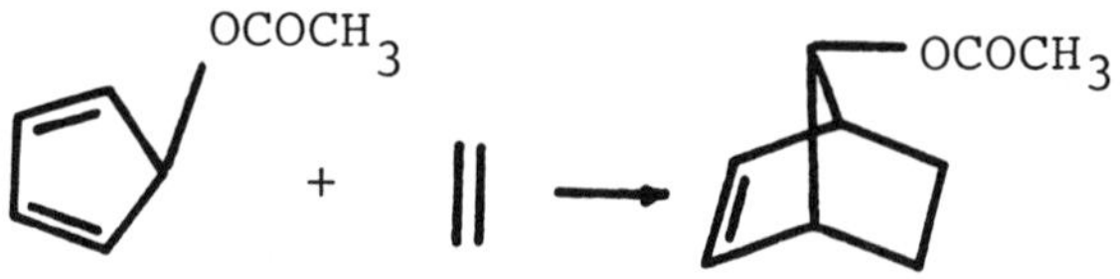

$OCOCH_3$ $OCOCH_3$

+ (5)

16 17 18

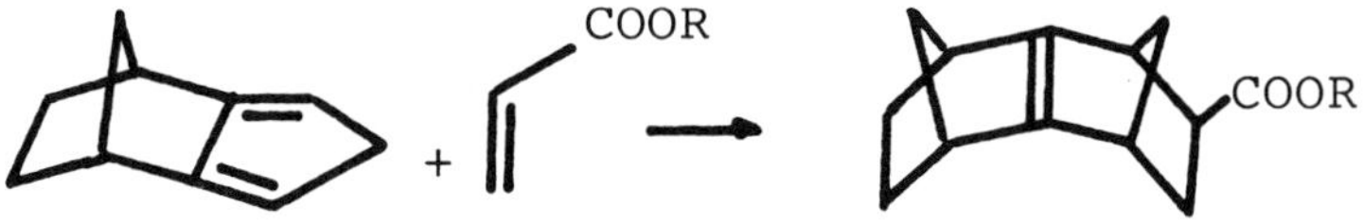

19 + 20 → 21 (6)

c. Cycloaddition Reactions Involving Polar Groups

An example for this type of reaction is given in eq. (7). It is shown that the syn product 24 is formed preferably [9], although the thioether corresponding to 22 yields mainly the anti product.

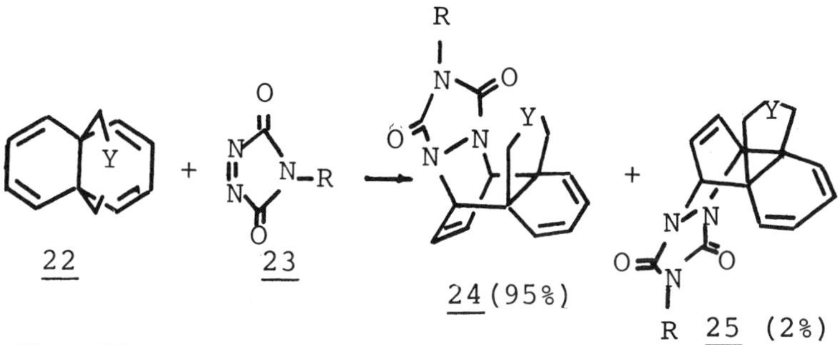

22 + 23 → 24 (95%) + 25 (2%) (7)

$Y = SO_2$

Perturbation Theory

In order to explain the stereospecificity and the regioselectivity of cycloaddition reactions it is necessary to reformulate and rearrange some basic expressions encountered in frontier orbital theory [5,10].

We divide the two reacting molecules into different regions: the Active Centers (AC), the Active Frame (AF), and the molecular fragments that are not involved in the reaction, ie, the Inactive Frame (IF). The Active Centers are those centers between which the new bonds are formed. The Active Frame consists of those atoms that are involved in the π-σ reorganization during cycloaddition but which are not Active Centers. To clarify this classification some examples are listed.

In contrast to most frontier orbital treatments which consider π orbitals only we consider also the σ orbitals at the Active Centers. This may lead to a π-σ interaction in the frontier orbital pairs. This π-σ interaction can be the result of interactions in the transition state (time dependent) or may originate in the topology of the Inactive Frame (time independent). Another role of the Inactive Frame is its capability of polarizing the π system.

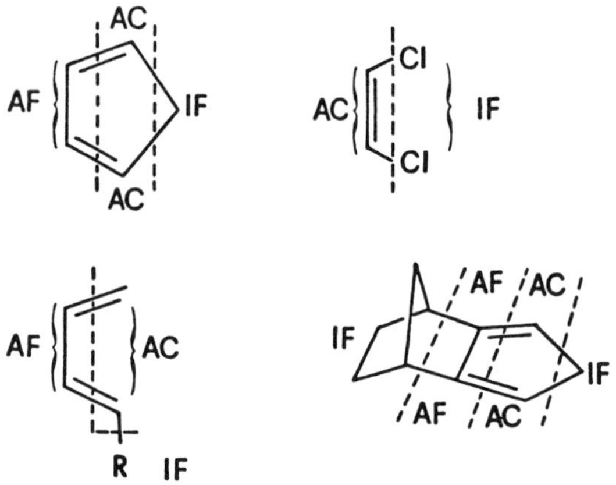

The total interaction energy of two reacting π systems can be divided on the basis of the aforementioned classification scheme. Let us consider reaction (2). In this Diels-Alder cycloaddition reaction we call the coupling between the π orbitals at the Active Centers (1 and 4 of the diene and 1' and 2' at the anhydride) as "first-order interaction". All other interactions between the two components that are not considered first-order interactions we call "second-order interactions". As we will see this category can be subdivided according to the contributions from AF and IF. Using second-order perturbation theory we can write an expression for the total energy lost or gained (ΔE_{tot}) when orbitals of one fragment interact with those of another:

$$\Delta E_{tot} = \Delta E_{(1)} + \Delta E_{(2)} + \Delta E_{(p)} + E_{(\pi/\sigma)} \qquad (8)$$

In this summation $\Delta E_{(1)}$ is expressed in terms of an electrostatic term, an expression of the frontier orbital interactions between occupied and unoccupied MOs and by a destabilizing component due to the interaction between occupied orbitals. This latter part is very often neglected in perturbational treatments.

$$\Delta E_1 = \underset{A,B \in \{AC\}}{\sum_A \sum_B} \frac{q_A q_B}{r_{AB}} + 2\underset{\substack{\mu\nu \in \{\pi\} \\ A,B \in \{AC\}}}{\left(\sum_i^{occ} \sum_j^{unocc} + \sum_i^{occ} \sum_j^{unocc}\right)} \frac{(C_{\mu i} C_{\nu i} S_{\mu\nu})^2}{\varepsilon_j - \varepsilon_i} + 4 \underset{\substack{\mu\nu \in \{\pi\} \\ A,B \in \{AC\}}}{\sum_i^{occ} \sum_j^{occ}} \frac{\left[(\varepsilon_{ij} - H_{ij}) C_{\mu i} C_{\nu j} S_{\mu\nu}\right]^2}{1 - S_{\mu\nu}^2} \qquad (9)$$

In eq. (9) the following abbreviations have been used:

q_A, q_B = net charges at centers A, B

r_{AB} = distance between centers A and B

$C_{\mu i}$ = AO coefficient of the μ'th AO at centers A and B

$S_{\mu\nu}$ = overlap integral

ε_i = orbital energy

ε_{ij} = $(\varepsilon_i + \varepsilon_j)^{1/2}$

H_{ij} = $k \cdot S_{ij}$ [11]

k = -39.7 eV

S_{ij} = group overlap integral

In the case of nonpolar cycloaddition reactions, ie, most Diels-Alder reactions, the first term can be neglected and thus we end up with the first order interaction between the Active Centers of the two fragments.

A similar expression can be written if we include the centers which do not belong to the AC. For the simple case involving two AOs μ and ν the following formula results.

$$\Delta E_2^{\mu\nu} = \sum_{A}\sum_{B} \frac{q_A q_B}{r_{AB}} + 2\left(\sum_{i}^{occ}\sum_{j}^{unocc} + \sum_{i'}^{occ}\sum_{j'}^{unocc}\right) \frac{(c_{\mu i} c_{\nu i} S_{\mu\nu})^2}{\varepsilon_j - \varepsilon_i} + 4\sum_{i}^{occ}\sum_{j}^{occ} \frac{\left[(\varepsilon_{ij} - H_{ij})\, c_{\mu i}\, c_{\nu j}\, S_{\mu\nu}\right]^2}{1 - S_{\mu\nu}^2} \quad 10)$$

$A, B \notin \{AC\}$ $\qquad$ $\mu\nu \in \{\pi\},\ A,B \notin \{AC\}$

$A \in \{AC\},\ B \notin \{AC\}$ $\qquad$ $\mu \in \{\pi\}, \nu \notin \{\pi\},\ A \in \{AC\},\ B \notin \{AC\}$

$\mu,\nu \notin \{\pi\},\ A, B \in \{AC\}$

ΔE_2 is given by

$$\Delta E_2 = \sum_{\mu}\sum_{\nu} \Delta E_2^{\mu\nu} \qquad \mu\ \nu \notin AC \tag{10a}$$

In this expression the first term contains electrostatic interactions between Active Centers from one reactant and Non Active Centers from the other and also electrostatic interactions between Non Active Centers. The second term sums up the contributions of AO interactions at the Active Frame between occupied and unoccupied MOs and the third term contains the same interaction between occupied MOs only.

The expression for $\Delta E_{(p)}$ for the polarization effects of electron-donating or electron-withdrawing groups is similar to eq. (10). Instead of the unperturbed LCAO coefficients we have to write

$$\Delta c_{\mu i} = c'_{\mu i} - c_{\mu i} \tag{11}$$

Rules to calculate these coefficients have been given by Libit and Hoffmann [12]. ΔE_p is a twofold summation where always one unperturbed LCAO coefficient ($C_{\mu i}$, $C_{\nu i}$) must be replaced by the

$C'_{\mu i}$th ($\Delta C_{\mu i}\ C_{\nu j}\ C_{\mu i}\ \Delta C_{\nu j}$; $\Delta C_{\mu i}\ \Delta C_{\nu j}$ elements are neglected). The term $\Delta E_{\pi,\sigma}$ is given by a perturbational summation where also the σ orbitals of the ACs are taken into account. The structure of $\Delta E_{(\pi,\sigma)}$ is completely equivalent to the orbital summation of eq. (10).

If we assume that the net charges at the Active Centers of both fragments are small, as in most Diels-Alder reactions, the electrostatic term in ΔE_1 can be set to zero and we can use the simple classification given in Figure 2.

FIRST ORDER ORBITAL INTERACTIONS

in phase relation between AO's of **Active Centers** in an unperturbed frame

SECOND ORDER ORBITAL INTERACTIONS

Secondary Orbital Effects	Substituent Effects	Polar Group Effects
in phase relation between AO's of **Active Frames**	a) polarization of π-systems b) σ/π mixing at **Active Centers**	interaction between **Active Centers** and **Non-Active Centers** and between **Non-Active Centers**

Figure 2. Classification of nonpolar Diels-Alder reactions according to eqs. (8)-(11).

In the following we will discuss all three cases illustrated by examples. For historical reasons we will start with "Secondary Orbital Interactions".

a. Secondary Orbital Interactions

According to Alder and Stein [13] diene and dienophile add to each other in such a way that an accumulation of double bonds is achieved. The most well-known example is the reaction between cyclopentadiene 1 and maleic anhydride 6 shown in eq. (2).

Further examples to be mentioned here are cycloaddition reactions of cyclic dienes with p-benzoquinone [14,15], furan with maleimide [16], and fulvene with maleic anhydride [17].

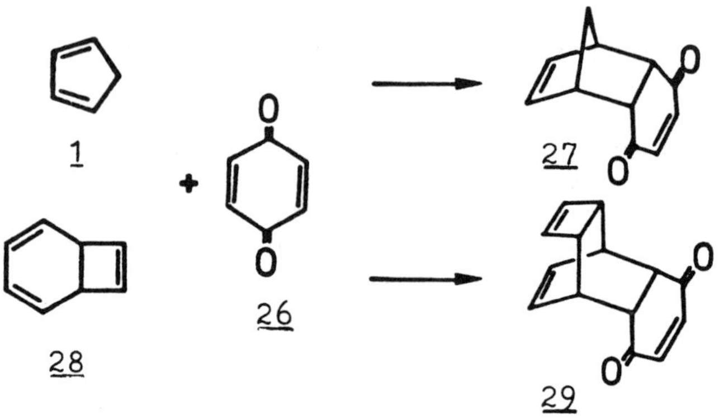

(12)

(13)

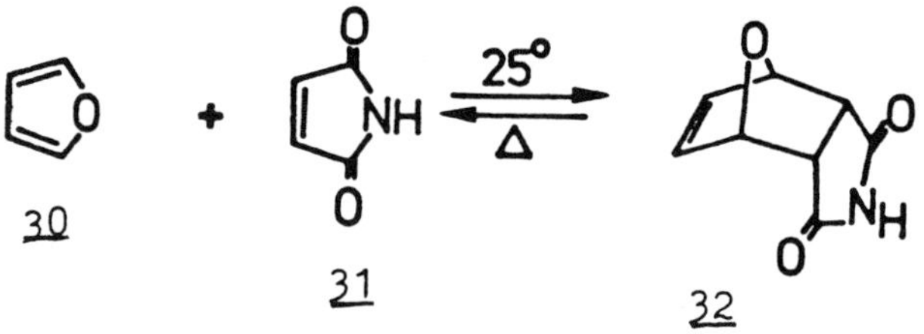

(14)

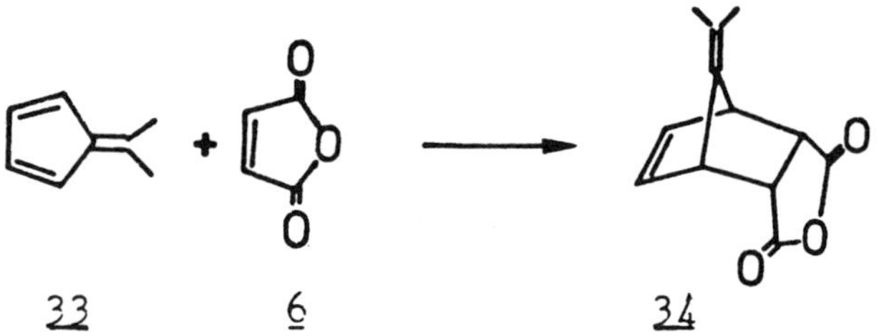

(15)

Although the observed ratio endo:exo amounts to 100:1, it should be noted that the energy difference in the corresponding endo or exo transition states is calculated to be 3 kcal/mol.

An MO-based rationalization of the Alder-Stein rule has been suggested by Woodward and Hoffmann [4]. If we consider again the important frontier orbitals of a diene and a dienophile as done in Figure 3, we find in addition to the first-order orbital interaction (in-phase relation between the centers 1 and 4 of the diene and the centers 1' and 2' of the dienophile) a further in-phase relation between the Active Frames of 1 and 6 in the endo transition state, ie, the AO at the centers 2 and 3 of the diene and the centers 3' and 5' of the maleic anhydride (see Figure 3).

Figure 3. HOMO of cyclopentadiene and LUMO of maleic anhydride.

Other examples which demonstrate nicely the importance of secondary orbital effects for the stereoselectivity are reactions of propellanes investigated by D. Ginsburg and co-workers [9,18,19].

X = O, S
35

+

23 (R = C_6H_5)

36

(16)

X = O, NH
37

+

23 (R=C_6H_5)

38

(17)

Reaction between the propellanes of the type 35 with triazolinedione 23 yields the anti product 36. This can be rationalized by invoking the steric effects of the bridge. Replacing the CH_2 groups by CO to yield 37 one observes only the syn products 38. Although the steric effect of the bridging unit is reduced in 37 compared to that in 35 it is not clear why there should be exclusively syn attack in 37 since the overall geometry is not that different between 35 and 37.

A simple explanation for the observed change is offered by secondary orbital interactions: The transition state for syn attack of 23 is stabilized by interactions between the n_- combination of the lone-pairs at the NN-bridge and the π^* orbital of the CO-X-CO bridge of 37 as shown below in 39.

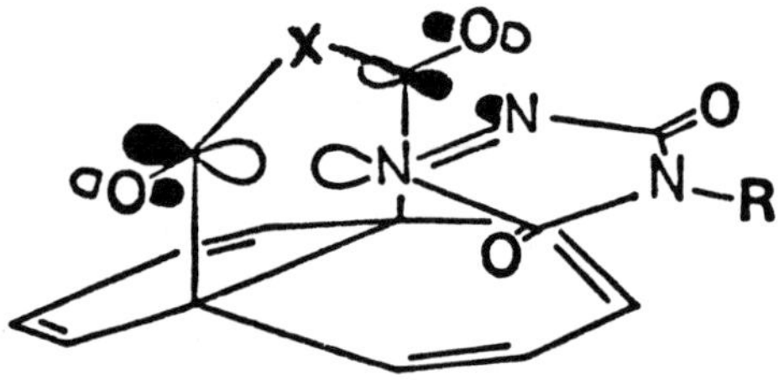

39

To pursue this idea further we have investigated a part of the potential surface for the reaction of 40 with TAD using the EH method [36].

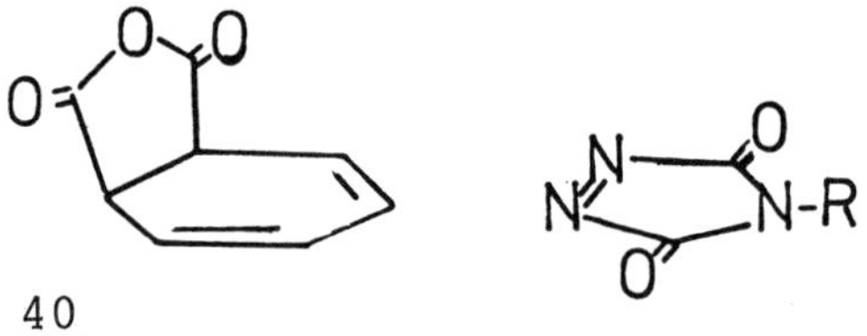

(18)

40 23

In Figure 4 we show an energy difference map for the syn approach of 23 to the butadiene plane (along the +z axis) and the anti approach (along the -z axis). The y component of the O-a vector is given along the abscissa.

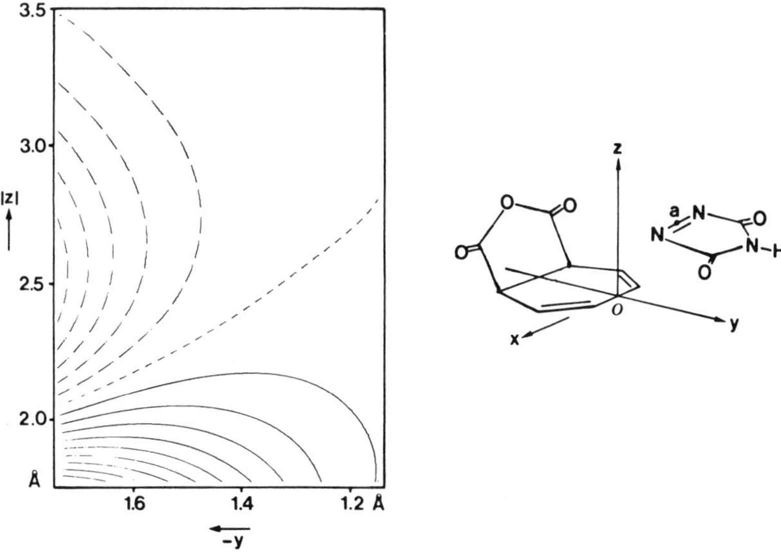

Figure 4. Contour diagram of the EH potential surface for the addition of 40 and 23. The contours are drawn every 4 kcal/mol and represent the difference in energy between addition syn (+z) and anti (-z) relative to the anhydride group. The full lines correspond to situations where syn addition is energetically favored; the broken lines indicate anti addition.

The full lines in Figure 4 indicate that syn approach is more favored than anti. The broken lines indicate the reverse. The result of the calculation suggests overall attraction in the region between y = -1.1 Å and z = +1.75 Å to 2.0 Å.

In principle, there are two effects which must be considered: (a) the first-order orbital interaction between the HOMO of the butadiene unit with π^* of 23 and (b) a secondary orbital interaction between the π system of the anhydride bridge and the n-orbitals of the azo group of 23.

The first effect is present whether 23 approaches from the syn or anti side. The second effect is only present for the syn approach. In Figure 5 the interaction between the lone pairs of the azo group and the π^* orbitals of the dicarbonyl system is shown. A strong interaction between n_- and π_A^* is encountered. This stabilizes the transition state for the syn approach.

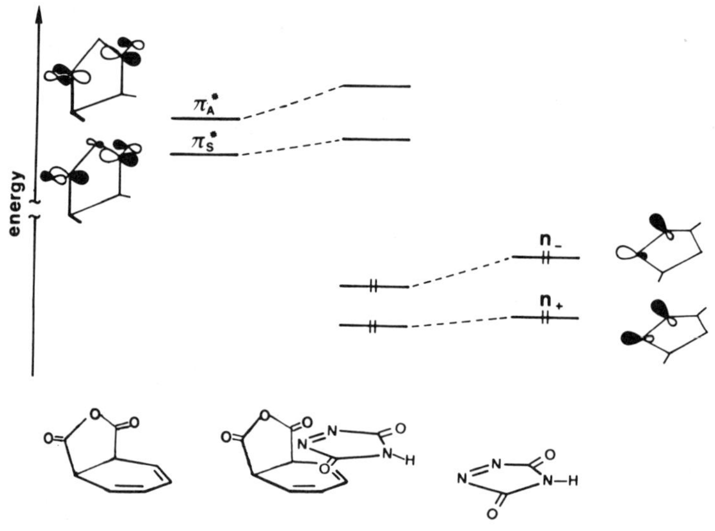

Figure 5. Qualitative interaction diagram for the syn approach of 40 and 23. Only the interaction between the lone pairs on the nitrogen atoms of 23 and the π^* carbonyl orbitals is shown.

In the case of an ethylene bridge in the propellane (eg, 41), the interaction between the π^* orbital of the ethylene and the n_- combination is minute. The dominant interaction is that between the π orbital of ethylene and the n_+ linear combination of the lone pairs of the azo-group. Since both of the latter MOs are occupied (see Figure 6), the net result is a destabilization of the transition state for the syn approach of 23 to 41. These predictions are in agreement with experimental findings: reaction between the [4.4.2]propellane (41) and PTAD yields the anti product 42.

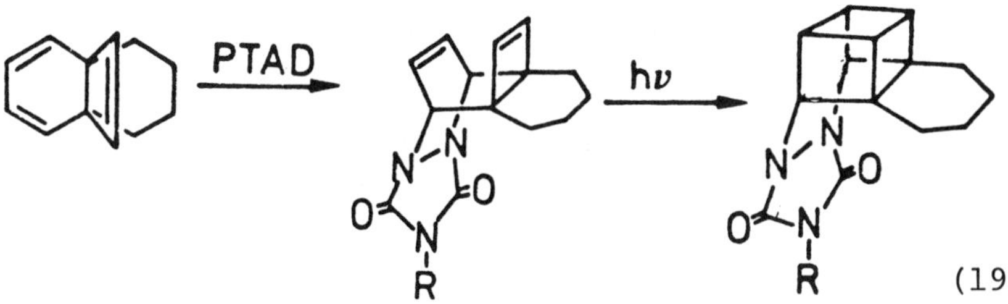

PTAD $h\nu$ (19)

41 42 43

The configuration of 42 is that shown, because irradiation of 42 leads to 43.

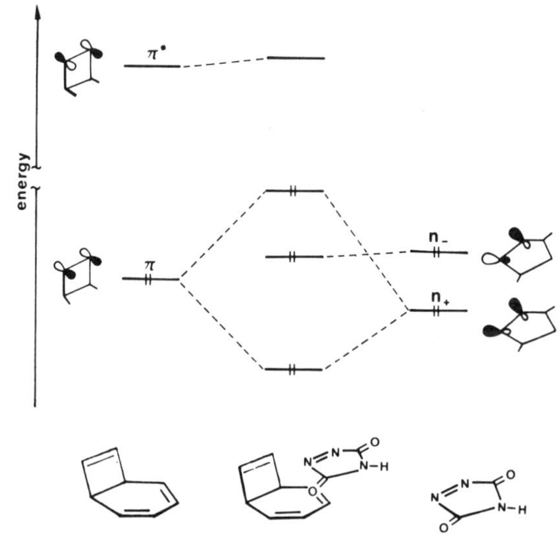

energy π^* π n_- n_+

Figure 6. Qualitative interaction diagram between the ethylene part of 41 and 23.

b. Substituent Effects

The substituent effects can be divided roughly into two groups:

1. Substituents which cause a polarization of the frontier orbitals of π character in such a way that the size of the amplitudes are changed, but the $2p_\pi$ lobes stay parallel

2. Substituents which cause a strong π-σ mixing, leading to rotation of the p_π lobes

Polarization Effects

To demonstrate the first effect let us go back to the Diels-Alder reactions described by eqs. (3) and (4). In these reactions which have been discussed extensively by Houk [20] and others [21] the energy difference between the HOMO(diene) and the LUMO(dienophile) is substantially smaller than that between the LUMO(diene) and the HOMO(dienophile). Therefore we can simplify our discussion by only considering the HOMO of the diene and the LUMO of the dienophile.

In Figure 7 we have shown the frontier orbitals of some dienes substituted in the 1 and 2 position and monosubstituted dienophiles [20]. With the simplifications just discussed eq. (10) simplifies to:

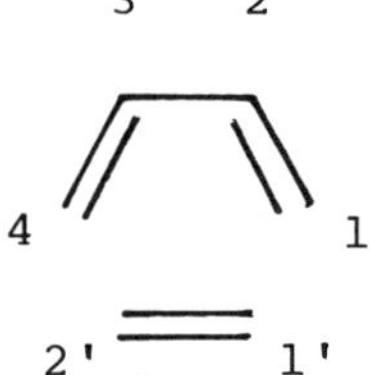

$$\Delta E \simeq \frac{\{\Delta c_1(\mathrm{HOMO}) \cdot c_1'(\mathrm{LUMO})\, S_{11'}\}^2 + \{\Delta c_4(\mathrm{HOMO}) \cdot c_2'(\mathrm{LUMO})\, S_{2'4}\}^2}{\varepsilon(\mathrm{HOMO}) - \varepsilon(\mathrm{LUMO})} \qquad (20)$$

$$+ \frac{\{\Delta c_1'(\mathrm{LUMO}) \cdot c_1(\mathrm{HOMO})\, S_{11'}\} + \{\Delta c_2'(\mathrm{LUMO}) \cdot c_4(\mathrm{HOMO})\, S_{2'4}\}^2}{\varepsilon(\mathrm{HOMO}) - \varepsilon(\mathrm{LUMO})}$$

ΔC_i is defined in eq. (11). By simple algebra one can show that the interaction between two large and two small AO coefficients (configuration A in Figure 7) is more efficient than between two times a large and small one (configuration B in Figure 7). From this it follows that the ortho arrangement in the case of a substituent at position 1 of the diene and the para product in case of a 2-substituent should result.

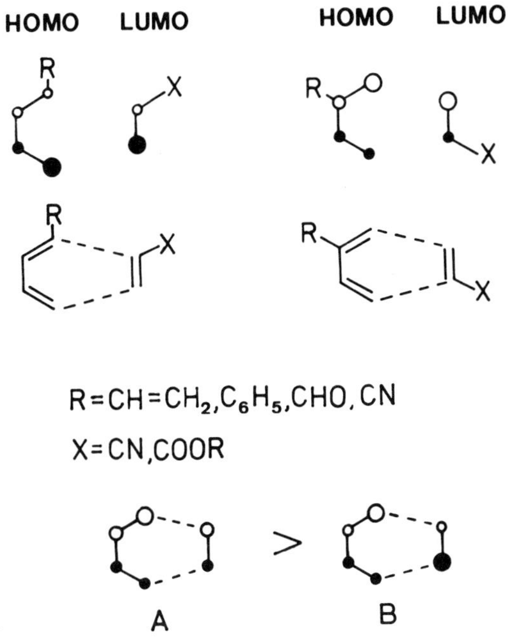

Figure 7. Highest occupied MOs of substituted dienes and lowest unoccupied MO of alkenes (top). Two arrangements (A,B) for the HOMO(diene) and LUMO(dienophile) are shown (bottom).

c. σ-π Interaction

Two examples of the influence of σ-π interaction are given in eqs. (6) and (7). Further examples are listed below [8b]. In order to explain the preference of the dienophile to attack anti to the CH_2 group we can ignore steric effects since the newly formed bonds are separated from the bridges by three σ bonds. We also can exclude any polar interactions since we are dealing with simple hydrocarbons with small dipole moments. Important for the understanding of the observed stereoselectivity shown in eqs. (21) and (22) are π-σ interactions. To demonstrate this let us consider

COOR

19 + 44 ⟶ 45 COOR (21)

46 + 26 ⟶ 47 (22)

the interactions that dominate in the boat conformation of cyclohexene summarized in Figure 8.

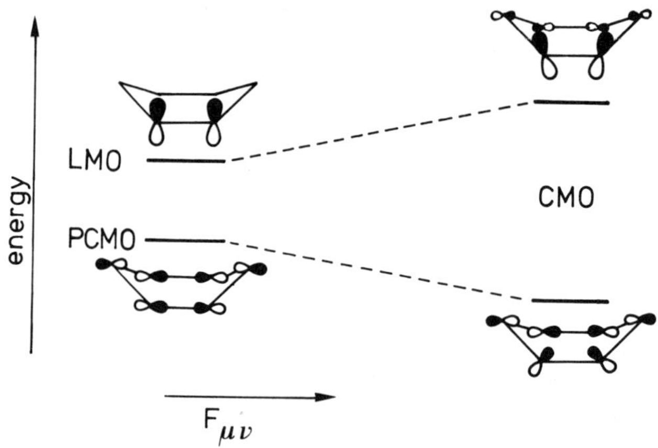

Figure 8. Mixing between the localized π orbital and the precanonical σ orbital in cyclohexene (tub conformation) with increasing interaction between both.

An analysis (transformation of the canonical MOs into the localized and precanonical MOs [22]) shows that the π orbital in cyclohexene interacts mainly with one ribbon-type σ orbital. The resulting changes in the π wave function depends to a large extent on two factors:

1. The relative position of the basis orbital energies of the interacting orbital wave functions

2. The size of the matrix element $F_{\mu\nu}$

In Figure 8 the basis energy of the localized (LMO) π orbital is situated above the basis energy of the precanonical (PCMO) σ orbital.

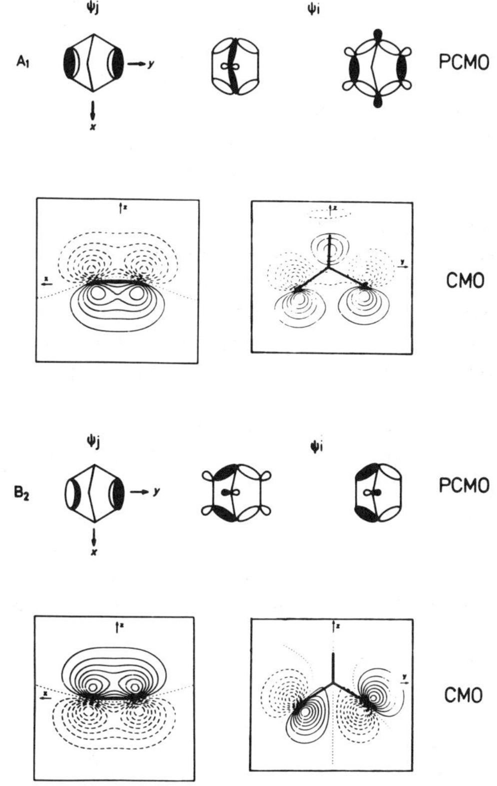

Figure 9. Contour diagrams of the canonical π MOs (CMOs) a_1 and b_2 of norbornadiene. The CMOs are due to an interaction between the localized π orbitals Ψ_j and the precanonical σ orbitals Ψ_i. Full and dashed lines distinguish between amplitudes of different signs. Nodes are shown in a plane through the atoms 2 and 3 (left) and in the yz plane (right).

With increasing $F_{\mu\nu}$ the interaction between both wave functions is increased and thus a disrotatory deformation of the π lobes follows. A further example for a strong σ-π interaction is found in case of norbornadiene [8b]. In Figure 9 we show the PCMOs which strongly interact with the symmetric $a_1(\pi_+)$ and antisymmetric $b_2(\pi_-)$ linear combination of the pure localized π orbitals. As a result of the π-σ interaction a rotation of the π amplitudes in the canonical MOs is obtained. The $a_1(\pi_+)$ linear combination is rotated with respect to the xz plane in a disrotatory manner so that the amplitude of the wave function on the side of the methylene group is decreased. In the case of the $b_2(\pi_-)$ level, the disrotatory motion increases the shape of the CMO syn to the methylene bridge.

Superimposed on the rotation in the xz plane is a second type of π deformation with respect to the xy plane. Importantly, the $a_1(\pi_+)$ linear combination is found to be rotated toward the methylene group. In case of $b_2(\pi_-)$ this rotation moves the two π orbitals away from the methylene group.

The different rotations just described are due to an admixture of the semilocalized p_π orbitals and the whole σ frame. This effect must be separated carefully from a nonequivalence due to hybridization [23], ie, a mixing between s and p orbitals. A detailed analysis of the corresponding CMOs of norbornadiene shows that sp mixing is unimportant and dominated by the p/p interaction of the molecular fragments.

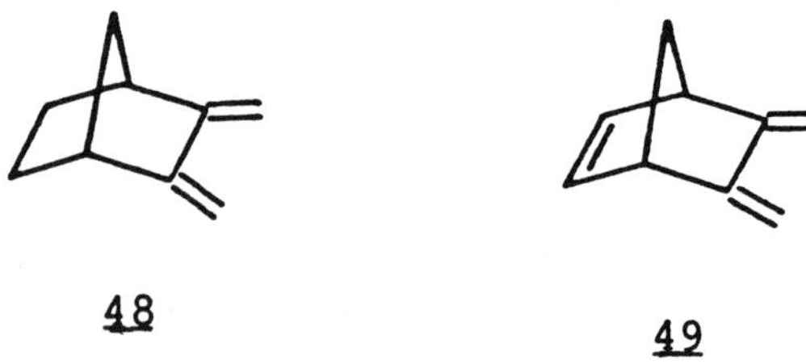

To rationalize the stereoselectivity observed in the Diels-Alder reactions of 19 and 46 we have carried out extensive calculations on 48 and 49 as simpler model systems. Semiempirical methods (ETH,

MINDO/3, SPINDO, a recently developed modified INDO) and ab initio calculations within the STO-3G basis set were employed. All methods applied to 48 and 49 predict a strong mixing between the lowest occupied π orbital (π_S) and high lying σ orbitals of proper symmetry. Due to the fact that STO-3G, SPINDO, INDO and EHT predict no symmetrical σ orbital above the lowest occupied π orbital the wave function predicted by these methods is rather different from the π MO predicted by MINDO/3. Both are shown schematically in Figure 10. The rotation of the terminal p_π lobes for π_S of 48 is shown in the contour diagram of Figure 10b.

It is seen that the rotation leads to significant differences in the electron distribution of the syn and anti side. The various theoretical models applied predict about 20-40% σ admixtures at the carbon centers of the diene moiety.

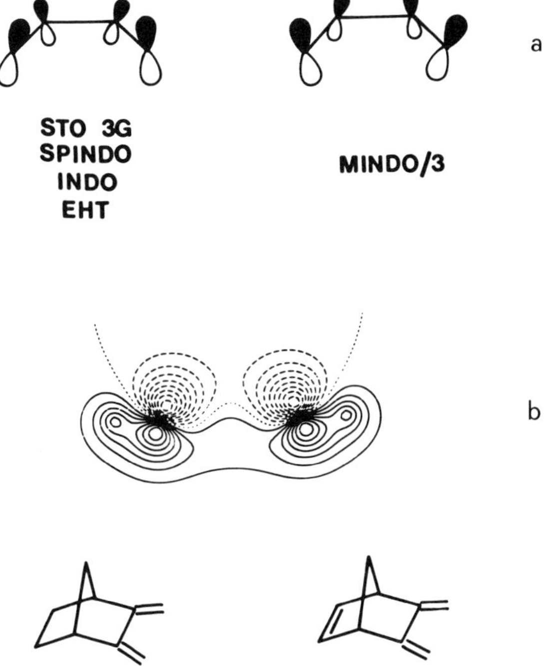

Figure 10. (a) Schematic representation of the π_S orbital of 48 and 49 as obtained with different theoretical methods. (b) Contour diagram for 49 showing the deformation of the two terminal π lobes.

The remarkable rotation of the lobes within the CMO π_S is a result of a strong interaction between the semilocalized π_S orbital and the precanonical σ orbitals of the same symmetry as shown in Figure 11. Consequently the precise sequence

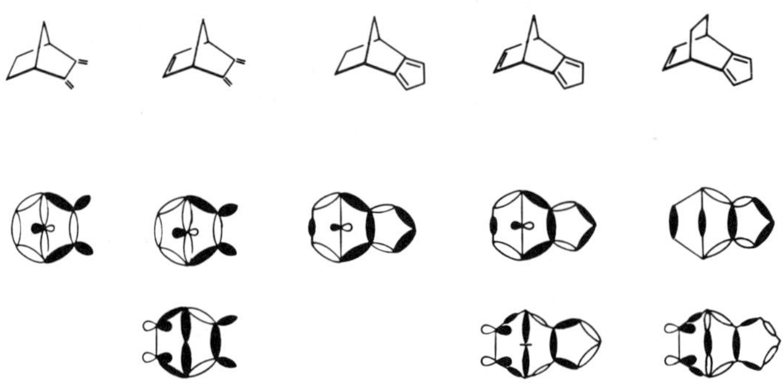

Figure 11. Schematic representation of the most important precanonical σ orbitals for the orbital mixing with π_S in 19, 46, 48, 49, and 50.

(π_S above or below σ_S) is particularly crucial for our argument. Indeed, the MINDO/3 calculations predict a rotation opposite that of the other theoretical procedures.

To judge the reliability of our model calculations on 48 and 49 we have recorded their PE spectra [8b]. Using Koopmans' theorem the experiment demonstrates that both π orbitals lie above the π orbitals. A comparison between the PE data and the computational results on 48 and 49 indicates that the INDO and SPINDO methods provide MO models capable of predicting the distortion of the orbitals due to the σ-π interaction. An analysis of the CMOs of 19 and 46 shows, that similar to 48 and 49, there is a negligible π-σ interaction in π_A (HOMO). The π_S orbital, however, interacts significantly with the σ ribbon.

As in 48 and 49 the π-σ interaction encountered in π_S leads to an enhancement of the amplitudes syn to the methano group. To understand the preferred addition of dienophile anti to the methano bridge let us consider Figure 12. If the dienophile adds from below, the antibonding interaction between π_S of the butadiene moiety and the HOMO of the dienophile is smaller than in the case of syn attack. This is due to the different overlap between the rotated $2p_\pi$ orbitals at the terminal carbon atoms of the butadiene fragment and the dienophile. To estimate the energy difference for exo and endo

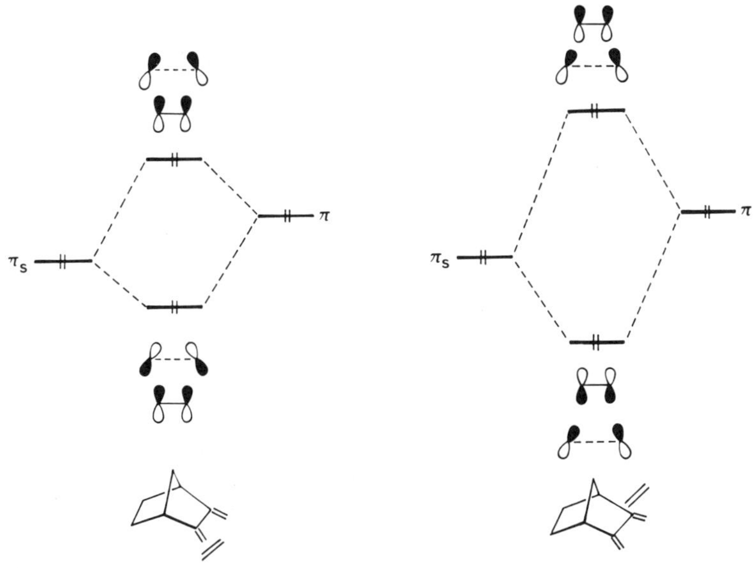

Figure 12. Qualitative diagram of the interaction between π_S of the butadiene unit of 48 and 49 with a π bond. On the left the situation of the approach of the ethylene anti to the methylene group, on the right the corresponding syn approach.

attack in the case of the dienes 19 and 44 we have calculated the four-electron destabilization energy

$$\Delta E_{ij} = \frac{4\,(\varepsilon_{ij}S_{ij} - H_{ij}S_{ij})}{1 - S_{ij}^2} \qquad (23)$$

[24] ΔE_{ij}(exo) and ΔE_{ij}(endo) between the canonical MOs π_s of the dienes in 19 and 44 and π of ethylene. The abbreviations used in eq. (23) are defined in eq. (9).

Assuming a mean distance of 2.18 Å for the transition state between butadiene and ethylene we obtain for 19 and 44 energy differences ($\Delta\Delta E_{ij}$) for the endo and exo addition in the order of 3 to 4.5 kcal/mol (see Table 1).

Table 1. Four-Electron Destabilization Energies ΔE_{ij} for 19 and 44

	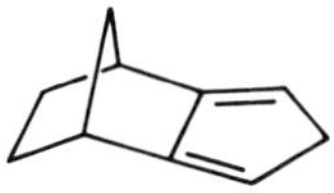 19	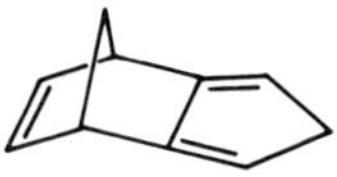 44
ΔE_{ij}^{exo} [kcal/mol]	9.9	7.0
ΔE_{ij}^{endo} [kcal/mol]	5.5	4.0
$\Delta\Delta E_{ij}$ [kcal/mol]	4.5	3.0

The results listed in Table 1 clearly show that in the case of 19 and 44 exo attack is favored. The relatively large energy differences explain why only the endo product is observed. In the case of 50, however, $\Delta\Delta E_{ij}$ is significantly reduced (-0.84 kcal/mol) and a slight preference for endo attack is predicted. This small value is in line with the observation of a mixture of isomers (51:52 = 21:79) when 50 is reacted with methyl propiolate as shown in eq. (24).

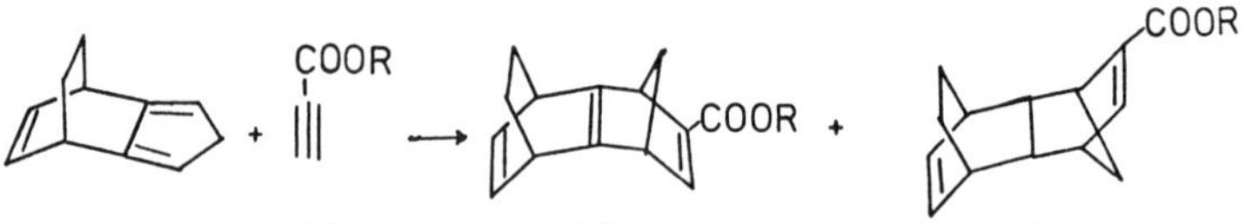

50 + 44 (COOR) ⟶ 51 (COOR) + 52 (COOR) (24)

Another reaction which must be mentioned here is the bromination of 1,6-methano[10]annulene (53). By means of NMR spectroscopy it has been shown that bromine adds syn to the methano bridge [25]. The intermediate (54) was trapped at low temperatures with PTAD [26] and the stereochemistry was established by an X-ray analysis of the product.

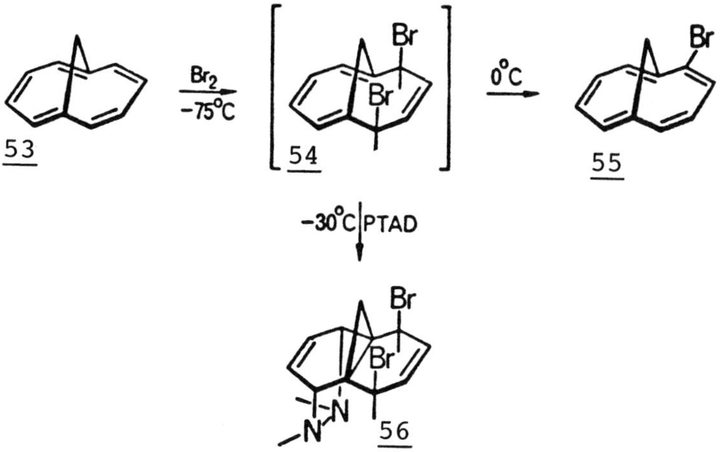

53 54 55 56 (25)

A similar course of bromine addition has been encountered with 1,6-oxido[10]annulene (57) and 2,7-methanoaza[10]annulene (60) as shown in eqs. (26) and (27) [26].

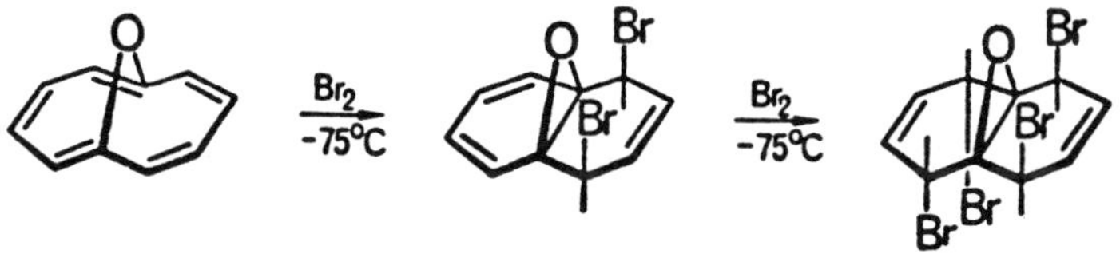

57 58 59 (26)

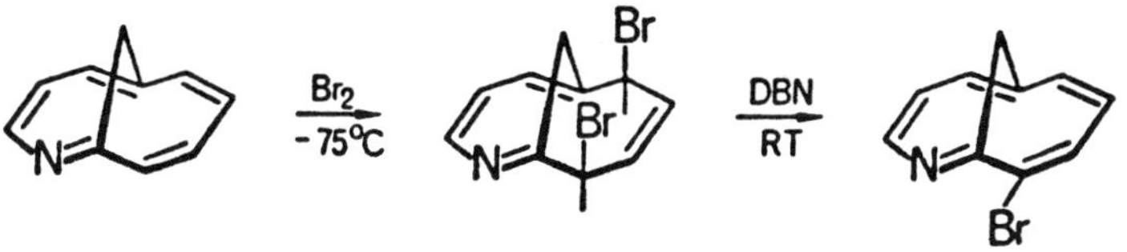

60 61 62 (27)

The explanation of this observation is straightforward if we consider the shape of the wave functions of the highest occupied MOs (see Figure 13).

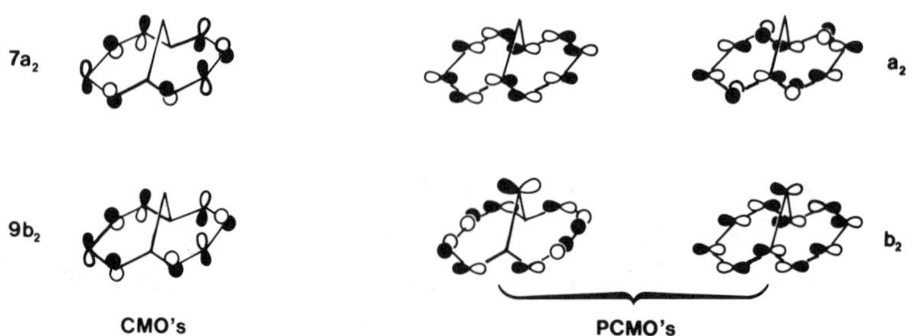

Figure 13. Highest occupied canonical π MOs of 53 (left). Schematic representation of the most important precanonical σ orbitals for the orbital mixing in 53 (right).

According to MO calculations [27] using a modified INDO version one finds a strong interaction between two high-lying pure π orbitals of the π perimeter and four precanonical σ orbitals (PCMOs) shown in Figure 13 at the right. The resulting canonical π orbitals (CMOs) $7a_2$ and $9b_2$ are shown in Figure 13 at the left. The strong π-σ interaction causes a rotation of the p_π AO's at the

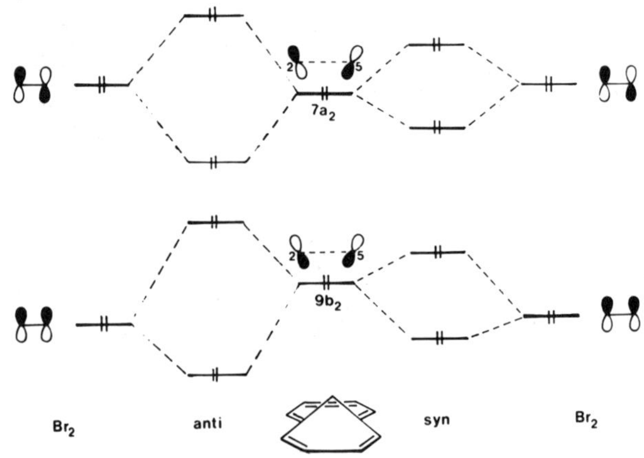

Figure 14. Qualitative diagram of the interaction between the occupied π and π* orbital of Br_2 and $9b_2$ and $7a_2$ syn (right) and anti (left) to the methano bridge.

centers C(2)/C(5) and C(7)/C(10) away from the methano bridge. Assuming a symmetrical transition state for the addition of Br_2 to 53, we find in analogy to Figure 12 that the antibonding interaction between the occupied π level of the halogen (π and π^* in Figure 14) and $9b_2$ and $7a_2$ of 51 is smaller if Br_2 adds syn to the methano bridge (right side of Figure 12) than in the case of an anti addition (Figure 14, left).

There are many other examples for which π-σ interaction causes a strong rotation of the p_π lobes at the Active Centers and thus dominates the stereoselectivity. If we restrict ourselves to [4+2] cycloaddition reactions, we have to mention here the work by P. Vogel et al. on 2-substituted 5,6-bis(methylene)norbornanes (63) as well as work by L. A. Paquette et al. on 64 and 65 and related compounds.

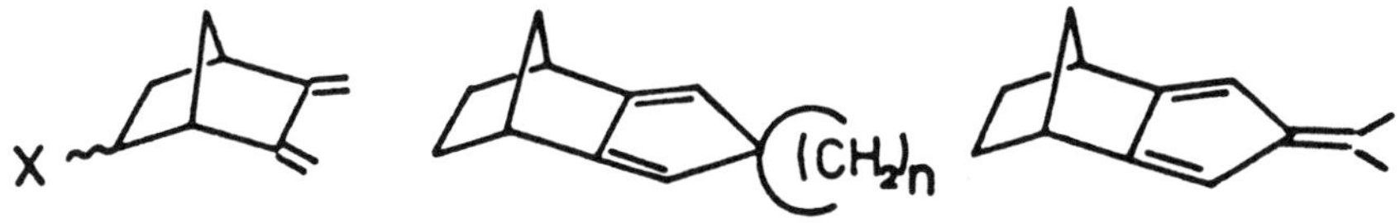

X

$(CH_2)_n$

Our preliminary calculations on 1-acetoxycyclopentadiene (16) suggests that the remarkable regioselectivity observed for the Diels-Alder reaction with ethylene as shown in eq. (5) [7] is due to π-σ interaction and rotation of the p_π-lobes at the Active Centers of 16. Here it should be mentioned that also other explanations have been put forward to rationalize the regioselectivity observed for eq. (5) [30].

The examples listed so far demonstrate that π-σ interaction is important in the discussion of the reactivity of nonplanar π systems. It is important to note that one has to consider the influence of the whole σ frame. To support this we have carried out model calculations using the EH model on 49 in which we varied the valence state ionization potential for the 2p AO [H_{ii}(2p)] at the centers 2, 3 and 7. The results of the model calculations for the lowest occupied π orbital for the diene moiety is summarized in Figures 15 and 16.

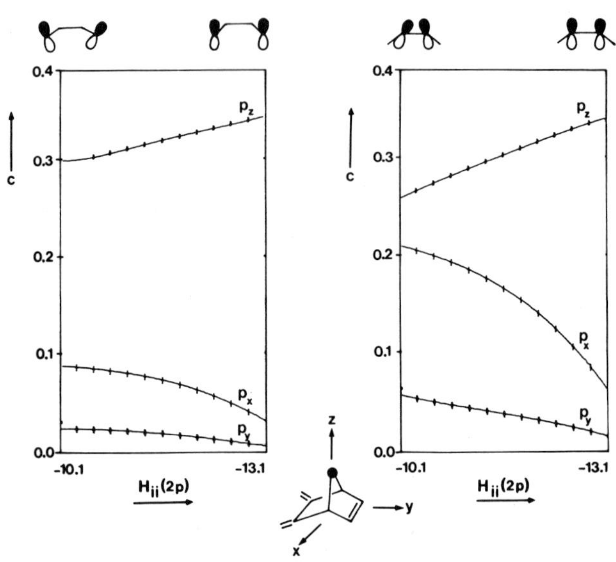

Figure 15. AO coefficients ($2p_z$, $2p_x$, $2p_y$) of π_S at the butadiene unit of 49 as a function of the variation of H_{ii} at center 7.

Let us first consider Figure 15, which summarizes the H_{ii} variation at center 7 of 49. Increasing H_{ii}(2p) from -10.1 eV to -13.1 eV reduces the - mixing at all centers.

The results displayed in Figure 16 are just opposite: Increasing H_{ii} from -10.85 eV to -12.35 eV reinforces the π-σ coupling and thus at larger H values a stronger coupling between $2p_z$ and $2p_x$ is encountered.

It should be mentioned here again that the predicted rotation is very sensitive to the relative position of π and σ orbitals and so the results in Figure 15 and 16 should only be taken as examples in which distant substituents may strongly influence the reactivity due to π-σ coupling.

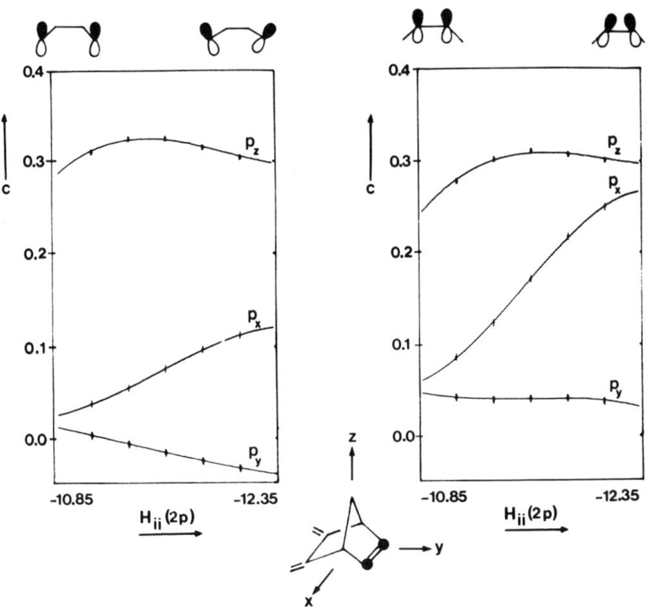

Figure 16. AO coefficients ($2p_z$, $2p_x$, $2p_y$) of π_S at the butadiene unit of 49 as a function of the variation of H_{ii} at centers 2 and 3.

So far a time-independent static picture has been used to explain the reactivity of several dienes by σ-π interaction in the ground state. It should be clear, however, that this static picture can be modified by means of the attacking dienophile. In an extended time-dependent (dynamic) approach the coupling between all components (π,σ diene and π,σ dienophile) as a function of the internuclear distance has to be investigated. Thus in certain cases the frontier orbitals of the dienophile may cause a strong direction dependent perturbation of the π amplitudes of the diene.

Polar Group Effects

A good example of the influence of polar groups on the regioselectivity of Diels-Alder reactions is the comparison of eq. (7) with eq. (16) (X = S). Replacing X=S in 35 by SO_2 changes the regioselectivity considerably [37]. At first one might attribute this effect to the 3d orbitals of sulfur, but the levels of those orbitals at an SO_2 group are comparable in energy with the π^* orbital of the ethylene bridge in 41. The reaction of 41 with PTAD, however, has demonstrated (see eq. (19)) how ineffective the π^*-n_- interaction is and thus an influence of frontier orbitals cannot be seen.

22

Since the SO_2 group is very polar one has to consider the electrostatic term in eq. (10).

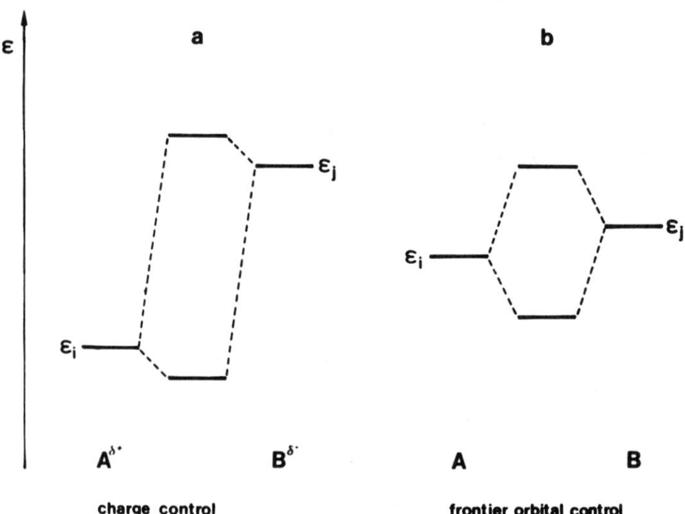

Figure 17. MO scheme for reactions in which (a) the reactants A and B carry a considerable net charge with a large HOMO-LUMO gap and (b) without considerable net charge and small HOMO-LUMO gap.

This term usually dominates if there is a large energy gap between the levels of the interacting fragments and if both centers carry a considerable net charge (see Figure 17a). In Figure 17 the limiting cases of charge control (left) and frontier orbital control are shown schematically.

To understand the directing effect operating in case of 22 and 66 it is necessary to have a physical term which rationalizes pictorially the electrostatic interaction between the reacting species. A quantum-chemical expectation value which meets this challenge is the electrostatic potential (EP) exerted by a molecule due to a test charge. With this definition of EP a positive point charge acts as a probe to test the potential around the molecule. We have calculated the EP in a modified central field monopole approximation [31] according to eq. (29).

$$EP(P) = \sum_A Z_A/R_{AP} - P_{ij}\int \frac{\Psi_i \Psi_j \, d\tau}{r_P} \quad (28)$$

$$EP(P) = \sum_A Q_A/R_{AP} \quad (29)$$

In these equations the following abbreviations are used:

$EP(P)$ = electrostatic potential acting on a point charge at P

Z_A = core charge

R_{AP} = distance between positive charge at P and atom A

r_P = distance between positive charge at P and the electron distribution $\Psi_i\Psi_j$

Q_A = net charge

P_{ij} = first-order density matrix

For a qualitative discussion, approximation (29) is sufficient [32]. The calculations are less time consuming than those associated with eq. (28) [33].

Analogously to recent investigations concerning protonation reactions [32,34], we have calculated the EP of 22, 23 and 35 (X = S) using the EH method. The results are displayed in Figure 18.

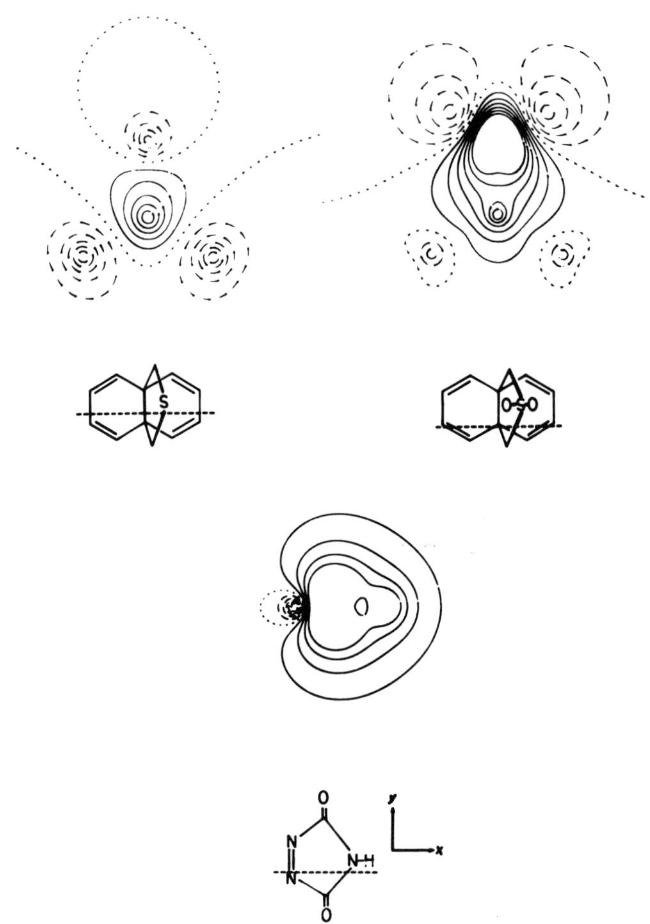

Figure 18. Contour diagrams of the calculated electrostatic potentials of 22, 23, and 35 (X = S). The maps are drawn in the plane parallel to the xz plane indicated by the dashed line in the formula. The interval between the contours is 15 kcal/mol in the case of 35 (X = S) and 30 kcal/mol in the case of 22 and 23. Positive potentials are indicated with full lines, negative potentials with broken lines. Nodes are indicated by short dashes.

Inspection of Figure 18 shows that the preferred syn attack in case of 22 can be rationalized as due to a stabilizing Coulomb attraction between the strongly electron-deficient S atom in the SO_2 group and the electron-rich N_2 group in PTAD.

In addition to the interaction just discussed there is still another interaction between the area of high electron density around the oxygen centers of the SO_2 fragment and the charge-deficient region above and below the π plane of PTAD (see 13). This interaction is sketched below. In connection with polar group effects experiments on the arylsubstituted 9-isopropylidenebenzonorbornenes 66 and 67 should be mentioned [35].

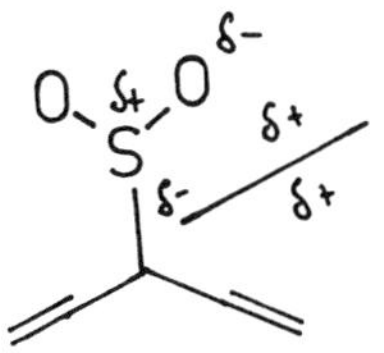

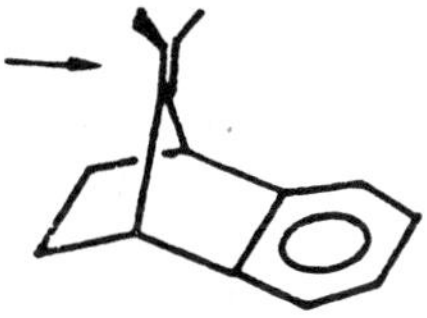

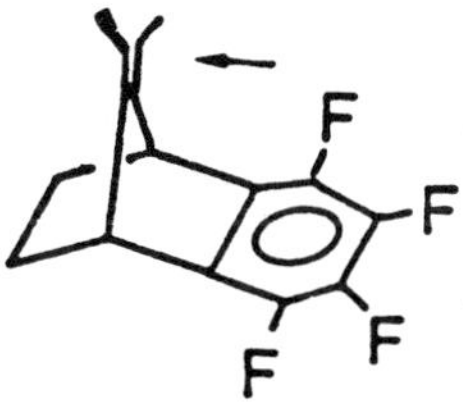

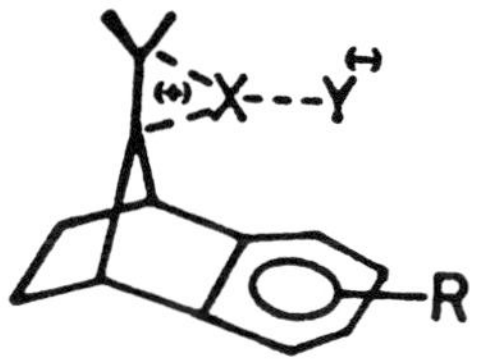

66 67 68

When weak electrophilic reagents such as singlet oxygen, mCPBA, NBS, and t-butylhypochlorite were studied the product distribution depended on the substitution patterns of the benzene ring. In the case of 66 anti attack is preferred; the amount of syn adduct, however, is enhanced if 67 is the reaction partner. The observed switch in the syn/anti ratios when the H atoms are replaced by F in 66 might be due to a strong reduction of electron density above and below the ring as shown in Figure 19. Accordingly, transition states as those shown in 68 are stabilized.

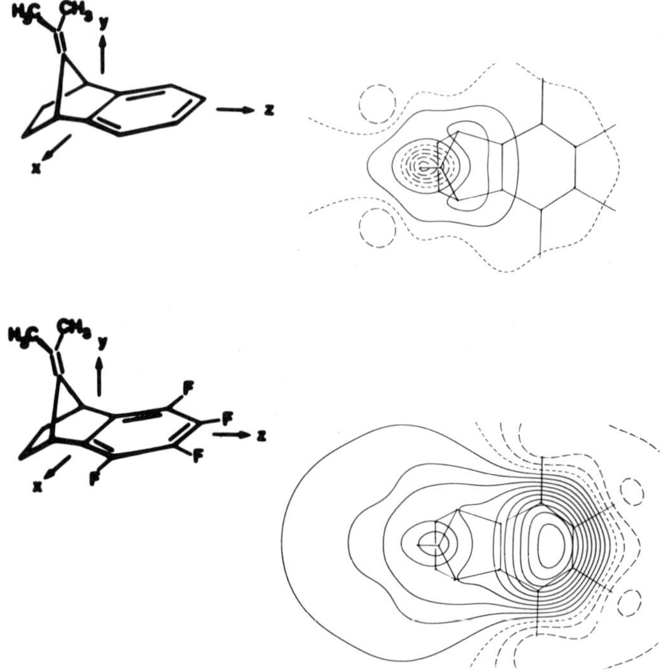

Figure 19. Contour diagram of the calculated electrostatic potentials of 66 and 67. The maps are drawn parallel to the xz plane 1.5 Å above the benzene ring. The gap between the contours is 5 kcal/mol in 67 and 2 kcal/mol for 66. Positive potentials are indicated by solid lines, negative ones with broken lines. Nodes are indicated by short dashes.

Final Remarks

The preceding discussion has shown that electronic effects are very important in order to rationalize the stereo- and regioselectivity of Diels-Alder reactions. It should be noted, however, that the energy difference between the possible transition states of a certain category listed in Figure 2 (eg, endo vs exo transition state) is in the order of 1-5 kcal/mol. This implies that other effects such as van der Waals forces, induced dipole-dipole interactions, and steric effects have to be considered.

I would like to stress again that most of our arguments are based on the electronic structure of the ground state of the molecule. The extrapolation to specific transition states should be checked by accurate calculations.

Acknowledgements. Our work on the regio- and stereoselectivity of cycloaddition reactions has been initiated by David Ginsburg and greatly stimulated by him, L.A. Paquette, and E. Vogel. It was a great pleasure to present our results at the conference on "Stereochemistry and Reactivity of π Systems" initiated and perfectly organized by P. D. Bartlett and W. H. Watson. We are grateful to the Deutsche Forschungsgemeinschaft, the Fonds der Chemischen Industrie and the BASF, Ludwigshafen, for financial support.

Discussion

Vogel: You have explained the high stereoselectivity of the Diels-Alder additions of strong dienophiles to [4.3.2]propella-3,5,11-triene and to [4.4.2]propella-3,5,8,12-tetraene [L. A. Paquette, R. E. Wingard and R. K. Russell, J. Am. Chem. Soc. 94, 4739 (1972); D. Ginsburg, Acc. Chem. Res. 7, 286 (1974)] by invoking electronic repulsions between the cyclobutene double bond and the dienophile that make the attack syn to the cyclobutene ring less favored than the anti attack. Are you sure the cyclohexa-1,3-dienes are planar in these systems? The face selectivity observed could be due to a distorted conformation of the diene.

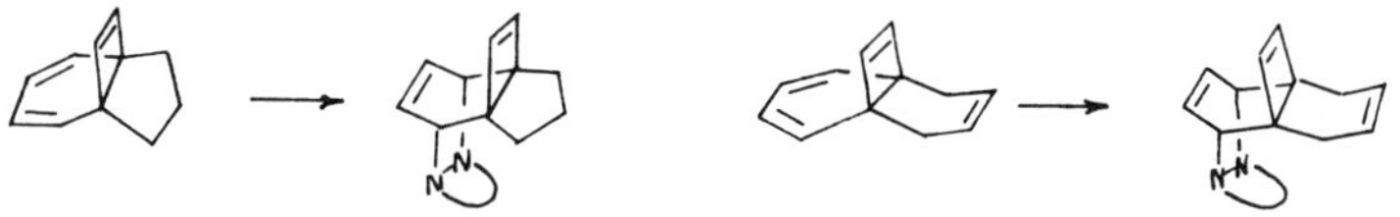

Gleiter: No. There are no X-ray data. I have asked Prof. Ginsburg about geometrical parameters but he doesn't have any.

Vogel: We have found the Diels-Alder additions of 5,6-bis(methylene)bicyclo[2.2.2]oct-2-ene to be endo face selective; the dienophiles prefer to attack onto the face of the diene syn to the endocyclic double bond. This is contrary to your predictions. Can you comment on that?

Gleiter: What are the ratios?

Vogel: We have obtained the following results:

+ dienophile +

R = Cl	+	TCNE	78	:	22
R = D	+	TCNE	75	:	25
	+	maleic anhydride	60	:	40
	+	N-phenyltriazoline-dione	>95	:	<5

It appears that the more the electrophilic character of the dienophile or the faster the cycloaddition, the more endo face selective the reaction.

Gleiter: We have done calculations on the species here. We have found that exo addition is favored by about 3 kJ/mol which is really not very much. This is really a tough case. So I cannot give you a good answer.

Paquette: Doesn't the answer depend on the energy of the olefinic double bond in question? What you are talking about is a cyclobutene bond in one case and a norbornene bond in the other. How different would you expect those energies to be?

Gleiter: As I pointed out the energy from the π bond does not do any good. We also looked for rotation, and we did not find any significant rotation about these π systems. But I do feel there is a big difference if you make a bond or you don't make a bond. We have done calculations on the system with this group removed over here and we see rotation. I think the energy difference from above and from below is very small if we carry out the calculation.

Herndon: In these strained bicyclic and tricyclic olefins I can imagine that there would be σ-π mixing in the highest filled MO. If that orbital had the same shape or extension in space that you write for the lowest filled MOs, then the prediction is that the HOMO-LUMO interactions with ethylene or a dienophile would lead to a result that is just the opposite of what one predicts when using the lowest filled MO.

Gleiter: We have carried out model calculations

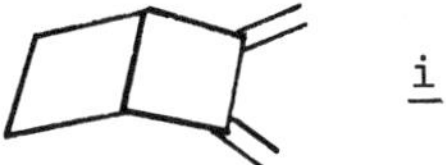

i

on molecule i which I have not discussed in my talk. In this compound we find a strong interaction between the σ frame and the lowest occupied π orbital. The interaction between the antisymmetric π orbital, the HOMO, and the σ frame is not very large. Concerning strong σ-π interactions very strained hydrocarbons should be checked. This was pointed out by Ken Houk in his talk. We have only calculated a few so far.

Houk: The closed shell repulsion arguments that you make to explain the stereochemistry of isodicyclopentadiene cycloadditions worry me. In spite of the fact that these effects certainly are an important cause of activation energies, I am particularly bothered that you give such a rather delicate geometrical argument. For the isolated geometries, the overlap of the lowest occupied π orbital is greatest with the ethylene filled orbital when attack occurs on the top side, but in the transition state of course the geometry is entirely different so this might change. Even for quite reactive Diels-Alder addends, I think the geometry will change. That there is a significant amount of geometrical change is going to damage these arguments.

Gleiter: What do you mean with different? In our simple frontier orbital model we have neglected geometrical modifications in the transition state. We have done a calculation with the unperturbed molecules assuming 2.18 Å as the distance in the transition state. I do not know but this value may be taken from your work. It is from a model calculation on butadiene and ethylene.

Houk: It is Salem's work. The distances are all right, but the carbon atoms undergoing bond formation are significantly pyramidalized in the transition state. This is also true in the fulminic acid, ethylene problem that I talked about earlier. In fact, it is true in every reaction we have studied. This pyramidalization is the thing that is going to cause the tilting of pi orbitals to be very different, ie, the alkenes will no longer be planar in the transition state and so all of the angles, etc, will change appreciably.

I do not have a question. That is just a worry. The question is what would happen if you stretched the alkene, or did a [6+4] cycloaddition. Now you have the termini attacking the diene much further apart. I believe that the overlap results will change and the prediction would be more attack from the exo face rather than the endo face.

Gleiter: We have not studied this problem in detail but there is no reason that the overlap arguments are invalid as the transition state geometry must be comparable with the [4+2] reaction.

Houk: It would be nice to calculate the overlap because I think it will be different and it would be a good test. A third thing is other electrophilic reactions. Your argument only applies to concerted four-center Diels-Alder reactions. Not to anything else. What about just adding electrophiles, bromination, attacking some electrophile on the diene system? What is the stereochemistry there? I would again think that whatever is going on is going to cause endo attack to be preferred by any electrophile. Your arguments would apply only to a concerted Diels-Alder reaction.

Gleiter: Now we are working on this question, but I cannot give you any answer.

References

[1] O. Diels and K. Alder, Justus Liebigs Ann. Chem. 460, 98 (1928).

[2] Reviews: J.G. Martin and R.K. Hill, Chem. Rev. 61, 537 (1961); J. Sauer, Angew. Chem. 78, 233 (1966); 79, 76 (1967); Angew. Chem. Int. Ed. Engl. 5, 211 (1966); 6, 16 (1967); J. Sauer and R. Sustmann, Angew. Chem. 92, 773 (1980); Angew. Chem. Int. Ed. Engl. 19, 779 (1980), and references therein.

[3] J.B. Lambert and J.D. Roberts, Tetrahedron Lett., 1457 (1965).

[4] R.B. Woodward and R. Hoffmann, "The Conservation of Orbital Symmetry", Verlag Chemie, Weinheim, 1970; R. Hoffmann and R.B. Woodward, Acc. Chem. Res. 1, 17 (1968); A.P. Marchand and A.P. Lehr, "Pericyclic Reactions", Vol. I and II, Academic Press, New York 1977.

[5] K. Fukui, Fortschr. Chem. Forsch. 15, 1 (1970); Acc. Chem. Res. 4, 57 (1971); W.C. Herndon, Chem. Rev. 72, 157 (1972); K.N. Houk, Acc. Chem. Res. 8, 361 (1975); R. Sustmann, Pure Appl. Chem. 40, 569 (1974); K. Fukui "Theory of Orientation and Stereoselection" in "Reactivity and Structure Concepts in Organic Chemistry", Vol. 2 Springer, Berlin 1975; I. Fleming, "Frontier Orbitals and Organic Chemical Reactions", John Wiley & Sons 1976 and references therein.

[6] K. Alder, G. Stein, F. Buddenbrock, W.v. Eckardt, W. Frercks, and S. Schneider, Justus Liebigs Ann. Chem. 514, 1 (1934); H. Stockmann, J. Org. Chem. 26, 2025 (1961).

[7] S. Winstein, M. Shavatsky, C. Norton, and R.B. Woodward, J. Am. Chem. Soc. 77, 4183 (1955).
[8] (a) T. Sugimoto, Y. Kobuke, and J. Furukawa, J. Org. Chem. 41, 1457 (1976), (b) M.C. Böhm, R.V.C. Carr, R. Gleiter, and L.A. Paquette, J. Am. Chem. Soc. 102, 7218 (1980), (c) W.H. Watson, J. Galloy, P.D. Bartlett, and A.A.M. Roof, J. Am. Chem. Soc. 103, 2022 (1981).
[9] R. Gleiter and D. Ginsburg, Pure Appl. Chem. 51, 1301 (1979).
[10] G. Klopman, J. Am. Chem. Soc. 90, 223 (1968); L. Salem, J. Am. Chem. Soc. 90, 543 and 553 (1968).
[11] N.D. Epiotis, W.R. Cherry, S. Shaik, R.L. Yates, and F. Bernardi, Top. Curr. Chem. 70, 1 (1977).
[12] L. Libit and R. Hoffmann, J. Am. Chem. Soc. 96, 1370 (1974).
[13] K. Alder and G. Stein, Angew. Chem. 50, 510 (1937).
[14] K.B. Wiberg and W.J. Bartley, J. Am. Chem. Soc. 82, 6375 (1960).
[15] R.C. Cookson, E. Crundwell and J. Hudec, Chem. and Ind., 1003 (1958); R.C. Cookson and E. Crundell, Chem. and Ind., 1004 (1958); G.O. Scherick and R. Steinmetz, Chem. Ber. 96, 520 (1963); R.C. Cookson, E. Crundwell, and R.R. Hill, J. Chem. Soc. (London), 3062 (1964).
[16] R.B. Woodward and H. Baer, J. Am. Chem. Soc. 70, 1161 (1948); H. Kwart and I. Burchuk, J. Am. Chem. Soc. 74, 3094 (1952).
[17] D. Craig, J.J. Shipman, J. Kiehl, F. Widmer, R. Fowler, and A. Hawthorne, J. Am. Chem. Soc. 76, 4573 (1954); R.B. Woodward and H. Baer, J. Am. Chem. Soc. 66, 645 (1944).
[18] D. Ginsburg, "Propellanes, Structure and Reactions", Verlag Chemie, Weinheim, 1975; D. Ginsburg, Acc. Chem. Res. 7, 286 (1974).
[19] P. Ashkenazi, D. Ginsburg, G. Scharf, and B. Fuchs, Tetrahedron 33, 1345 (1977).
[20] K.N. Houk, J. Am. Chem. Soc. 95, 4092 (1973); K.N. Houk, J. Sims, R.E. Duke, Jr., R.W. Strozier, and J.K. George, J. Am. Chem. Soc. 95, 7287 (1973).
[21] J. Feuer, W.C. Herndon, and L.H. Hall, Tetrahedron 24, 2575 (1968); N.D. Epiotis, J. Am. Chem. Soc. 95, 5624 (1973); O. Eisenstein, J. M. Lefour, and N.T. Anh, Chem. Comm., 969 (1971); O. Eisenstein, J.M. Lefour, N.T. Anh,

and R.F. Hudson, Tetrahedron 33, 523 (1977).
[22] E. Heilbronner and A. Schmelzer, Helv. Chim. Acta 58, 936 (1975).
[23] S. Inagaki, H. Fujimoto, and K. Fukui, J. Am. Chem. Soc. 98, 4054 (1976); S. Inagaki and K. Fukui, Chem. Lett., 509 (1974).
[24] K. Müller, Helv. Chim. Acta 53, 1112 (1970); N.C. Baird and R.M. West, J. Am. Chem. Soc. 93, 4427 (1971); N.D. Epiotis and R.L. Yates, ibid. 98, 461 (1976).
[25] E. Vogel, W.A. Boll and M. Biskup, Tetrahedron Lett., 1569 (1966).
[26] E. Vogel, T. Scholl, and J. Lex, Angew. Chem. 94, 924 (1982).
[27] R. Gleiter, M.C. Böhm and E. Vogel, Angew. Chem. 94, 925 (1982).
[28] P.-A. Carrupt, M. Avenati, D. Quarroz, and P. Vogel, Tetrahedron Lett. 45, 4413 (1978); M. Avenati, J.-P. Hagenbuch, C. Mahaim, and P. Vogel, Tetrahedron Lett. 21, 3167 (1980); J.-P. Hagenbuch, P. Vogel, A.A. Pinkerton, and D. Schwarzenbach, Helv. Chim. Acta 64, 1818 (1981).
[29] L.A. Paquette, T. Kravetz, M.C. Böhm, and R. Gleiter, J. Org. Chem. 48, 1250 (1983); L.A. Paquette, P. Charumlind, M.C. Böhm, R. Gleiter, L.S. Bass, and J. Clardy, J. Am. Chem. Soc. 105, 3136 (1983). L.A. Paquette, P. Hayes, P. Charumlind, M.C. Böhm, R. Gleiter, and J.F. Blount, J. Am. Chem. Soc., 105, 3148 (1983).
[30] N.T. Anh, Tetrahedron 29, 3227 (1973).
[31] E.U. Condon and G.H. Shortley, "The Theory of Atomic Spectra", Cambridge University Press, 1970.
[32] J. Almlöf, E. Haselbach, F. Jachimowicz, and J. Kowalewski, Helv. Chim. Acta 58, 2403 (1975).
[33] C. Giessner-Prettre and A. Pullman, Theor. Chim. Acta 25, 83 (1972).
[34] E. Scrocco and J. Tomasi, Topics Curr. Chem. 42, 95 (1973); J. Almlof, A. Hendriksson-Enflo, J. Kowalewski, and M. Sundbohm, Chem. Phys. Lett. 21, 560 (1973).
[35] L.A. Paquette, L.W. Hertel, R. Gleiter, M.C. Böhm, M.A. Beno, and G.G. Christoph, J. Am. Chem. Soc. 103, 7106 (1981).
[36] M.C. Böhm and R. Gleiter, Tetrahedron 36 3209 (1980).

[37] J. Kalo, E. Vogel, and D. Ginsburg, *Tetrahedron* **33**, 1183 (1977).

5. EXOCYCLIC DIENES, TETRAENES AND HEXAENES. DIELS-ALDER REACTIVITY, REGIOSELECTIVITY AND STEREOSELECTIVITY

Pierre Vogel

Institut de Chimie Organique, 2, rue de la Barre,
CH 1005 Lausanne, Switzerland

Introduction

The spectroscopic [1] and chemical properties of an exocyclic s-cis-butadiene moiety grafted onto a rigid skeleton can be modified by remote substitution of the bicyclic skeleton [2-8]. The 2,3,5,6-tetrakis(methylene)bicyclo[2.2.n]alkanes (1-5) and [2.2.2]hericene [9] (6) are very attractive starting materials for the preparation of polycyclic,

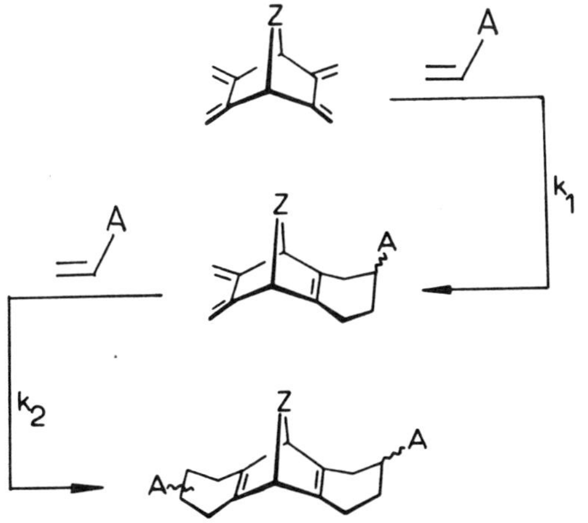

Table 1. Kinetic data for cycloadditions of tetraenes 1, 2, 4, pentaenes 3, 5 and [2.2.2]-hericene 6 and their adducts to ethylenetetracarbonitrile (TCNE) in toluene.

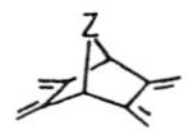	$\Delta H^{\ddagger}$ kcal/mol	$\Delta S^{\ddagger}$ cal/molK	k_1(298K) $mol^{-1}s^{-1}$	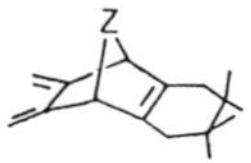	$\Delta H^{\ddagger}$	$\Delta S^{\ddagger}$	k_2(298K)	k_1/k_2'(298K)
Z = O: 1	14.8 ±0.7	-26.4 ±2.3	1.5×10^{-4}	14	17	-30	4×10^{-7}	376
Z = CH_2: 2	12.2 ±0.5	-25 ±1.5	255×10^{-4}	15	14.1 ±1.0	-30.1 ±3.4	0.7×10^{-4}	362
Z = $C{=}C(CH_3)_2$: 3	11.3 ±0.6	-25.6 ±2.1	850×10^{-4}	16	14.8 ±0.9	-24.6 ±2.8	3.4×10^{-4}	250
Z = CH_2-CH_2-: 4	10.6 ±0.4	-24 ±1.4	5.9×10^{-1}	17	10.6 ±0.6	-30 ±2	0.334×10^{-1}	17
Z = CH=CH: 5	11.3 ±0.8	-24 ±2.8	1.72×10^{-1}	18	13.0 ±0.2	-29.5 ±1	6.1×10^{-4}	282
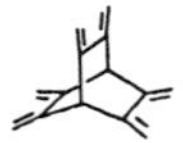 6	11.3 ±0.2	-24.5 ±0.8	1.36×10^{-1}	19	10.9 ±0.5	-30 ±1.5	156×10^{-4}	9
			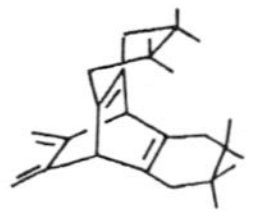	20	15	-30	0.5×10^{-4}	k_2/k_3 = 300

Scheme 2

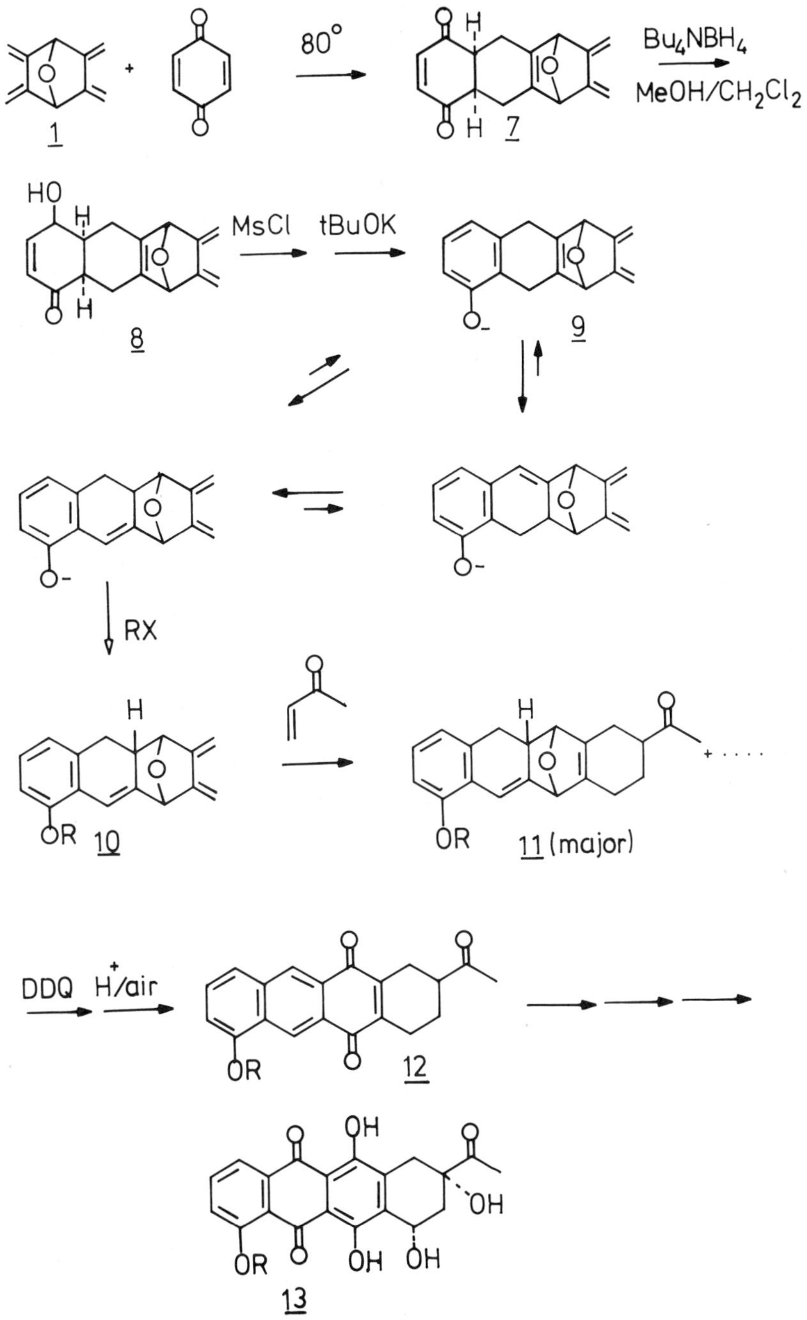

1

80°

7

Bu_4NBH_4

$MeOH/CH_2Cl_2$

HO

8

MsCl tBuOK

9

RX

10

11 (major)

DDQ H^+/air

12

13

polyfunctional systems by two successive Diels-Alder additions with different dienophiles. 2,3-5,6-Tetrakis(methylene)-7-oxabicyclo[2.2.1]heptane [10] (1), readily obtained in four steps from the inexpensive furan and maleic anhydride, can be used to prepare various anthracycline derivatives [11,12]. The principle of our strategy rests upon the fact that the rate of the Diels-Alder addition of the tetraene 1 (k_1) is much higher than that (k_2) of the corresponding monoadduct (Scheme 1). This is true for dienophiles such as benzoquinone, methyl acetylene dicarboxylate, tetracyanoethylene (TCNE) (See Table 1), methyl vinyl ketone, acrylic esters, etc ($k_1/k_2>100$). With benzyne, the rate ratio k_1/k_2 is no greater than 5 [12]. It depends also upon the nature of the bridge Z of the exocyclic polyenes 1-6. The various possible factors that can affect k_1/k_2 will be discussed.

In the synthesis of daunomycinone (13), a key intermediate is the exocyclic diene 10 perturbed by a homoconjugated exocyclic double bond [13] (Scheme 2). The latter causes the diene to add to methyl vinyl ketone (in the presence of BF_3, -80°) with good regioselectivity (11 being the major adduct). The remote substituent effect on the Diels-Alder regioselectivity of s-cis-butadienes will be discussed with reference to the popular PMO theory. It will be shown that a carbonyl group homoconjugated with a π system can act as an electron-donating substituent, a hyperconjugative interaction (polarizability effect) overriding the "normal" field effect of the oxo group.

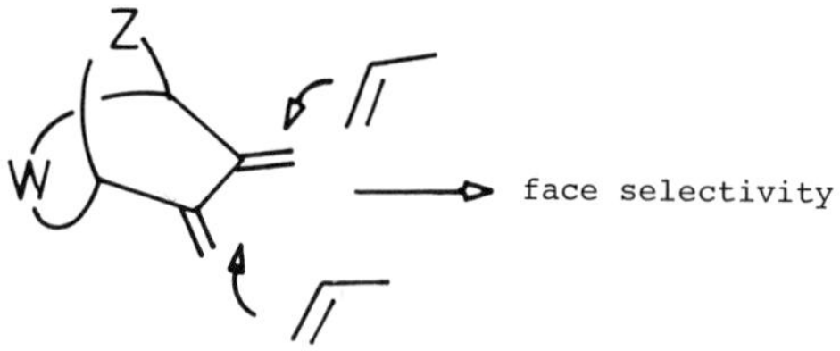

Z

W

face selectivity

Scheme 3.

The two faces of the exocyclic dienes grafted onto a norbornane or a substituted bicyclo[2.2.2]-octane skeleton are not equivalent and can lead to face selectivity for the reactions at the π systems (Scheme 3). The possible factors that can intervene in the Diels-Alder face selectivities of our dienes will also be discussed briefly.

1. The Diels-Alder Reactivity of 2,3,5,6-tetrakis-(methylene)bicyclo[2.2.n]alkanes.

The Diels-Alder reactivities of the polyenes 1-6 and of the corresponding monoadducts 14-19 and bis adduct 20 toward TCNE are reported in Table 1. One observes that the rate constant ratio k_1/k_2 is large for the olefins 1 [10], 2 [8], 3 [14] and for the monoadducts 19 of [2.2.2]hericene 6 [9] (k_2/k_3); it is much smaller for 4 [15,16] and 6 [9,15]. The main contribution to k_1/k_2 comes from a change of the $\Delta H^{\ddagger}$ term.

A priori, the Diels-Alder reactivity of TCNE in toluene to exocyclic dienes can be affected by the following possible factors: (a) geometry of the diene (1,4-distance between the exocyclic methylidene C-atoms [17], conformation of the diene [18]), (b) polarizability of the diene [19] (ionization potentials, IPs [20]), and (c) exothermicity of the reactions (the Dimroth [21], Bell-Evans-Polanyi principle [22] can be followed!).

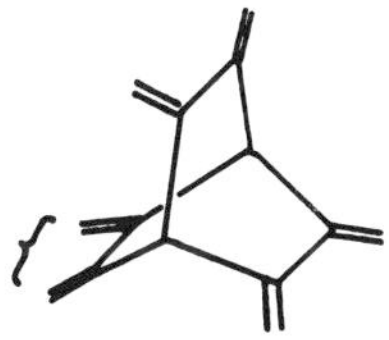

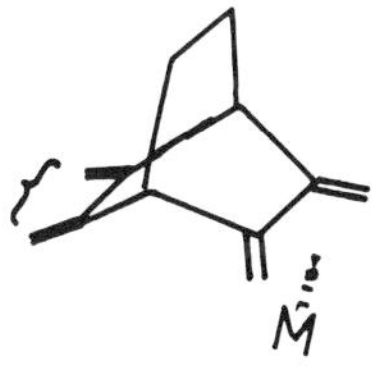

[23] $d_{1,4}$ = 3.066 Å 3.037 Å [24]

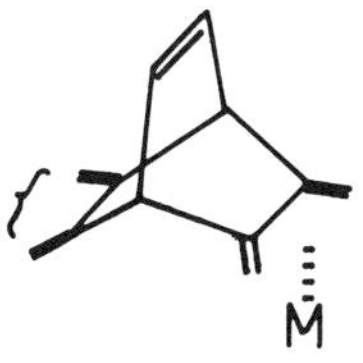

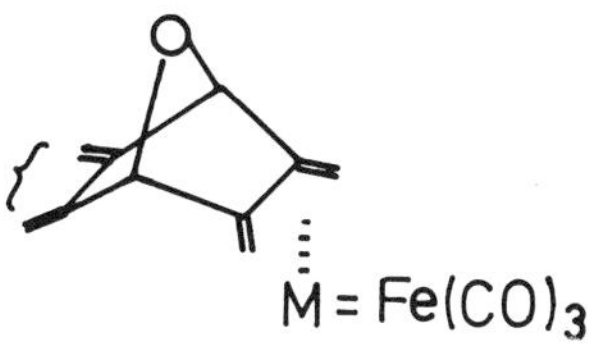

[16b] $d_{1,4}$ = 3.055 Å 3.159 Å [25]

X-ray diffraction studies on the [2.2.2]hericene 6 [23], and the irontricarbonyl complexes 21-23 demonstrate that the 1,4-distance between the methylidene C-atoms of the exocyclic s-cis-butadiene does not vary significantly when one bridge of the bicyclic skeleton varies from an exocyclic diene to an endocyclic double bond. MINDO/3 [26], MNDO [27] as well as force-field calculations [28] (see Figure 1) suggest that 2,3-bis(methylene)-norbornane and 2,3-bis(methylene)bicyclo[2.2.2]-octane systems have single minimum energy hypersurfaces independent of the degree of unsaturation of the bicyclic skeleton. The s-cis-butadiene moiety prefers to be perfectly planar. Examination of models of 2,3-bis(methylene)bicyclo[2.2.2]-octane suggests that the gauche interactions between the hydrogen atoms of the methylidene groups syn to C-2,3 should make the diene deviate from planarity. Accordingly, it was thought that in 6, where we cumulate three of these van der Waals repulsive effects, we should have had the best possible opportunity to observe nonplanar dienes. Surprisingly, the X-ray crystal structure of 6 [23] shows a twisting about the C_3 axis not larger than 0.3° and a dihedral angle for the exocyclic dienes of 1.34°. These values show that 6 must be considered as a D_{3h} species in the crystalline state. High thermal motion prevented any observation of a twist about the C_3 axis as observed for bicyclo-[2.2.2]octane-1,4-dicarboxylic acid [30]. This suggests that the dienes 1-6 and 14-20 can be considered planar and the more so for norbornane and 7-oxanorbornane derivatives 1-3 and 14-16 (see Figure 1). Therefore, we think that the rate ratio k_1/k_2 cannot be attributed to differences in geometry between the polyenes 1-6 and their monoadducts 14-19 and bis-adduct 20.

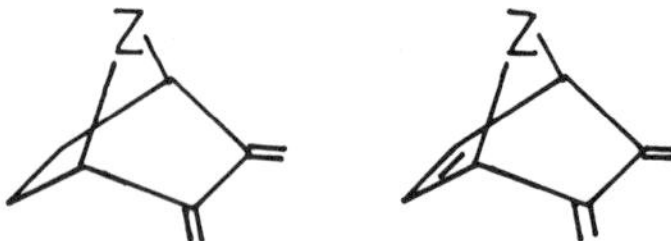

Z = CH_2 24 IP_1 = 8.41 eV 25 IP_1 = 8.48 eV

Z = O 26 IP_1 = 8.79 eV 27 IP_1 = 8.87 eV

The Diels-Alder reactivity of strong dienophiles toward the dienes 24 and 26 is higher than that toward the corresponding trienes 25 and 27 [6,8]. This can be attributed to a homoconjugative interaction (bishomoaromaticity?) in the trienes 25 and 27 that is not present in the dienes 24 and 26. The LUMO localized on the endocyclic double bond could stabilize the HOMO of the homoconjugated diene.

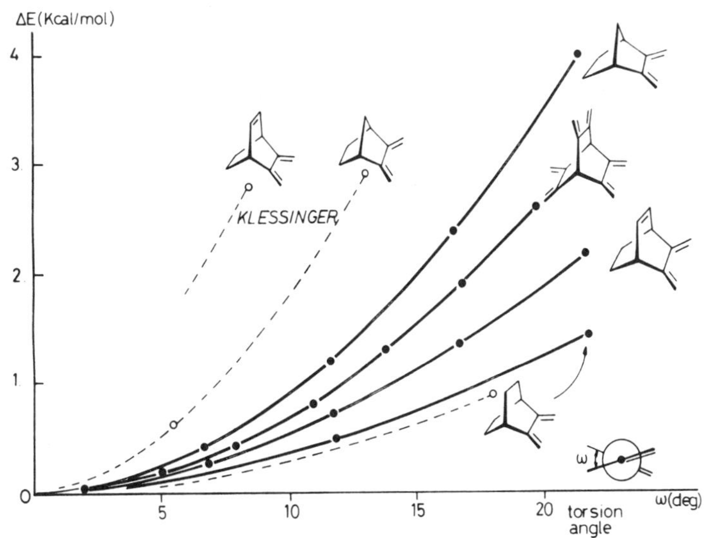

Figure 1. Force-field calculations (MM1/MMPI(QCPE Program No. 318)) on exocyclic dienes grafted onto bicyclic skeletons [28] (dotted lines: calculations reported by Klessinger et al. [29])

This seems to be verified when comparing the IPs of the trienes 25 and 27 with those of the dienes 24 and 26 and the tetraenes 2 and 1, respectively (see Figure 2). However, the measured IP differences are very small and might not be significant if there were an experimental error of ±0.05 eV in the IP (see also [31]). Furthermore, for the series of dienes grafted onto bicyclo[2.2.2]octanes, the observed IPs do not vary with the degree of unsaturation and with the type of homoconjugated π systems (exocyclic diene or endocyclic double bond, see Figure 2). The fact that the rate ratio k_1/k_2 (and k_2/k_3 for 19) is large for Z = O, CH_2, $C{=}C(CH_3)_2$, and CH=CH but small for Z = CH_2-CH_2 and

$H_2C{=}C{-}C{=}CH_2$ which because of the small variations makes it difficult to explain the differences in Diels-Alder reactivity between an exocyclic tetraene and its corresponding monoadduct based on a change of the IP of these systems.

The four cyano groups in the monoadducts 14-19 could retard their Diels-Alder additions because of the field effect of these remote substituents. If such a factor should intervene, it should not affect the rate constant ratio k_1/k_2 more than by a factor of 5-10 judging from the k_1/k_2 value measured for the tetraene 4 and the hexaene 6. The lower reactivity of the 7-oxanorbornane derivatives compared with that of the hydrocarbon analog is probably due to the inductive effects of the ether bridge, as indicated by the IP (see Figure 2).

To a first approximation, assuming the strain energy of an unsaturated bicyclic system to be a linear function of the number of sp^2 hybridized carbons it contains, we estimate there should be no change of strain energy between a tetraene, its monoadduct, and its bis-adduct since the number of sp^2 carbon atoms is the same. Accordingly, we predict no significant changes in the exothermicities between the Diels-Alder addition of a tetraene and that of its monoadduct. We shall see that this assumption is incorrect. Let us consider first the cycloadditions of the bicyclo[2.2.2]-octane derivatives 4-6 and 17-20.

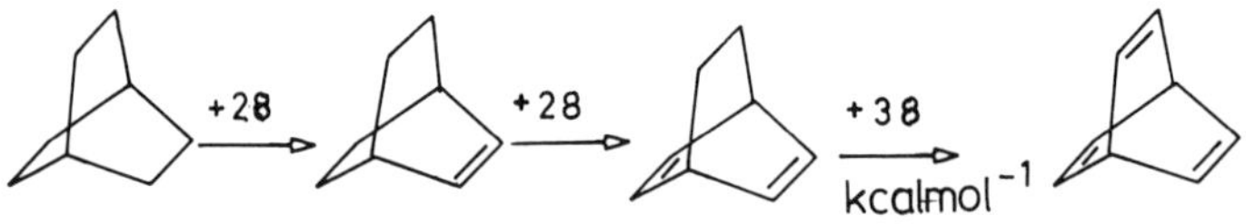

+28 → +28 → +38 kcalmol^{-1}

It is well known that barrelene [32] has an excess strain energy of about 10 kcal/mol compared with that of 2,5-bicyclo[2.2.2]octadiene and 2-bicyclo[2.2.2]octene [33]. It has been attributed to a repulsive electronic interaction between the π electrons of the three double bonds [34]. Such repulsive interactions are not predicted for the polyenes 4-6 and 17-20. They must be present only in the bis-adducts of the pentaenes 5 and 19.

Thus, we predict that the exothermicities of the Diels-Alder additions of 18 and 20 should be lower than those of the cycloadditions of 4-6, 17 and 19 since the former reactions generate barrelene derivatives that are more strained than the adducts formed in the latter reactions. Accordingly, and assuming the validity of the Dimroth [21], Bell-Evans-Polanyi principle [22], the Diels-Alder additions of 4, 6, and 19 should have similar rates to those of the corresponding monoadducts, whereas in the case of the 2-bicyclo[2.2.2]octene derivatives 5 and 19, the tetraenes are expected to be more reactive than their corresponding monoadducts.

We present now data that suggest that the monoadducts of the tetraene grafted onto bicyclo-[2.2.1]heptanes are less strained than the corresponding bis-adducts. The X-ray structure [35] of the 5,6-bis(methylene)-7-oxabicyclo[2.2.1]hept-2-ene derivative 8 (see Scheme 2) shows that the substituents at the endocyclic double bond C-2,3 deviate from the C-1,2,3,4 plane toward the endo face by 10° whereas the exocyclic s-cis-butadiene C-5,5',6,6' must be considered to be in the same plane as C-1,4,5,6. The oxygen bridge O(7) of 8 tilts away from the endocyclic double bond showing that the π-exoC(2,3) electronic repulsion is more important than the π exo-C(5,6) electronic repulsion. This is due to the exo-face polarization of the π-electron density of the 2-norbornene double bond, as predicted by Fukui [36]. Similar structural features have been recorded recently [37] for the maleic anhydride adduct (28) to 7,7-dimethyl[2.2.1]hericene (3). Thus, when adding a dienophile to the monoadducts 8, 14 and 16 gen-

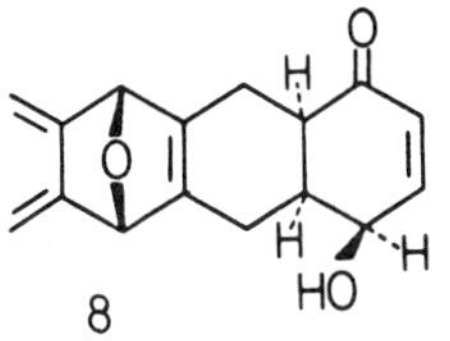

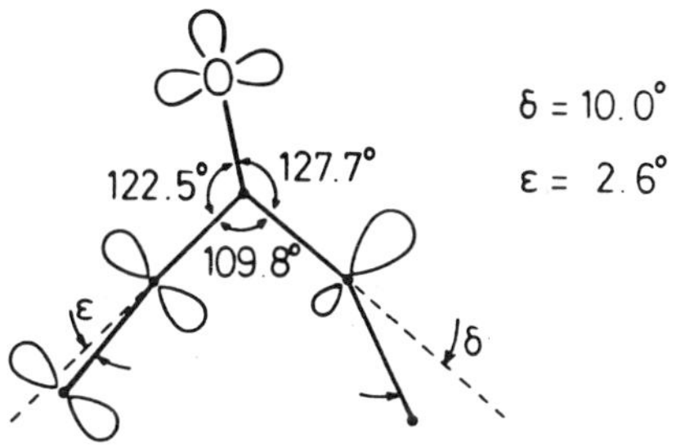

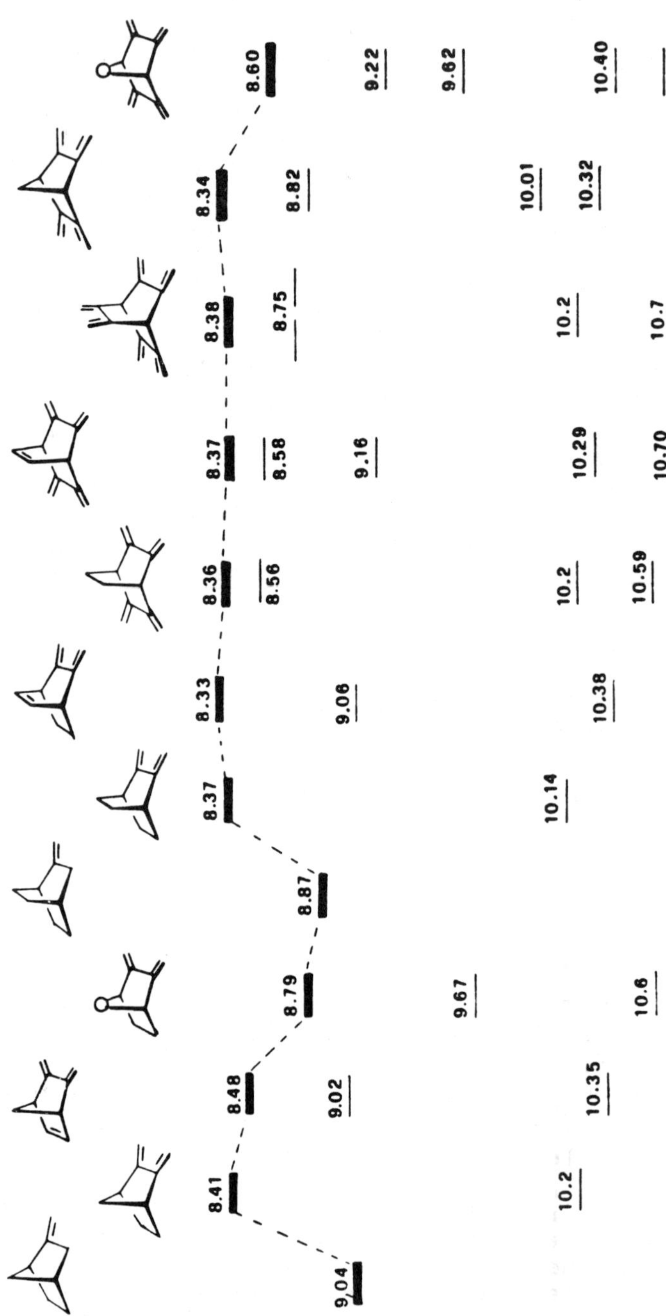

Figure 2. Ionization potentials (eV) of exocyclic dienes, tetraenes, and hexaene (by PES [1a,b]).

erating norbornadiene derivatives, we expect the new endocyclic double bond to push the ethereal and isopropylidene bridges "back" to a "more symmetrical" position. This gives a picture of the "extra" strain expected for the bis-adducts of the bicyclo-[2.2.1]heptane series. It makes the Diels-Alder additions of the tetraenes 1 and 3 more exothermic than those of their corresponding monoadducts. Consequently, the cycloadditions of 1 and 3 must be more rapid than those of their corresponding monoadducts. X-ray data on a monoadduct of 2 (Z = CH_2) are not yet available. Nevertheless, there are data

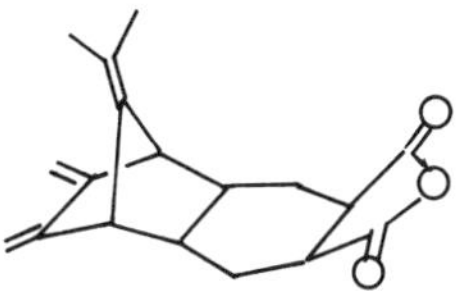

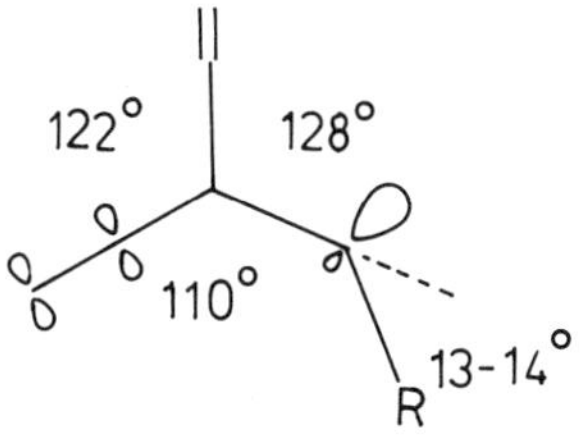

122° 128° 110° 13-14° R

in the literature [38a,39], and data that we have obtained on 2-norbornene derivatives (see below) that show that 14 and related 5,6-bis(methylene)-norborn-2-ene systems must present structural features analogous to those of 8 and 28; therefore, we propose that the Diels-Alder reactivity of 2 is higher than that of its monoadduct 15 because the cycloaddition of 2 is more exothermic than that of 15.

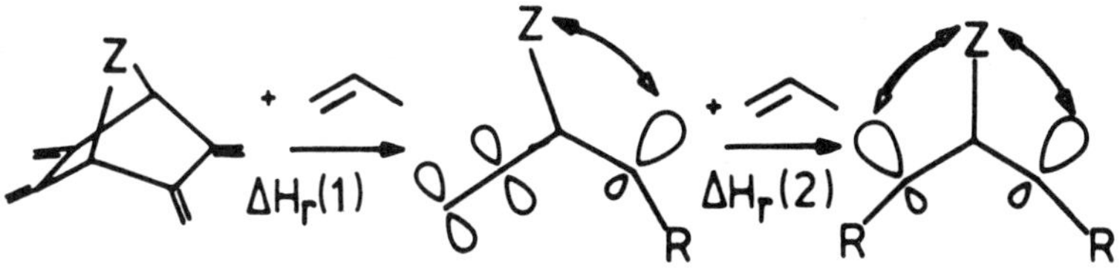

Z + $\Delta H_r(1)$ Z R + $\Delta H_r(2)$ Z R R

$|\Delta H_r(1)| > |\Delta H_r(2)|$ for Z = O, CH_2, $(CH_3)_2C{=}C$

and CH=CH

The out-of-plane deformation of the π-C(2,3) systems in 8 and 28 cannot be attributed to homoconjugative interactions between the endocyclic double bond and the exocyclic *s-cis*-butadiene moieties since no deviation from planarity was observed [37] in the X-ray diffraction study of compounds 29 (maleic anhydride adduct to 5,6-bis-(methylene)bicyclo[2.2.2]oct-2-ene [40] and 30 (maleic anhydride monoadduct to 4). The π-anisotropy of the endocyclic double bond in 8 and 28 (and other 2-norbornenes, see below) must be attributed to specific σ-π interactions of the bicyclo[2.2.1]-hept-2-ene system. We think that torsional effects involving the bridgehead H-atoms [41] as well as rehybridization of the olefinic C-atoms due to bond angle deformations [42] would have led to similar out-of-plane distortions for both the exocyclic diene and the R-substituents.

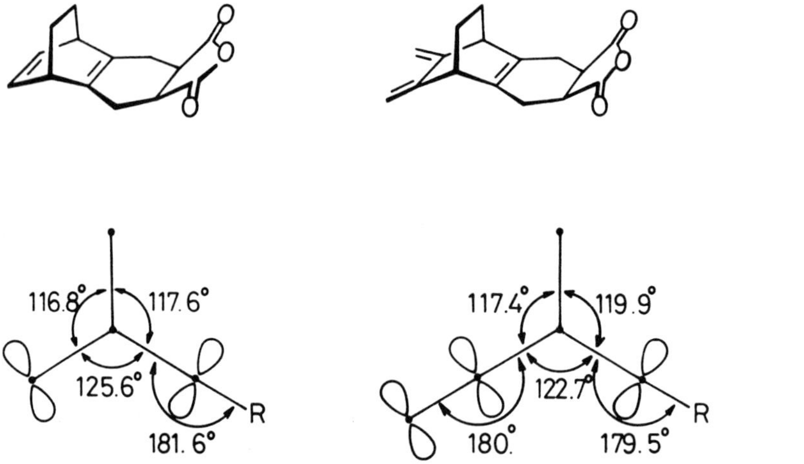

29 30

In apparent agreement with Fukui [36], we attribute the nonplanarity of the 2-norbornene double bond to a polarization of the π-electrons toward the exo face. We suggest that MO calculations should include electron correlation to show this effect and to give minimized geometries with relatively large out-of-plane deformation as observed by X-ray crystallography of 2-norbornene derivatives. Qualitative predictions of these effects have been presented by us earlier [47]. Our analysis considers the electron distribution in a cyclopentane ring (this corresponds more closely to a cyclopropane than to a cyclobutane or cyclohexane ring as far as its polarizability (hyperconjugative interactions) is concerned) in the C_s symmetry field of an interacting olefenic system such as in 2-norbornene and 2,3-bis(methylene)norbornane [47]. Under these circumstances, the a' MOs of the C-1,7,4,5,6 five-membered ring "concentrates" the electron density along the mirror plane and in the endo face whereas the a" MOs "concentrate" it away from the mirror plane and in the exo face of these systems (analogy with the e_s and e_a Walsh orbitals of cyclopropane). Considering σ-π electron repulsions, a simple explanation is thus found for the nonplanarity of 2-norbornene, the planarity of the exocyclic butadiene in 2,3-bis(methylene)norbornane derivatives, the exo preference of radical additions to syn-sesquinorbornene double bond and the loss of this exo face selectivity for the electronically excited syn-sesquinorbornene [38b].

2. The Face Stereoselectivity of Diels-Alder Additions of Exocyclic Dienes Grafted onto Bicyclic Skeletons.

The stereoselectivity of the Diels-Alder additions to "isodicyclopentadiene" 31 has been investigated by several authors [43,44]. In the cycloaddition of methyl acrylate, methyl propynoate [43b,44], N-methyltriazolinedione [39], methyl acetylenedicarboxylate, phenyl vinyl sulfone, benzoquinone, and benzyne [43b] endo attack was found to be preferred.

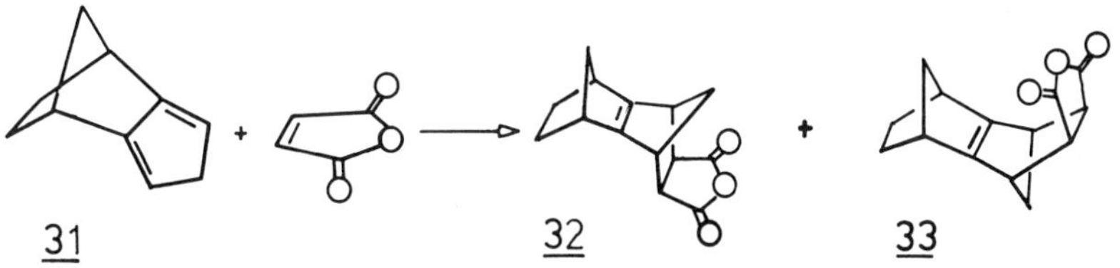

31 32 33

The photooxidation of 31, however, proceeded only with moderate endo stereoselectivity [45]. With maleic anhydride, the endo/exo-face selectivity (product ratio 32/33) varied between 55:45 and 35:65 depending upon the temperature and the solvent [38]. Paquette et al. attributed the endo face selectivity of the cycloaddition of 1 to a kinetic stereoelectronic control involving secondary orbital interactions between the dienes and the dienophiles [43b,45]. It was assumed that the adducts resulting from the endo-face or from the exo-face attack of 1 should have similar stabilities and should not affect the relative rate of their formation.

We have prepared the (2-norborneno)[c]furan (34), and oxa analog of 31 [47]. Because of the aromaticity of the furan ring, 34 can be equilibrated with its Diels-Alder adducts even at room temperature where strong dienophiles are used. This, in principle, allows one to test whether the cycloadditions give different face selectivities under kinetic and thermodynamic control. With methyl acetylenedicarboxylate and maleic anhydride we have found that the endo-face attack is always preferred.

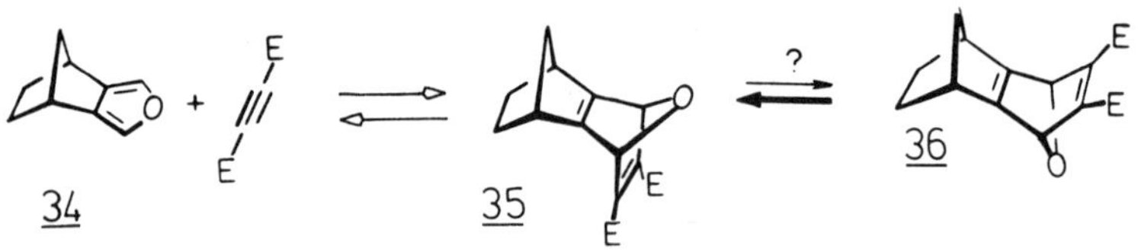

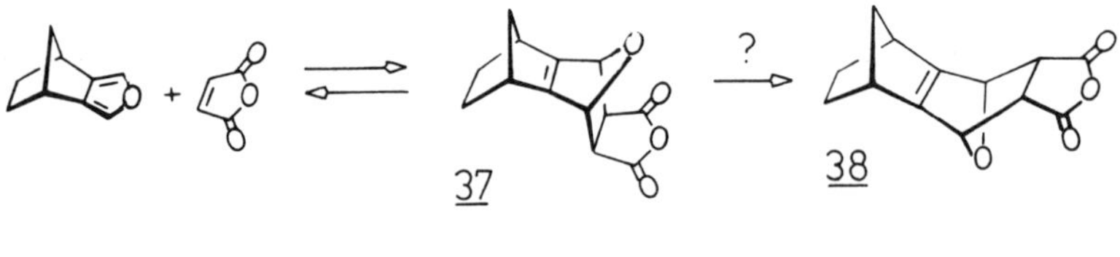

34 35 36

37 38

\+ 2 other isomers

A priori, four isomeric adducts (37, 38 + two other isomers) can be formed in the addition of 34 to maleic anhydride. At -60° ($CDCl_3$) or at higher temperatures, only 37 was observed. The equilibrium constants $K_{21^\circ} \simeq 2.2$ and $K_{77^\circ} \simeq 0.6$ L/mol^{-1} were measured for the reaction 34 + maleic anhydride $\rightleftarrows$ 37. No trace of the other isomeric adducts (or other products) could be detected even after prolonged heating in $CDCl_3$ or benzene at 150°. Considering a rate constant k > 1/60 L/mol s^{-1} at 77° for the endo attack, we estimate a $\Delta G^{\ddagger}$ < 23.5 kcal/mol and $\Delta H^{\ddagger}$ < 13 kcal/mol with $\Delta S^{\ddagger}$ = -30 eu. At 150° and after 10 days, we can consider k < 1/200 3600 L/mol s^{-1}, thus giving $\Delta G^{\ddagger}$ > 36.5 kcal/mol and H 21 kcal/mol for the exo attack, if the exo isomers should be as stable as 37. This is a much higher activation enthalpy than 13 kcal/mol and makes it difficult to postulate that 38 and the other possible isomeric adducts had no time to be generated at 150° after 10 days. It is more reasonable to postulate that 37 is the most stable adduct, and because of that, the products arising from exo-face attack are not observed under our equilibrium conditions. Considering a detection limit of 1% one can estimate 37 to be at least 2.5 kcal/mol more stable than any other isomer. Similarly, methyl acetylenedicarboxylate added to 34 and is equilibrated only with the adduct 35 arising from the endo-face attack. At 21°, the equlibrium constant K = 280 $Lmol^{-1}$ (in $CDCl_3$) was reached in about 12 h. No other adduct could be observed after prolonged heating to 150°. The structure of 37 was deduced from its spectral data and by its reactions. Air oxidation of 37 gave the epoxide 39; catalytical hydrogenation furnished 40, which underwent a thermoneutral dyotropic transfer of hydrogen at 130° (40 ⇄ 41) [48].

37 —air→ 39   40 ⇄(130°) 41

E = COOMe

The structure of the adduct 37 was assigned by X-ray diffraction studies. It showed a nonplanar endocyclic double bond, the out-of-plane deviation being 17° as in the case of syn-sesquinorbornenes [38] (see Figure 3). Repulsive interactions between the H-atoms H-C(4,5 endo) and H-C(9,10 endo) in 37 (separated by 2.35 Å) would tend to reduce the bending of the C(2,7) double bond of this 11-oxa-syn-sesquinorbornene. If one assumes a 10°

deviation due to the 7-oxanorbornene subsystem (see structure 8 [47]) and a similar deviation due to the norbornene subsystem [38], we understand now why the adducts 35 and 37 arising from the exo-face attack to the furan 34 are more stable than their isomers arising from the exo-face attack: in the 11-oxa-syn-sesquinobornenes there is the possibility of a "synergic" effect of the double π anisotropy of the norbornene and 7-oxanorbornene joined together by the C(2,7) double bond. This

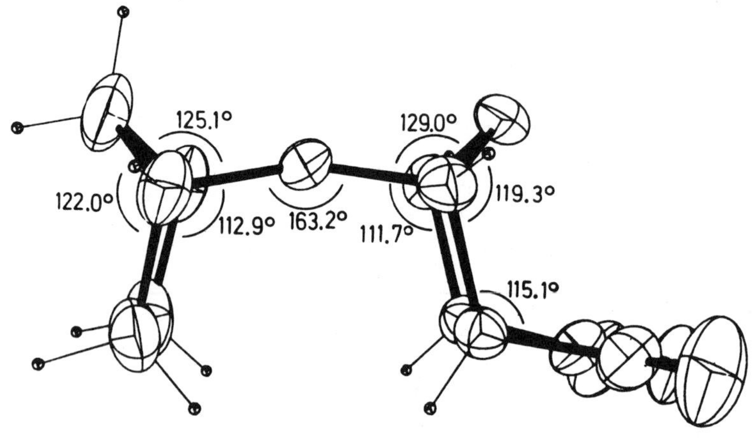

Figure 3. X-Ray structure of the adduct 37. Note that the CH_2 bridge as the oxygen bridge is repelled by the endocyclic double bond whose π electron density must be polarized toward the exo face.

"synergic" effect is not possible in the 11-oxa-anti-sesquinobornene derivatives 36 and 38. The apparent kinetic endo-face selectivity of the cycloadditions of 34 (at low temperature) parallels the relative stability of the isomeric adducts: the Dimroth [21] Bell-Evans-Polanyi principle [22] might well be followed with the Diels-Alder additions of 34 (as for the cycloadditions of the tetraenes 1-6 and their corresponding monoadducts, see above) and perhaps also with those of the "isodicyclopentadiene" 31. This possibility is not demonstrated; it is not ruled out yet.

The exocyclic dienes grafted onto 7-oxabicyclo-[2.2.1]heptanes 42, 43 [49], and 44 added to strong dienophiles preferentially onto the exo-face [50], in contrast to the endo-face selectivity observed generally for the Diels-Alder additions of 31 and 34. The TCNE additions to the (Z)- and (E)-chlorodienes 42 and 43, respectively, show the same face selectivity (exo vs endo attack, 80:20 at 130° in C_6H_5Cl). In the first case, kinetic face stereoselectivity has been proved not to be governed by the stability of the adducts. It should be noted, however, that there is no prerequisite for the same kinetic face selectivity in the cycloadditions of (Z)- and (E)-substituted s-cis-butadienes grafted onto bicyclic skeletons, even though the reactions should exhibit the same exothermicity. The addition of N-phenyltriazolinedione to 44 was exo-face selective at >97%.

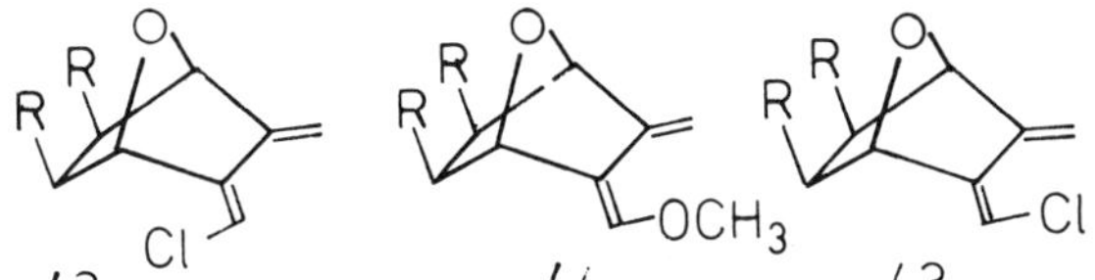

R = CH_2Cl

The substituted tetraene 45 added TCNE to the methoxy-diene much faster than to the unsubstituted diene moiety, also with good exo-face selectivity (85:15 at 20° in acetone or benzene [50]).

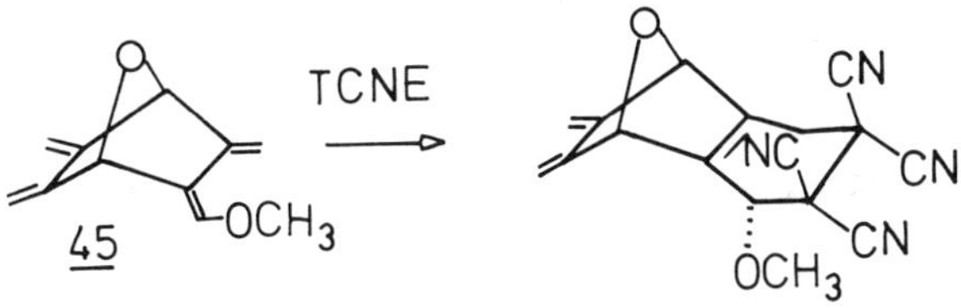

The simplest explanations for the observed exo-face selectivity in the Diels-Alder additions of 42-45 is to assume a coordination (formation of a charge-transfer complex) of the dienophiles by the O(7)-atom (entropy and/or enthalpy effects). The ethereal bridge could assist the cycloaddition onto the exo face.

exo face preferred

5,6-Bis((E)-deuteriomethylene)bicyclo[2.2.2]oct-2-ene (46) (obtained by Zn/Cu-dioxane/D_2O reduction of 5,6-bis((E)-chloromethylene)bicyclo[2.2.2]oct-2-ene (47)) adds to strong dienophiles with endo face selectivity [51]. In this case one cannot invoke a

46 + maleic anhydride + COOMe–C≡C–COOMe + TCNE + N-phenyltriazolinedione

60/40 70/30 75/25 95/5

difference in the stability of the adducts arising from the endo- or exo-face attack. The stereoselectivity increases with the electronic demand of the dienophile; the faster the cycloaddition, the higher the endo-face preference. It can be attributed to a differential attractive steric effect of the two bridges of the bicyclic skeleton. Since the unsaturated bridge C-2,3 is more polarizable than the saturated bridge C-7,8, the strong dienophiles prefer to attack onto the endo face. It is possible also that the endo-face selectivity is due to differential repulsive steric effects between the endo and exo faces. We think, however, that this steric argument is not valid, at least not in the case of the Diels-Alder additions of the dichlorodienes 48-50 [52], where the two faces of the dienes are expected to offer the same bulk to the approaching dienophile. The endo face was preferred also for the TCNE additions to 47-50, probably because the unsaturated or substituted C-2,3 bridge is more polarizable than the $H_2C(7,8)$ bridge. The ketone 51 and the alcohol 52 added to TCNE with exo face selectivity [52].

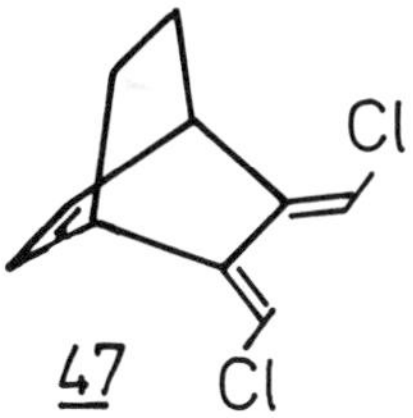

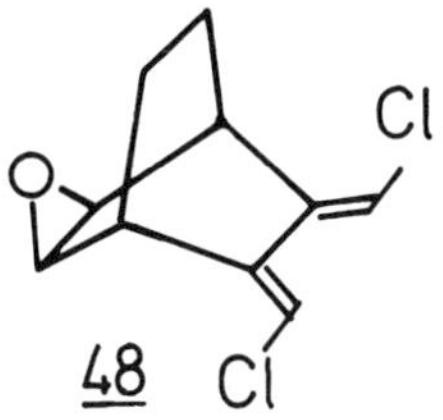

endo/exo	47	48	49
	78/22	80/20	68/32

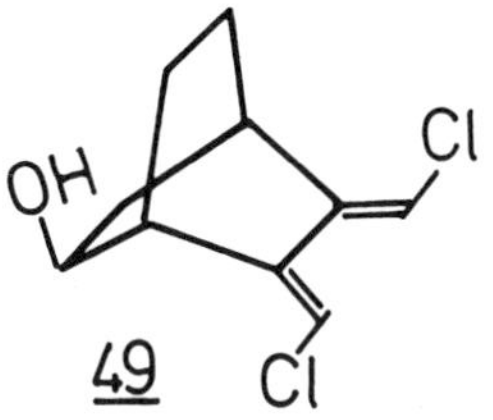

	50	51	52
	60/40	30/70	3/96

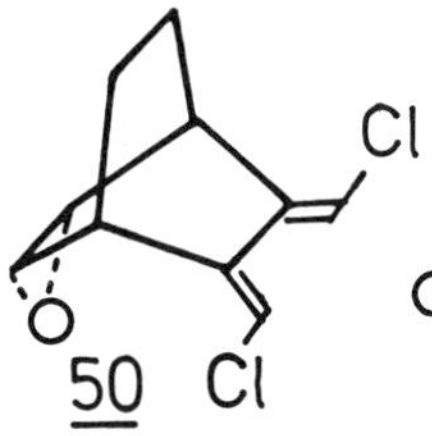

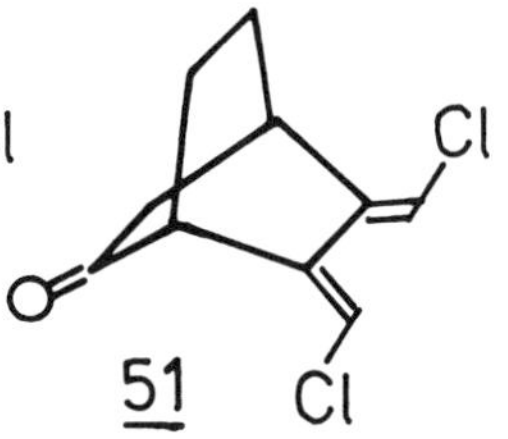

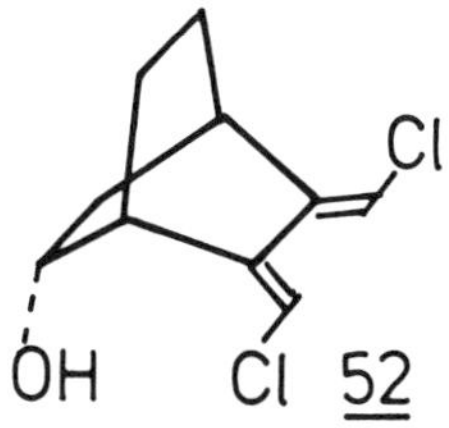

Analysis of the maleic anhydride adducts to 46 showed that the anti-Alder-rule isomers (Scheme 4) were slightly preferred over those of the Alder-rule additions (52:48). In the former mode of addition this implies, in principle, a smaller steric demand than in the latter. The endo face is preferred, thus confirming the hypothesis that the endo face of 46 and 47 can better assist a Diels-Alder transition state than the exo face because of its higher polarizability.

The endo face selectivity of the Diels-Alder additions of 46 and 47 contrasts with the exo-face selectivity reported for the additions of methyl propynoate and acetylenedicarboxylate to the cyclopentadiene analog 53 [46b,53]. However, it parallels the endo preference observed for the *N*-methyltriazolinedione addition to 53 [39].

We think more experimental data must be collected before a general theory can be developed of the face selectivity of the Diels-Alder additions of dienes annelated to bicyclic skeletons. For the moment we must retain the steric factor, the differential polar and polarizability effect of the

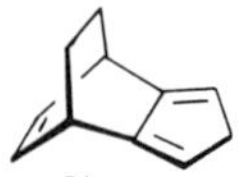

53

two bridges of the bicyclic systems. When the isomeric adducts arising from the exo and endo face attacks have different stabilities it is possible that it can affect the kinetic face stereoselectivity.

Scheme 4

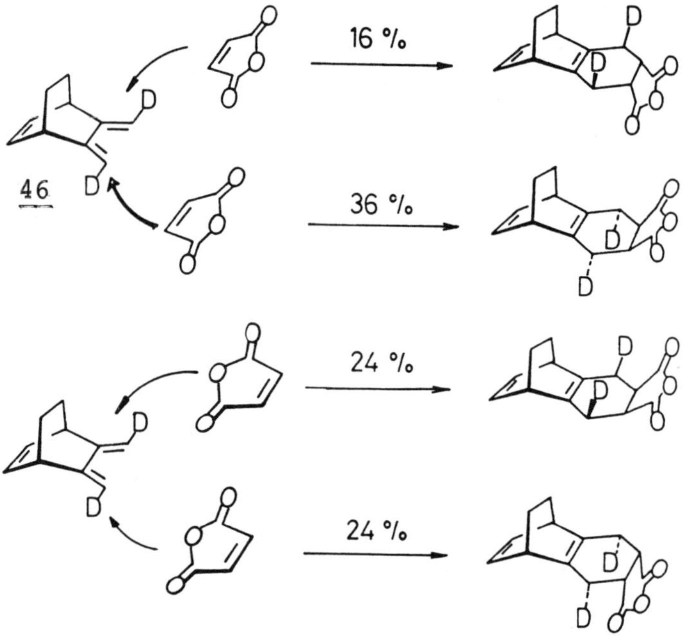

46

16 %

36 %

24 %

24 %

3. The Regioselectivity of Diels-Alder Additions of Exocyclic Dienes Grafted onto Bicyclic Skeletons Remotely Perturbed.

Our synthesis of daunomycine [13] employs an intermediate 10 (Scheme 2), that is, a 2,3-bis-(methylene)-7-oxanorbornane homoconjugated with an "exocyclic" double bond. The latter appears to induce a regioselective Diels-Alder addition of methyl vinyl ketone and readily allows the generation of the daunomycine skeleton with the correct regiochemistry (control of the substitution pattern of rings A and D). We show now that remote substitution of the bicyclic skeleton of a 2,3-bis-(methylene)norbornane or 7-oxanorbornane can lead to regioselective cycloadditions of the dienes.

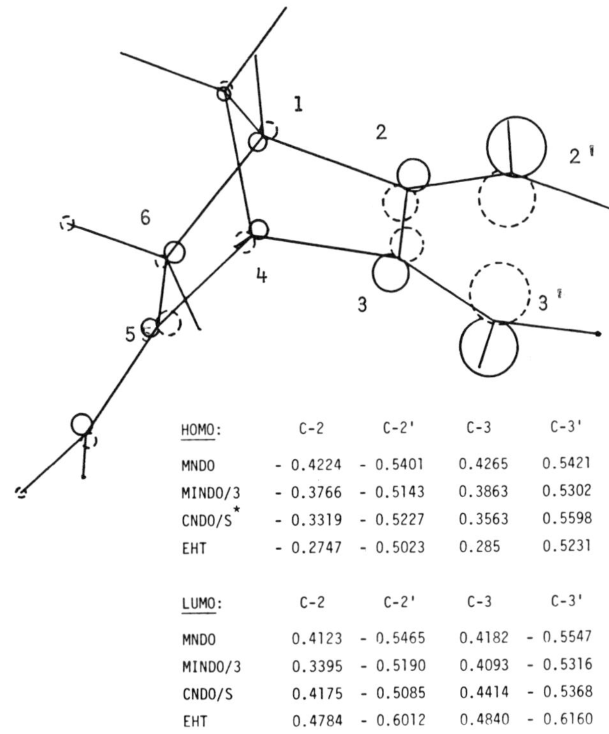

HOMO:	C-2	C-2'	C-3	C-3'
MNDO	- 0.4224	- 0.5401	0.4265	0.5421
MINDO/3	- 0.3766	- 0.5143	0.3863	0.5302
CNDO/S*	- 0.3319	- 0.5227	0.3563	0.5598
EHT	- 0.2747	- 0.5023	0.285	0.5231

LUMO:	C-2	C-2'	C-3	C-3'
MNDO	0.4123	- 0.5465	0.4182	- 0.5547
MINDO/3	0.3395	- 0.5190	0.4093	- 0.5316
CNDO/S	0.4175	- 0.5085	0.4414	- 0.5368
EHT	0.4784	- 0.6012	0.4840	- 0.6160

Figure 4. Molecular orbital coefficients for the triene 54.

The PMO theory [54] has been very successful in rationalizing the Diels-Alder reactivity [19,20], stereoselectivity, and regioselectivity [55]. Accordingly and to a first approximation, the regioselectivity of the Diels-Alder addition of 10 to methyl vinyl ketone should be given by the shape (p coefficients) of the HOMO of 10. The HOMO of the model triene 54 calculated by the MNDO, MINDO/3, CNDO/S, and EHT techniques (minimized geometries) shows comparable p coefficients for the two methylene groups of the *s-cis*-butadiene moiety (see Figure 4). A possible homoconjugative interaction between the exocyclic diene and monoolefin does not manifest itself by the shape of this HOMO. Thus, the PMO theory would predict little regioselectivity for the Diels-Alder additions of 54, contrary to experiment that gave with various dienophiles a good "para" regioselectivity [56] as in the case of 10.

("méta") ("para")

54 (minor) (major)

The Woodward-Katz visualization of Diels-Alder transition states as diradicaloids [57] seems to be more appropriate for predicting the "para" regioselectivity observed with 54. If one writes the possible configurations of a diradicaloid which is supposed to represent the transition state of the "para" mode of addition of 54 to a strong dienophile, one sees that a charge-transfer configuration putting a negative charge α to the electron-withdrawing substituent A and a positive charge homoconjugated with the exocyclic double bond can profit from the homoconjugative stabilization. This situation is not possible in the transition state of the "meta" mode of addition. Following this model, we predict 5-(dicyanomethylene)-2,3-*bis*(methylene)norbornane (55) to add preferentially with "meta" regioselectivity. This was observed experimentally for various strong dienophiles [56]. If one now replaces the 5-methylene group of 54 by a carbonyl group which is supposed to be as electron-withdrawing as the dicyanomethylene group [58] in 55, one would predict also "meta" regioselective Diels-Alder additions for 5,6-*bis*(methylene)-2-norbornanone 56. The opposite "*para*" regioselectivity was actually observed for the additions of methyl vinyl ketone and methyl propynoate [59].

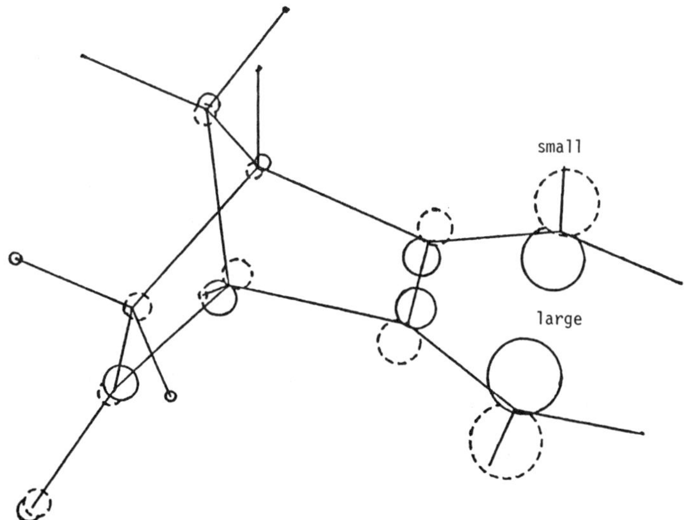

Figure 5. HOMO (EHT) of 5,6-bis(methylene)-2-norbornanone (56).

The HOMO of 56 (see Figure 5) shows a significant difference in the size of the p coefficients of the two methylene group of the diene. According to the PMO theory and considering a Diels-Alder addition with normal electronic demand the "para" regioselectivity is predicted and observed [59]. The shape of the HOMO of 56 suggests also that there is an important hyperconjugative interaction of the type n(CO)↔σC(1,2)↔πC(5,6) in 56 that makes the carbonyl group act as an electron donating homoconjugated substituent. If one writes the diradicaloid species that represent the transition state of the Diels-Alder additions of 56 in the "para" mode, one sees that the "usual" charge-transfer configuration places the positive charge homoconjugated with the carbonyl group. This configuration will be stabilized by the oxo function if it can be considered as electron-donating rather than as electron-withdrawing substituent.

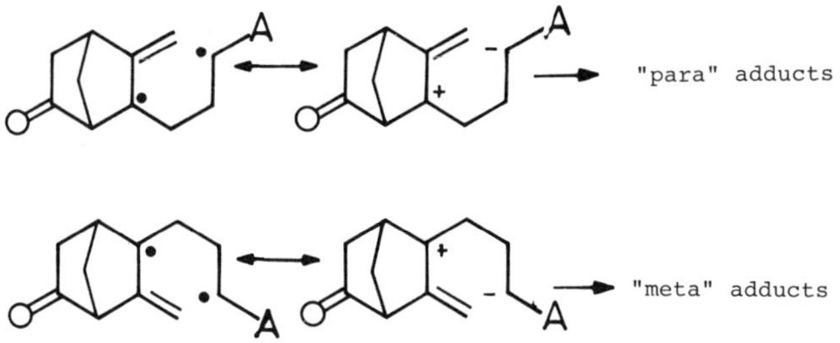

"para" adducts

"meta" adducts

MINDO/3 and MNDO calculations [60] of substituted 2-norbornyl cations predicted the 6-oxo-2-norbornyl cation 57 to be more stable than the 5-oxo-isomer 58.

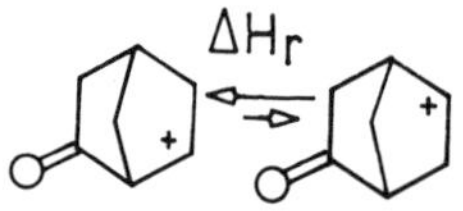

ΔH_r

57 58

MINDO/3 ΔH_r = +8 kcal/mol

MNDO ΔH_r = +2.8 kcal/mol

The stabilizing n(CO)↔σC(1,6)↔pC(2) interaction overides the -I and -M effects of the homoconjugated carbonyl group in 57. Accordingly we predict the electrophilic additions (E^+) of the double bond of 2-norborn-5-enone (59) to be regioselective in the sense that the counterion X^- should attack C(6) rather than C(5). This has been confirmed experimentally [61].

Benzeneselenylchloride (PhSeCl) and bromide (PhSeBr), 2-nitro-(NBSCl), and 2,4-dinitrobenzenesulfenyl chloride (DNBSCl) add to olefins [62,63] and generate bridged seleniranium and sulfeniranium ion intermediates whose stability and reactivity can be influenced by homoconjugated endocyclic [63,64] or exocyclic double bond [65] and by the medium [63]. With PhSeCl and PhSeBr in $CHCl_3$, CH CN, or AcOH, 59 gave the adducts 60 (93%) and 61 (95%,isolated), respectively. At 20°, the reactions were instantaneous. At -78° in THF, the addition 59 + PhSeBr gave 61 quantitatively in ca. 2-5 h. The arenesulfenyl chlorides added more slowly ($CHCl_3$, 20°, 12-16 h). With NBSCl only 62 was found (93%,

EX =		
PhSeCl:	60	64
PhSeBr:	61	65
NBSCl :	62	66
DNBSCl:	63	67

(E = DNBS)

isolated). In CH_3CN or AcOH, the addition of DNBSCl gave 63 (60-70%) together with 68, isolated as the acid 69 (20-30%). In AcOH + 2 eq of $LiClO_4$, 68 was the major product (80%). No trace of the isomeric adducts 64-67 could be detected (HPLC, ^{1}H NMR (360 MHz)) in the mother liquors after crystallization of 60-63. Prolonged heating in CH_3CN (90°, 10 h) of the *a priori* most labile adduct 61 [68] led to the slow formation of 65 and polymers, thus confirming that the adducts were formed under kinetic control. The exo preference for the electrophile (E^+) attack of 2-norbornenes was expected [69]; the high preference of the endo-C(6) vs endo-C(5) attack by the counterion X^- is most simply attributed to the higher polarizability of the C(6)-E^+ bond rather than of the C(5)-E^+ bond in the cationic intermediate 71; the limiting structure 70 is more stable than 72 because of the hyperconjugative interaction n(CO)$\leftrightarrow\sigma$C(1,2)$\leftrightarrow$pC(6). The carbonyl group behaves as an electron-donating group (polarizability) rather than as an electron-withdrawing group (-M, -I). The "harder" the electrophile and the stronger the ionizing power of the medium, the more favored is the n(CO), σC(1,2) participation and the σC(1,2) leakage (frangomeric effect [70]).

Steric hindrance of the quenching of X^- with 71 could be larger at C(5) than C(6) and thus explain the observed regioselectivity. Another hypothesis would be to assume an addition of X^- to the carbonyl group followed by a simple 1,3-transfer as shown in 73. The additions of PhSeCl and PhSeBr to 2-bicyclo[2.2.2]oct-5-enone 74 gave the adducts 75+76 and 77+78, respectively, in high yield [66] (20°, $CHCl_3$ or CH_3CN, 6-12 h). Since the same high regioselectivity was observed for the endo and exo

73 74 +EX

EX = PhSeCl: 75 (~70%) 76 (~30%)

PhSeBr: 77 (~70%) 78 (~30%)

attacks of the C(5,6) double bond of 74 the two latter hypotheses are not valid, at least not for 74 + PhSeCl → 76 and 74 + PhSeBr → 78.

As expected for olefins perturbed by -I substituents such as CN and Cl, the addition of PhSeCl to the 2-chloronorborn-5-ene-2-carbonitriles were much slower (48 h, 20°, $CHCl_3$) than those of norbornene and 2-norborn-5-enone (59). They were highly stereo- and regioselective giving the corresponding adducts 79 and 80 [66]. Similarly, the 2-chlorobicyclo[2.2.2]oct-5-ene-2-carbonitriles added PhSeCl very slowly (20% of adducts formed after 72 h at 20°, $CHCl_3$, excess of PhSeCl) giving the adducts 81 and 82, respectively, with the same high regioselectivity (>95%) but opposite to that observed with the enones 59 and 74.

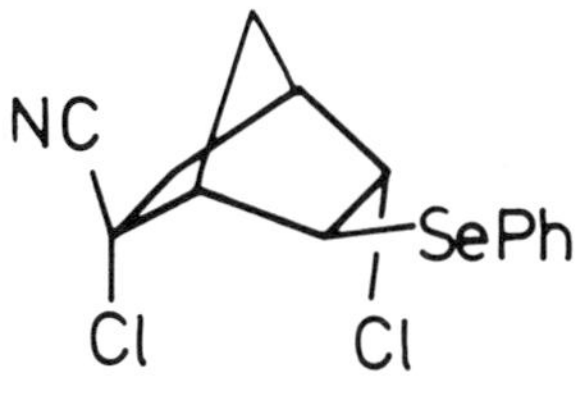

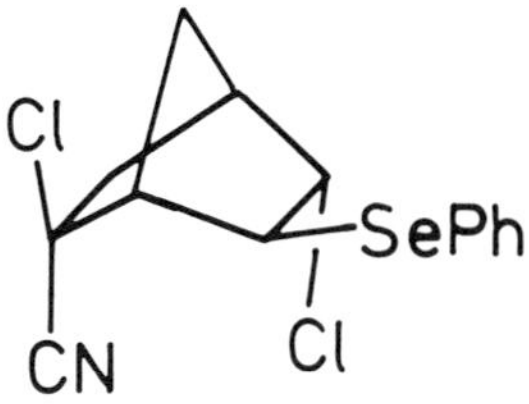

This can be attributed to the field effect of the CN and Cl substituents that makes the C(6)-E bond in the bridged seleniranium ion intermediate 83 less prone to nucleophilic displacement than the C(5)-E^+ bond. The high preference of the endo-C(5) vs endo-C(6) attack by X^- could also be attributed to steric hindrance in its approach to the endo face of 83.

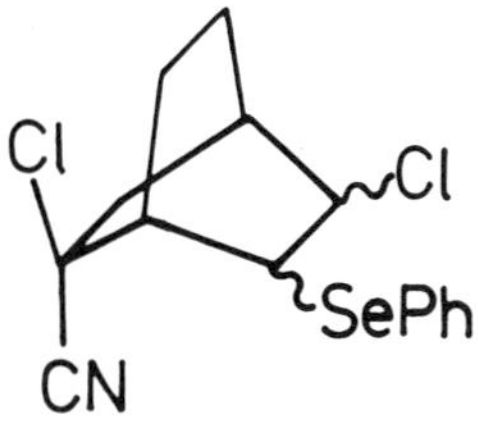

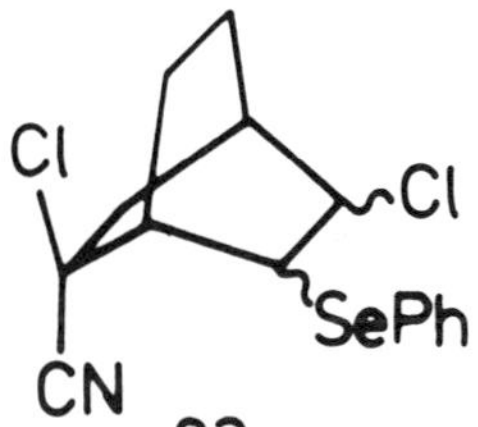

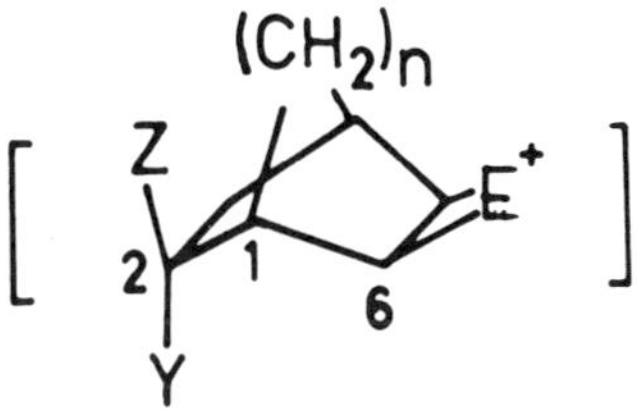

The dimethylacetal 84 reacted with PhSeCl in $CHCl_3$ (20°, ca. 10s) and gave 85 + 86. Similarly, the adducts 87 + 88 were formed on treatment with PhSeBr ($CDCl_3$, 20°, 10s). The large rate enhancement of these reactions compared with those of the addition of PhSeCl to the 2-chloronorborn-5-ene-2-

carbonitriles was unexpected for MeO substituents which have a field effect and a bulk analogous to those of Cl and CN. With the "harder" electro-

EX =			
PhSeCl:	85 (~15%)	86 (~85%)	-
PhSeBr:	87 (~15%)	88 (~85%)	-
NbSCl :	89 (~90%)	90 (< 5%)	93 (~ 5%)
DNBSCl:	91 (~90%)	92 (< 5%)	94 (~10%)

philes NBSCl and DNBSCl (in $CHCl_3$) an opposite regioselectivity was observed, the adducts resulting from the Cl^- attack onto C(5) were not found (90, 92) and 93 and 94 were formed in small amounts together with 89 and 91, respectively. As in the case of 71, this suggests that the σC(1,2) bond can participate in stabilizing the cationic intermediates 83 and orienting their reactions. The larger the electronic demand, the more the participation intervenes and competes with the steric effect of the substituent Y at C(2) in directing the quenching of the counter-ion. With +M substituents at C(2), the leakage of the σC(1,2) bond can become a favorable process. This interpretation was confirmed by the addition of DNBSCl (CD_3CN) to the ethyleneacetal 95 that gave only the cyclopentenylacetic acid derivative 96. In 95, the n(O) orbitals are forced to stay well aligned with the σC(1,2) bond, thus enhancing the electron-donating effect of the n(O)↔σC(1,2)↔C(6) interaction.

The relatively fast electrophilic additions of the C(5,6) double bond of 59 and 74 and their high regioselectivity opposite to that of the addition of the corresponding chlorocarbonitrile derivatives confirm that the homoconjugated carbonyl group in 59 and 74 can act as an electron-donating substituent because of the hyperconjugative n(CO)↔σC(1,2)↔πC(6) interaction (the polarizability of the O=C(2)-C(1) bonds overrides the field effect of the carbonyl group).

There are several examples of substituents that can be electron withdrawing or electron donating depending upon the sign and magnitude of the electronic demand (±q) [71]. This phenomenon can be explained by the simple electrostatic Field Model [72] that considers charge-dipole (μ) interactions ($V_C = \pm q\mu\cos\Theta/\varepsilon r^2$) and the charge-induced dipole interactions ($V_I = -q^2\alpha/2\varepsilon r^4$), the latter being always stabilizing. The polarizability can be evaluated by MO theory [54]. This has been recognized for instance for alkyl substituents [73], the cyano group [74], the α-carbonyl group [75], and the diene-$Fe(CO)_3$ group [76].

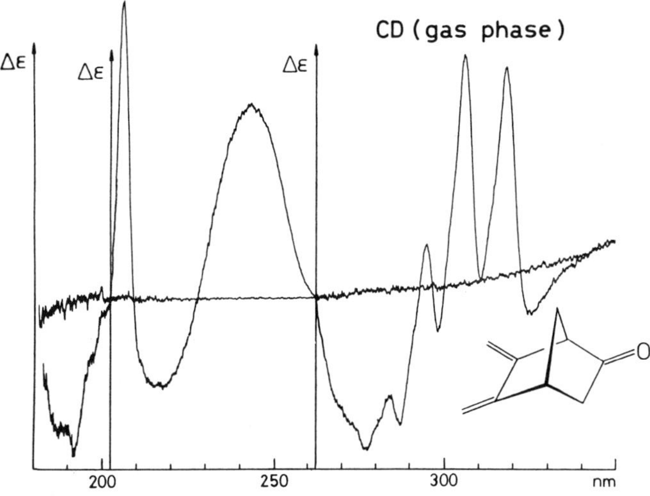

Figure 6. Gas phase CD spectrum of (+)-56.

The CD spectrum of (+)-(1R)-5,6-dimethylidene-2-norbornanone ((+)-56) in isooctane [77] or in the gas phase (see Figure 6) displays unexpected features between 265 and 340 nm. Instead of showing a strong Cotton Effect for the β,γ-unsaturated carbonyl [78], it presents the superposition of two Cotton Effects of opposite sign. The low-energy signal has a fine Franck-Condon contour typical of a "delocalized" n(CO)$\leftrightarrow\pi^*$(CO) transition. The second is a broad band suggesting a charge-transfer diene-skeleton-carbonyl transition. CNDO/S calculations [79] including all singly and doubly excited configuration interactions on 56 reproduced the observed CD spectrum. We calculated also the electronic spectra of the rigid bicyclic ketones 96, 59, and 97-99 (using MNDO minimized geometries) and predicted that an "extra" band between those attributed to the carbonyl and olefinic chromophores must be present in the UV and CD spectra of all these systems (see Table 2). This was confirmed experimentaly for 2-deuterio-7-norbornenone [80] and for 56, 59, 74, and 96 [81]. According to the CNDO/S calculations, the "extra" band in the β,γ-unsaturated ketones of Table 2 is a mixed transition with substantial charge-transfer components of the form π(olefin)$\rightarrow\pi^*$(CO) and n(CO),$\sigma\rightarrow\pi^*$((olefin).

Conclusion

A great variety of exocyclic dienes, tetraenes, and the [l.m.n]hericenes are now readily available compounds. They are potential precursors in the synthesis of polycyclic, polyfunctional molecules because their Diels-Alder additions can be stereo- and regioselective. This has been demonstrated by the development of a highly versatile synthesis of anthracyclinones starting with 2,3,5,6-tetrakis-(methylene)-7-oxanorbornane. The substitution of bicyclic skeletons affects the physical and chemical properties of the exocyclic dienes grafted onto them. The study of these transannular interactions has just begun and has delivered already a wealth of information on the reactivity of π-systems, the role of homoconjugative and hyperconjugative interactions, and the electron donating ability of homoconjugated methylene and carbonyl groups. Critical evaluation of models such as the popular PMO theory of stereo- and regioselectivity of cycloadditions is now possible. The Woodward-Katz representation of a Diels-Alder transition state merits reconsideration. A second thought

Table 2. Calculated (CNDO/S) and experimental (in parentheses, by CD) transitions (nm) in [81]

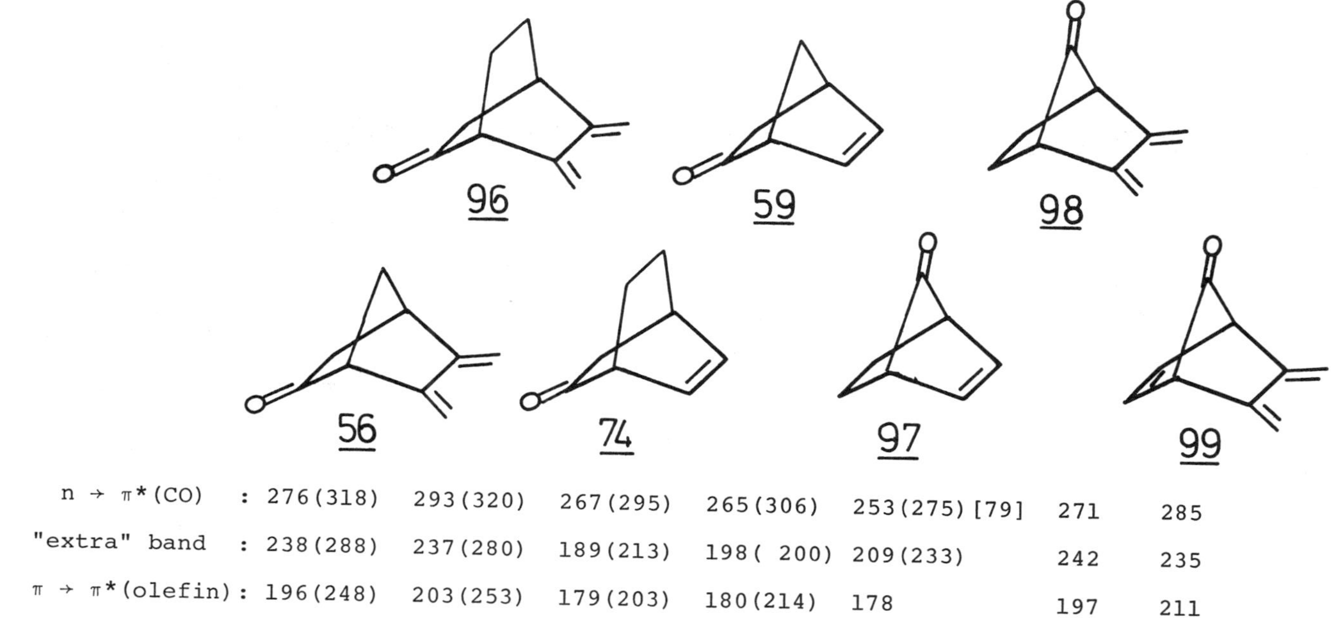

n → π*(CO) :	276(318)	293(320)	267(295)	265(306)	253(275)[79]	271	285
"extra" band :	238(288)	237(280)	189(213)	198(200)	209(233)	242	235
π → π*(olefin):	196(248)	203(253)	179(203)	180(214)	178	197	211

should be given to the role of exothermicity and the Dimroth, Bell-Evans-Polanyi principle on the Diels-Alder reactivity although this reaction is assumed to have early transition state and to deviate from this principle because of the Alder endo-rule.

Our polyenes give access to a large variety of olefins with two different faces in which the σ-π mixing can have dramatic consequences on their properties, a question difficult to comprehend. New avenues are ahead of organic synthesis as well as boulevards leading to fascinating basic problems in the reactivity of π systems.

Acknowledgements. This work would have been impossible without the enthusiastic collaboration of my co-workers whose names are given in the references. I wish to thank all of them as well as Hoffmann-La Roche & Co. (Basel), the Swiss National Science Foundation and the Fonds Herbette (Lausanne) for generous financial support.

Discussion

Traylor: I have a question about the last picture you have drawn. That, as I understand it, is a hyperconjugative picture of the carbonyl group. If you replace the carbonyl with a silicon it should be even better. It is electron withdrawing and has

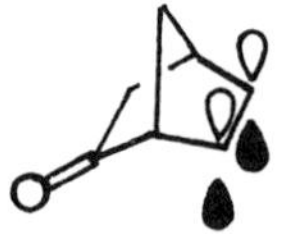

an empty p orbital just like the carbonyl, but it is very electron donating in that situation. If you make it you can assume it will be extremely stabilized. That always asks for the nucleophile to come in on the trans face and that is what bothers me. As far as I know hyperconjugative additions always prefer the trans arrangement, but you get cis.

Vogel: With norbornene and bicyclo[2.2.2]octene derivatives, we get trans additions and/or leakage of the C(1,2) bond. The electrophile attacks onto the exo face and the nucleophile adds from the endo side as shown for the intermediate i.

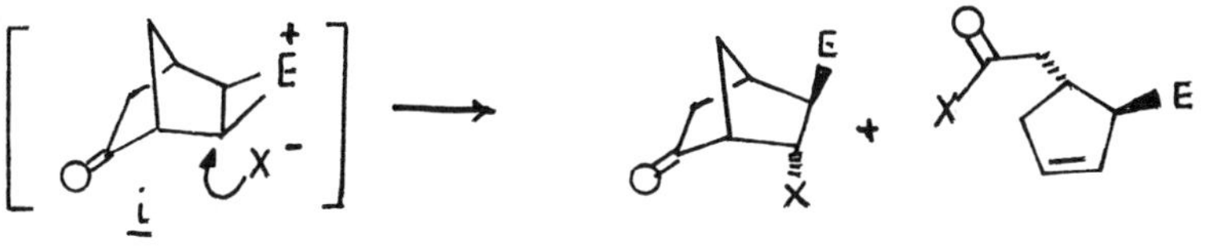

Traylor: Yes, but that is cis with respect to the hyperconjugative group. Norbornene likes exo and hyperconjugation likes exo. So why does it come on endo? If the carbonyl is not there will it come in exo?

Vogel: Electrophiles such as arenesulfenyl halides and areneselenyl halides add to olefins generating epi-sulfonium and epi-selenonium ion intermediates i. Thus, the situation here is different from that of a S_N1 solvolysis of a 2-norbornyl derivative. The nucleophile (halide) attacks the most electrophlic carbon of i or the carbon which bears the weakest $C-E^+$ bond (the most polarizable carbon). We have proposed that the $C(6)-E^+$ bond in i is more labile than the $C(5)-E^+$ bond because of the hyperconjugative interaction with the n(CO), σC(1,2) system. With a harder electrophile such as $AcOH + H_2SO_4$, norbornenone is reported to yield two volatile acetates ii and iii in small amounts together with polymeric material [H. Krieger, Ann. Acad. Scient. Fennicae, Ser. AII 109, 7 (1961)].

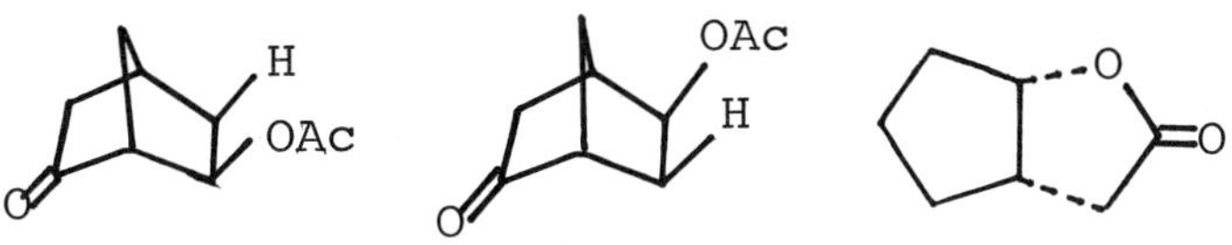

ii iii iv

In this case, the electrophile H^+ generates cationic intermediates of different types than i. We have also studied the acid additions to norbornenone. Contrary to Kriege, we do not detect any trace of acetate ii by treatment of norbornenone with CF_3SO_3H or HSO_3F in AcOH. Beside trace amount of iii, one obtained lactone iv whose formation can be explained by invoking the following mechanism:

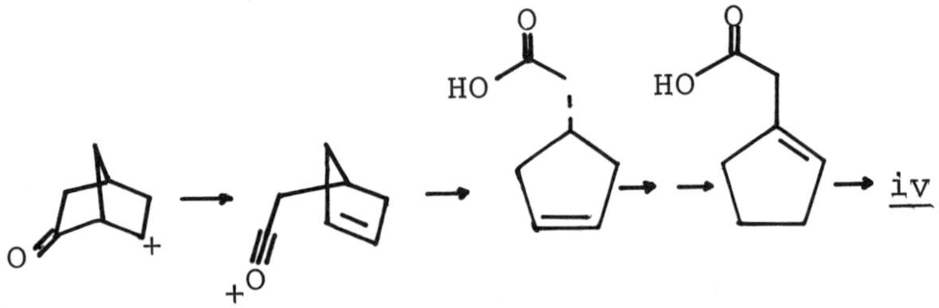

Houk: Let us suppose your electrophile is equally bonded on both sides, or more or less so. You do have an asymmetric system that is polarized on one side more than the other. If you were a nucleophile or anion, on which side of the bottom face would you attack? You would choose the side next to the dipole where the carbonyl is located. You don't necessarily have to propose hyperconjugation.

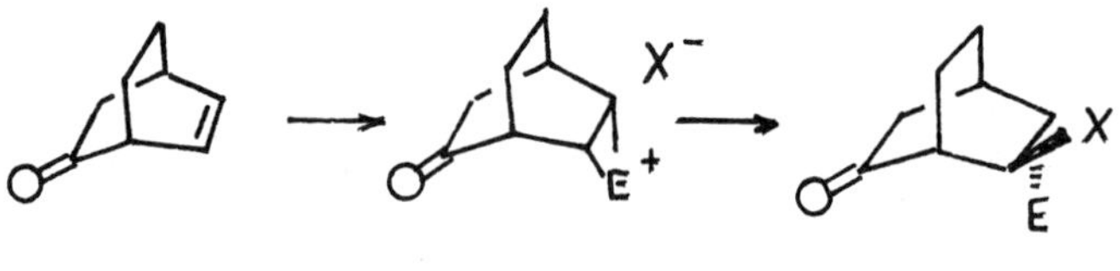

Vogel: You are perfectly right. Your hypothesis could explain our results with norbornenone. Nevertheless, we have an experiment for which, we think, it is difficult to maintain it. In the electrophilic additions to the C(5,6) double bond of bicyclo[2.2.2]oct-5-en-2-one both the exo and endo faces react with the same high regioselectivity. At least, in the case of the endo attack by the electrophile, the nucleophile must be trapped in the exo face in a position far removed from the carbonyl group. We favor the hyperconjugation interpretation because it allows us to explain also the regioselectivity of the electrophilic additions to the double bond of norbornenone acetals. This hypothesis can also explain the regioselectivity of the Diels-Alder additions to 5,6-bis(methylene)-2-norbornanone.

Gleiter: You are telling us that an electron-pushing effect of your carbonyl group explains your effect. This is a ground state effect and to support it you consider n,π* excited states. You argue that the carbonyl group is donating electron density via the lone pair orbitals. You say that CNDO results support this prediction of a n,π* charge-transfer state. You are mixing arguments valid for ground state with those for the excited state. This is dangerous.

Vogel: This drawing interprets our suggestion

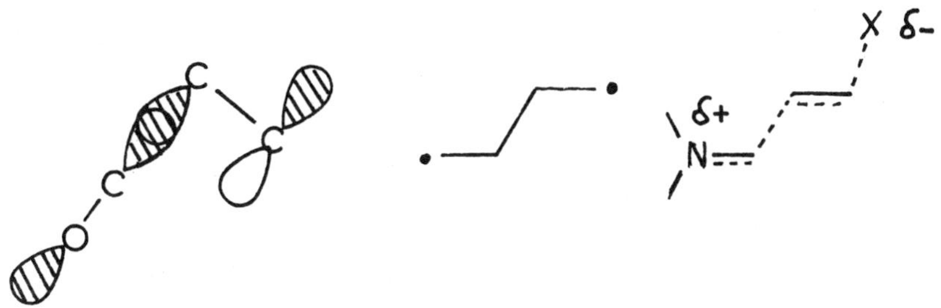

that the regioselectivity of the electrophilic additions to norbornenones are related to the 1,4-diradical problem and more specifically to the Grob fragmentation. We say that a homoconjugated electron-withdrawing function such as the carbonyl group can act as an electron donating group under special circumstances, i.e. if the electron demand is strong enough. There are many known cases where this situation occurs for functions directly attached to a charge center. The polarizability effect of the CO group overrides its charge-dipole effect, and we have interpreted this polarizability effect by invoking a typical hyperconjugative mechanism.

Gleiter: Another comment I would like to make. You said the decrease of intensity of the charge-transfer band was somehow proof that the lone pair donates electrons. The decrease in intensity may be a vibrational coupling.

Herndon: When you talked about the cycloaddition reactions you ruled out homoconjugation in essence as a good explanation for why the second reaction is much slower than the first. Yet you have shown here that homoconjugation is a very important part of describing the reactivity of all of these compounds. I was wondering why homoconjugation would not manifest itself in the cycloadditions where it does exist because I think your PES work shows the homoconjugative interactions are very large.

Vogel: No, they are not large.

Herndon: Large enough. They are on the order of 0.2 to 0.4 eV.

Vogel: No, the effects are small in the norbornane series and nonexistant in the bicyclo[2.2.2]-octane series. For instance the first IPs for 2,3-bis(methylene)norbornane, 5,6-bis(methylene)-2-norbornene, and 2,3,5,6-tetrakis(methylene)norbornane are 8.41, 8.48 and 8.34 eV, respectively. I do not want to say that homoconjugation is not important in our polyenes. I do not exclude a small contribution from minor variation in the IPs of our dienes. I say that the available PES data do not give a rationalization of all our kinetic data in terms of variation of the polarizability of our dienes. Transannular interactions between exocyclic dienes and endocyclic double bonds affect the stability of our bicyclic systems. We say that homoconjugative and hyperconjugative interactions modify the exothermicity of our Diels-Alder additions, and we propose that this might be a factor responsible for the reactivity differences observed between the cycloadditions of a tetraene and those of the corresponding monoadduct (rate constant ratio k_1/k_2 can be small or large depending upon the bridge of the bicyclic system).

Paquette: We certainly all agree with your sentiments that the simpler the explanation and the more general you can make it, the more acceptable it will be to everyone. But within that limitation you must accommodate all of the experimental facts. I would like to hear what you have to say about the experiments we reported on yesterday where we perturbed the carbon atom in the cyclopentadiene ring with alkyl groups in such a way that the steric demands are not changed too much, yet the face stereoselectivity is altered. Certainly, that would cause steric arguments to be considerably weakened in favor of some alternative explanations. That experiment leaves us with an electronic explanation, but I don't know where it leaves you. I would like to know what your explanation might be.

Vogel: Our results suggest that face selectivity in Diels-Alder additions of dienes grafted onto bicyclic skeletons is due to several factors. A priori, one can list:

(1) differential steric effects
(2) differential dipole effects
(3) differential polarizability effects
(4) non-equivalent extension of the π-electron densities (π-anisotropy)
(5) coordination of the dienophile with a substituent or a function of the bicyclic framework
(6) differences in the exothermicities of the cycloadditions (assuming validity of the Dimroth, Bell-Evans-Polanyi principle).

You attribute the face selectivity of the Diels-Alder additions that you have studied essentially to an electronic factor of the type (3) + (4) since the face selectivities you have observed appear to be correlated with the shapes of the subHOMO of your dienes. When applied to our reactions, the PMO approach that you have used together with Prof. Gleiter led to incoherent predictions. In our hands, the shapes and energies of what you consider the important sub HOMO's were dependent upon the calculation techniques and very small geometric variations of these molecules. This was true for your and our dienes. We think this is of no predictive value. All six factors listed above might intervene simultaneously or not. This depends upon the type of dienes and dienophiles involved.

Let us consider now the cycloadditions that give sesquinorbornene derivatives. The endo attack generate the syn-sesquinorbornenes which we propose are more stable than the anti-sesquinorbornenes.

If one assumes validity of the Dimroth, Bell-Evans-Polanyi principle, the endo attacks would thus be preferred because they are more exothermic than the exo additions giving the syn-sesquinorbornenes. Volumes of activation suggest that Diels-Alder transition states are in fact product-like, thus for a limited series of cycloadditions like yours, it is possible that exothermicity differences have something to do with the relative stabilities of the transition states. I am ready to recognize the existence of a correlation between norbornene π anisotropy and the face selectivity of the Diels-Alder additions of a diene grafted onto norbornane unless factors (1), (2) or/and (5) dominate. What makes the norbornene double bond polarized toward the exo face might also be responsible for the endo face selectivity of most of your dienes. The deviant behavior, I think, can be attributed to steric factors that can be repulsive or attractive. If one considers the following retro-Diels-Alder reaction and assumes the Woodward-Katz representation of its transition state, one of its charage-transfer configurations is reminiscent of a S_N2' reaction, in which the double bond of norbornene pushes preferentially from the exo face, thus making the dienophile to prefer the endo face.

Paquette: Let us get back to the stability of anti- and syn-sesquinorbornene. Ken Houk calculates 2 kcal/mol. I understand your argument is based on a reaction in which you see one product and not the other. Are you not basing your arguments on a non-observable result?

Vogel: You are perfectly right. We have not made the anti-oxasesquinorbornene and looked at how it would equilibrate with the other isomers. I agree, it is a weak point. However, under our conditions, if the anti-oxasesquinorbornene had the same stability as that of its syn isomer, we would have seen it. I do not see why the energy barrier to its formation should be higher than that of the *syn*-oxasesquinorbornene by more than 8 kcal/mol! Because the *anti*-oxasesquinorbornene is not seen in equilibrium with the cycloaddents and *syn*-oxasesquinorbornene at 150° after several days, we propose the anti isomers to be at least 2.5 kcal/mol less stable than the syn derivatives. I agree, we have not proven it.

Paquette: It could be much larger than that.

Houk: I would like to know a little more in detail what you mean by polarization of the double bond. You have done a calculation and you know there is no s mixing, but you seem to be saying that s mixing is important. Polarization is an even more vague term than calculation.

Vogel: I mean there is rehybridization from 2s character to being mixed with 2p character.

Houk: You know you do not get that in the calculation.

Vogel: You get a little bit. If you would allow electron correlation effects in the calculations, you would increase the effect. It would give you a larger admixture of 2s orbitals.

Houk: Citing undone calculations is even more dangerous than citing those done. How does the spontaneous pyramidalization stabilize the double bond? Why don't all double bonds pyramidalize? You said that *syn*-sesquinorbornene is stabilized because of polarization of the double bond.

Vogel: We assume norbornene to have its π electrons polarized toward the exo face because of repulsive interactions between the π system and the C-1,7,4,5,6 five membered ring. This hypothesis is consistent with the X-ray observations which show the C(7) and O(7) bridges to be repelled by the endocyclic double bond of norbornene and 7-oxanorbornene, respectively. In the *anti*-sesquinorbornene, this synergic effect is absent and thus the olefin adopts a planar double bond.

Houk: You are making them work in concert.

Vogel: Yes.

Houk: You know the way we would look at it. I think if we combine the staggering arrangement and hyperconjugation I think we may ultimately get to the explanation of the whole phenomenon.

Perrin: At last, a combining of forces.

Houk: What really bothers me about this is that you do the calculation and there is no s mixing of any significance in the π bond. There is extension of the π density on the exo face. You do not have to do a CI calculation to get Fukui's plot. The problem is that the exo π extension doesn't come from a mixing. Fukui is correct that the π density is bigger on the exo face, but it doesn't come from s-mixing. There isn't any need to use s-mixing on the exo face. It has to do with the antibonding interactions with the C-C orbitals. This is a destabilizing effect, not a stabilizing one.

Vogel: Oh yes, but if you were to allow this phenomenon to relax it would go even larger than it is.

Houk: The exo density is larger. The point I am making is it doesn't have anything to do with s mixing.

I would like to say one other thing. It is true that when you do any kind of an addition to a double bond that it wants to bend in a trans fashion. It is possible whatever the reasons for the endo bending in norbornene, that when you begin to add to the diene that in the Diels-Alder transition state there is some trans bending along both of the alkene linkages. If you bend the alkene in a trans fashion and have the normal propensity to bend endo at the fusion with the norbornene, then the attack of the diene termini is endo.

Traylor: I just want to ask what happens to a simple electrophile when you add it to that sesquinorbornene.

Vogel: It always adds from the top.

Houk: It adds from the top in the ground state. Professor Bartlett observed an excited state reaction that adds the other way.

Perrin: I want to get back to this quick dismissial of the torsional effects based upon the X-ray structure with a double bond on one side and a bismethylene on the other. The bending was much greater on the double bond side which is exactly what you would expect from torsional effects. Because the *bis*-methylene should be eclipsed with the bridgehead hydrogen, you would expect far smaller distortion on the *bis*-methylene side.

Vogel: You are saying the C-H wants to be aligned with the methylene. OK.

Greene: Gassman has presented some evidence recently, I think it was solvolyses experiments, that the cyano group can be electron donating toward carbenium ions. Following your line of reasoning with respect to an oxygen of a carbonyl being electron donating, I would have expected to see some effect in your dicyanomethylene substituent. I do not believe you observed any effect.

Vogel: This is a very good point. Here are our results with 5-dicyanomethylene-2-norbornene:

ENu =			
$PhSeCl$	68	:	32
$PhSeBr$	65	:	35
$2(NO_2)\text{-}C_6H_4Cl$	55	:	45
$2,4(NO_2)_2\text{-}C_6H_3SCl$	58	:	42

If you are willing to take the above regioselectivities as significant, you are right, the dicyanomethylene substituent acts also as an electron donating group in these electrophilic additions. But you recognize, the regioselectivities are much smaller than those observed with norbornenone. We calculate the following heats of formation for the corresponding 2-norbornyl cations:

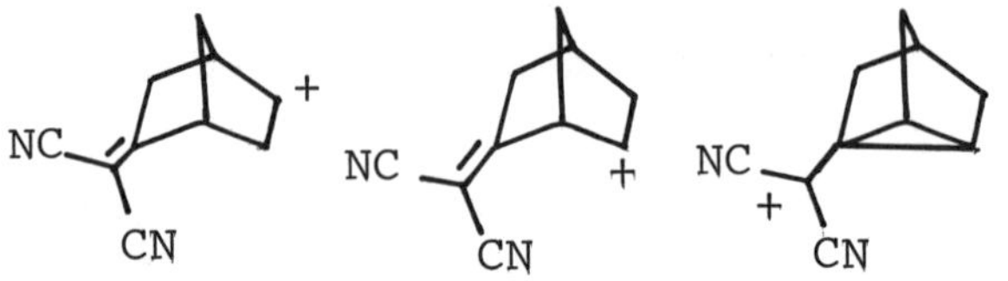

ΔH_f^o(MINDO/3)	280.0	280.0	271.1 kcal/mol
ΔH_f^o(MNDO)	307.1	308.6 kcal/mol	

I do not think that the above experimental regioselectivities and calculated heats of formation allows any conclusion to be drawn at this moment. Certainly, the cyano group can be electron donating when directly attached to a positively charged center. The problem is a more subtle one in our case. We have attributed the electron-donating effect of a homoconjugated carbonyl group to a hyperconjugative mechanism. This is obviously an oversimplification. I do not exclude some homoconjugative participation of the carbonyl double bond. However, I think this contribution is not very important judging from the results obtained with 5-dicyanomethylene-2-norbornene in which the dicyanomethylene group imitates the dipole and homoconjugative properties of the carbonyl group.

Schmid: This effect should also show up in the rate-determining transition state of electrophilic addition as well as in the product determining transition state.

Vogel: It does. We have preliminary kinetic data (NMR tubes) which show norbornene to add more rapidly than 5-dicyanomethylene-2-norbornene and 2-chloronorborn-5-ene-2-carbonitrile. In contrast, 2-norborn-5-enone reacts faster than norbornene. The regioselectivities reported correspond to products formed under kinetic conrol. This was checked.

References

[1] (a) M. Mohraz, C. Batich, E. Heilbronner, P. Vogel, and P.-A. Carrupt, Recl. Trav. Chim. Pays-Bas 98, 361 (1979). (b) M. Mohraz, W. Jian-gi, E. Heilbronner, P. Vogel and O. Pilet Helv. Chim. Acta 63, 568 (1980). (c) W.T. Borden, S.D. Young, D.C. Frost, N.P.C. Westwood, and W.L. Jorgensen, J. Org. Chem. 44, 737 (1979); P. Bischof, R. Gleiter, and R. Haider, Angew. Chem. 89, 112 (1977); P. Bischop, R. Gleiter, and R. Haider, J. Am. Chem. Soc. 100, 1036 (1978); L.A. Paquette, R.V.C. Carr, M.C. Böhm, and R. Gleiter, J. Am. Chem. Soc. 102, 1186 (1980). (d) H.-U. Pfeffer, M. Klessinger, G. Erker, and W.R. Roth, Chem. Ber. 108, 2923 (1975). (e) H.-U. Pfeffer and M. Klessinger, Org. Mag. Res. 9, 121 (1977); D. Quarroz, T.-H. Sonney A. Chollet, A. Florey, and P. Vogel, Org. Mag. Res. 9, 611 (1977); V. Gergely, Z. Akhavin and P. Vogel, Helv. Chim. Acta 58, 871 (1975). (f) P. Asmus and M. Klessinger, Tetrahedron 30, 2477 (1974). (g) P.V. Alston and R.M. Ottenbrite, J. Org. Chem. 41, 1635 (1976); ibid. 40, 322 (1975).

[2] D.N. Butler and R.A. Snow, Can. J. Chem. 53, 256 (1975); D.N. Butler and R. A. Snow, Can. J. Chem. 52, 447 (1974); D.N. Butler and R.A. Snow, Can. J. Chem. 50, 795 (1972); W.T. Borden and A. Gold, J. Am. Chem. Soc. 93, 3830 (1971); W.L. Jorgensen and W.T.Borden, J. Am. Chem. Soc. 95, 6649 (1973); L.W. Jelinsky and E.F. Kiefer, J. Am. Chem. Soc. 98, 281 (1976); H. Hogeveen, W.F.T. Huurdeman, and D.M. Kok, J. Am. Chem. Soc. 100, 871 (1978).

[3] M. Avenati, J.-P. Hagenbuch, C. Mahaim, and P. Vogel, Tetrahedron Lett., 3167 (1980).

[4] R. Gabioud and P. Vogel, Tetrahedron 36, 149 (1980).

[5] A. Chollet, C. Mahaim, C. Foetisch, M. Hardy, and P. Vogel, Helv. Chim. Acta 60, 59 (1977).

[6] M. Hardy, P.-A. Carrupt, and P. Vogel, Helv. Chim. Acta 59, 936 (1975).

[7] O. Pilet, A. Chollet and P. Vogel, Helv. Chim. Acta 62, 2341 (1979).

[8] O. Pilet and P. Vogel, Helv. Chim. Acta 64, 2563 (1981).

[9] O. Pilet and P. Vogel, Angew. Chem. Int. Ed. Engl. 19, 1003 (1980).
[10] P. Vogel and A. Florey, Helv. Chim. Acta 57, 200 (1974); C. Mahaim, P.-A. Carrupt, J.-P. Hagenbuch, A. Florey, and P. Vogel, Helv. Chim. Acta 63, 149 (1980).
[11] P.-A. Carrupt and P. Vogel, Tetrahedron Lett. 4533 (1979).
[12] Y. Bessière and P. Vogel, Helv. Chim. Acta 63, 232 (1980).
[13] J. Tamariz and P. Vogel, in preparation.
[14] L. de Picciotto, P.-A. Carrupt, and P. Vogel, J. Org. Chem. 47, 3796 (1982).
[15] O. Pilet and P. Vogel, in preparation.
[16] (a) A. Chollet, M. Wismer and P. Vogel, Tetrahedron Lett., 4271 (1976). (b) Ph. Narbel, T. Boschi, R. Roulet, P. Vogel, A.A. Pinkerton, and D. Schwarzenbach, Inorg. Chim. Acta 36, 161 (1979).
[17] H.-D. Scharf, H. Plum, J. Fleischhauer, and W. Schleker, Chem. Ber. 112, 862(1979); R. Sustmann, M. Böhm and J. Sauser, Chem. Ber. 112, 883 (1979).
[18] D. Craig, J.J. Shipman, and R.B. Fowler, J. Am. Chem. Soc. 83, 2885 (1961); P.D. Bartlett, G.E.H. Wallbillich, A.S. Wingrove, J. S. Swenton, L.K. Montgomery, and B.D. Kramer, J. Am. Chem. Soc. 90, 2049 (1968); Ch. Rucker, D. Lang, J. Sauer, H. Friege and R. Sustmann, Chem. Ber. 113, 1663 (1980).
[19] G. Klopman, J. Am. Chem. Soc.90, 223 (1968); L. Salem, J. Am. Chem. Soc. 90, 543, 553 (1968); K. Fukui, Acc. Chem. Res. 4, 57 (1971); Bull. Chem. Soc. (Japan) 39, 498 (1966); I. Fleming, "Frontier Orbitals and Organic Chemical Reactions," Wiley, 1976; R. Sustmann, Pure and Applied Chem. 40, 569 (1974).
[20] K.N. Houk and L.L. Munchhausen, J. Am. Chem. Soc. 98, 937 (1976); J. Sauer and R. Sustmann, Angew. Chem. Int. Ed. 19, 779 (1980).
[21] O. Dimroth, Angew. Chem. 46, 571 (1933).
[22] A. Pross and S.S. Shaik, J. Am. Chem. Soc. 104, 1129 (1982) and references therein.
[23] O. Pilet, P. Vogel, A.A. Pinkerton, and D. Schwarzenbach, in preparation.
[24] A.A. Pinkerton, G. Chapuis, P. Vogel, H. Hañish, Ph. Narbel and R. Roulet, Inorg. Chim. Acta 35, 197 (1978).

[25] E. Meier, O. Cherpillod, T. Boschi, R. Roulet, P. Vogel, C. Mahaim, A.A. Pinkerton, D. Schwarzenbach, and G. Chapuis, J. Organomet. Chem. 186, 247 (1980).
[26] R.C. Bingham, M.J.S. Dewar and D.H. Lo, J. Am. Chem. Soc. 97, 1287, 1307 (1975).
[27] M.J.S. Dewar and W. Thiel, J. Am. Chem. Soc. 99, 4899, 4907 (1977).
[28] J.-L. Birbaum and P. Vogel, unpublished calculations.
[29] H.-U. Pfeffer and M. Klessinger, Chem. Ber. 112, 890 (1979).
[30] O. Ermer and J.D. Dunitz, Helv. Chim. Acta 52, 1861 (1969).
[31] E. Heilbronner and A. Schmelzer, Nouv. J. Chim. 4, 23 (1980).
[32] H.E. Zimmerman and R.M. Paufler, J. Am. Chem. Soc. 82, 1514 (1966).
[33] (a) R.B. Turner, J. Am. Chem. Soc. 86, 3586 (1964). (b) R.B. Turner, W.R. Meador and R.E. Winkler, J. Am. Chem. Soc. 79, 4116 (1957); P.v.R. Schleyer, J. Am. Chem. Soc. 80, 1700 (1958).
[34] E. Haselbach, E. Heilbronner, and G. Schroeder Helv. Chim. Acta 54, 153 (1971).
[35] A.A. Pinkerton, D. Schwarzenbach, J.H.A. Stibbard, P.-A. Carrupt, and P. Vogel, J. Am. Chem. Soc.103, 2095 (1981).
[36] S. Inagaki and K. Fukui, Chem. Lett. 509 (1974); S. Inagaki, H. Fujimoto, and K. Fukui, J. Am. Chem. Soc. 98, 4054 (1976).
[37] A.A. Pinkerton, D. Schwarzenbach, P.-A. Carrupt, and P. Vogel, in preparation.
[38] (a) W.H. Watson, J. Galloy, P.D. Bartlett, and A.A.M. Roof, J. Am. Chem. Soc. 103, 2022 (1981) and references therein. (b) P.D. Bartlett, A.A.M. Roof, and W.J. Winter, J. Am. Chem. Soc. 103, 6520 (1981).
[39] L.A. Paquette, R.V.C. Carr, P. Charumilind and J.F. Blount, J. Org. Chem. 45, 4922 (1980), Supplementary Material.
[40] M. Avenati, dissertation, University of Lausanne, June 1982.
[41] P.v.R. Schleyer, J. Am. Chem. Soc. 89, 701 (1967).
[42] T. Kawamura, T. Koyama and T. Yonezawa, J. Am. Chem. Soc. 95, 3220 (1973); see also: H.-U. Wagner, G. Szeimies, J. Chandrsekar, P.v.R. Schleyer, J.A. Pople, and J.S. Binkley, J. Am.

Chem. Soc. 100, 1210 (1978).
[43] (a) P.D. Bartlett, A.J. Blakeney, M. Kimura, and W.H. Watson, J. Am. Chem. Soc. 102, 1383 (1980). (b) L.A. Paquette, R.V.C. Carr, M.C. Bohm, and R. Gleiter, J. Am. Chem. Soc. 102. 1186 (1980); M.C. Böhm, R.V.C. Carr, R. Gleiter, and L.A. Paquette, J. Am. Chem. Soc. 102, 7218 (1980). (c) K. Alder, F.H. Flock, and P. Janssen, Chem. Ber. 89, 2689 (1956).
[44] T. Sugimoto, Y. Kobuke, and J. Furukawa, J. Org. Chem. 41, 1457 (1976).
[45] L.A. Paquette, R.V.C. Carr, E. Arnold, and J. Clardy, J. Org. Chem. 45, 4907 (1980).
[46] J.-P. Hagenbuch and P. Vogel, Tetrahedron Lett., 561 (1979).
[47] J.-P. Hagenbuch, P. Vogel, A.A. Pinkerton, and D. Scharzenbach, Helv. Chim. Acta 64, 1818 (1981).
[48] J.-P. Hagenbuch, B. Stampfli, and P. Vogel, J. Am. Chem. Soc. 103, 3934 (1981).
[49] M. Avenati, J.-P. Hagenbuch, C. Mahaim, and P. Vogel, Tetrahedron Lett., 3167 (1980).
[50] C. Mahaim and P. Vogel, Helv. Chim. Acta 65, 866 (1982).
[51] M. Avenati and P. Vogel, in preparation.
[52] M. Avenati and P. Vogel, Helv. Chim. Acta 65, 204 (1982).
[53] L.A. Paquette, F. Bellamy, M.C. Böhm, and R. Gleiter, J. Org. Chem. 45, 4913 (1980).
[54] M.J.S. Dewar and R.C. Dougherty, "The PMO Theory of Organic Chemistry," Plenum Press, New York, 1975.
[55] R.F. Hudson, Angew. Chem. Int. Ed. 12, 36 (1973); N.D. Epiotis, J. Am. Chem. Soc. 95, 5624 (1974); K.N. Houk, Acc. Chem. Res. 8, 361 (1975); O. Eisenstein, J. M. Lefour, N.T. Anh, and R.F. Hudson, Tetrahedron 33, 523 (1977); P.V. Alston, R.N. Ottenbrite, and J. Newby, J. Org. Chem. 44, 4939 (1979); V. Bachler and F. Mark, Tetrahedron 33, 2857 (1977); P.V. Alston and D.D. Shillady, J. Org. Chem. 39, 3402 (1974); P.V. Alston, R.M. Ottenbrite, and D.D. Shillady, J. Org. Chem. 38, 4075 (1973); T. Cohen, R.J. Ruffner, D.W. Shull, W.M. Daniewski, R.M. Ottenbrite and P. V. Alston, J. Org. Chem. 43, 4052 (1978); P.V. Alston, R. M. Ottenbrite, and T. Cohen, J. Org. Chem. 43, 1864 (1978); V. Bachler and F. Mark, Theoret. Chim. Acta 43, 121 (1976); I.

Fleming, J.P. Mitchel, L.E. Overman, and G.F. Taylor, Tetrahedron Lett., 1313 (1978); W.B.T. Cruse, I. Flemming, P.T. Gallagher, and O. Kennard, J. Chem. Res. (S), 372 (1979); B.M. Trost, D. O'Krongly, and J.L. Belletire, J. Am. Chem. Soc. 102, 7595 (1980); I.-M. Tegmo-Larson, M.D. Rozeboom, N.G. Rondan, and K.N. Houk, Tetrahedron Lett., 2047 (1981); M.C. Böhm, and R. Gleiter, Tetrahedron 36, 3209 (1980).

[56] L. Schwager and P. Vogel, in preparation.

[57] R.A. Firestone, Tetrahedron 33, 3009 (1977) and references therein.

[58] K. Wallenfels, K. Friedrick, J. Riser, W. Ertel, and H.K. Thiene, Angew. Chem. Int. Ed. 15, 261 (1976).

[59] M. Avenati, P.-A. Carrupt, D. Quarroz and P. Vogel, Helv. Chim. Acta 65, 188 (1982).

[60] P.-A. Carrupt, dissertation, University of Lausanne, 1979.

[61] P.-A. Carrupt and P. Vogel, Tetrahedron Lett. 23, 2563 (1982).

[62] G.H. Schmid and D.G. Garratt, "The chemistry of Functional Groups. Suppl. A: The Chemistry of Double-bonded Functional Groups," S. Patai, Ed., J. Wiley and Sons, London 1977, pp.828-854.

[63] N.S. Zefirov, N.K. Sadovaja, L.A. Novgorodtseva, R.Sh. Achmedova, S.V. Baranov, and I.V. Bodrikov, Tetrahedron 35, 2759 (1979); N.S. Zefirov, A.S. Koz'min, and V.V. Zhdankin, Tetrahedron 38, 291 (1982).

[64] N.S. Zefirov, N.K. Sadovaja, R.Sh. Achmedova, I.V. Bodrikov, T.C. Morrill, A.M. Nersisyan, V. Rubakov, N.D. Saratseno, and Yu. T. Struchkov, Zh. Org. Khim. 16, 580 (1980); Chem. Abstr. 93, 9459z (1980); D.G. Garratt and P.L. Beaulieu, J. Org. Chem. 44, 3555 (1979); D.G. Garratt and A. Kabo, Can. J. Chem. 58, 1030 (1980).

[65] D. Heissler and J.J. Riehl, Tetrahedron Lett., 4707, 4711 (1980); 3957 (1979); M. Przybylska and D.G. Garratt, Can. J. Chem. 59, 658 (1981).

[66] The structures of the adducts were determined by proton NMR (360 MHz) with the help of double irradiation experiments and by C-13 NMR. They were confirmed by oxidative-elimination ($NaIO_4/NaHCO_3$ in $MeOH/H_2O$, 20°)

that yielded the corresponding substituted olefins. Details will be given in a full paper.
[67] J.L. Huguet, Adv. Chem. Ser. 76, 345 (1967); D.N. Jones, D. Mundy, and R.D. Whitehouse, J. Chem. Soc. Chem. Comm., 86 (1970); K.B. Sharpless, M.W. Young, and R.F. Lauer, Tetrahedron Lett., 1979 (1980); H. Reich, Acc. Chem. Res. 12, 22 (1979).
[68] S. Raucher, J. Org. Chem. 42, 2950 (1977).
[69] R. Huisgen, Pure Appl. Chem. 53, 171 (1981) and references therein.
[70] A. Eschenmoser and A. Frey, Helv. Chim. Acta 35, 1660 (1952); W. Fischer and C. Grob, Helv. Chim. Acta 61, 1588 (1978); C.A. Grob, Angew. Chem. 88, 621 (1976); cf. also: R. Hoffmann, Acc. Chem. Res. 4, 1 (1971); R. Hoffmann, A. Imamura, and W.J. Hehre, J. Am. Chem. Soc. 90, 1499 (1968); R.Gleiter, Angew. Chem. 86, 770 (1974); E. Heilbronner and A. Schmelzer, Helv. Chim. Acta 58, 936 (1975).
[71] J.I. Brauman and L.K. Blair, J. Am. Chem. Soc. 90, 6561 (1968); J.I. Brauman and L.K. Blair, J. Am. Chem. Soc. 92, 5986 (1970).
[72] C.S. Yoder and C.H. Yoder, J. Am. Chem. Soc. 102, 1245 (1980).
[73] R.W. Taft, M. Taagepera, J.L.M. Abboud, J.F. Wolf, D.J. DeFrees, W.J. Hehre, J.E. Bartmess, and R.T. McIver, Jr., J. Am. Chem. Soc. 100, 7765 (1978); R.L. Woodin and J.L. Beauchamp, J. Am. Chem. Soc. 100, 501 (1978).
[74] P.G. Gassman, K. Saito and J.J. Talley, J. Am. Chem. Soc. 102, 7613 (1980); P.G. Gassman and K. Saito, Tetrahedron Lett., 1311 (1981).
[75] X. Creary, J. Am. Chem. Soc. 103, 2463 (1981).
[76] C. Barras, L.G. Bell, R. Roulet, and P. Vogel, Helv. Chim. Acta 64, 2841 (1981).
[77] J.-M. Sonney and P. Vogel, Helv. Chim. Acta 63, 1034 (1980); See also: Ch. Barras, R. Roulet, E. Vieira, P. Vogel, and G. Chapuis, Helv. Chim. Acta 64, 2328 (1981).
[78] D.J. Sandman, K. Mislow, W.P. Giddings, J. Dirlam, and G.C. Hansom, J. Am. Chem. Soc. 90, 4877 (1968).
[79] J. Del Bene and H. Jaffe, J. Chem. Phys. 48, 1807 (1968).
[80] D.A. Lightner, J.K. Gawronski, A.E. Hansen and T.D. Bouman, J. Am. Chem. Soc. 103, 4291 (1981).
[81] P.-A. Carrupt and P. Vogel, Tetrahedron Lett.

4721 (1981).

6. FINDINGS, NEW AND OLD IN OLEFIN-ELECTROPHILE REACTIONS

Frederick D. Greene

Department of Chemistry, Massachusetts Institute of Technology, Cambridge, Massachusetts 02139

My talk today is about some electrophilic reactions of olefins. To begin in a straightforward fashion, our own entry into this area was rather indirect. We were (indeed, are) interested in making some four-membered cyclic azo compounds, diazetines. We had found a way to make this ring system and had shown that the products of decomposition were dinitrogen and the corresponding olefin [1]. We were much interested in making derivatives with different substituents at the ring carbons in order to get information on the question of "one-bond" versus "two-bond" breaking at the transition state. Of particular interest to us was 1,2-diphenyldiazetine (eq. 1) in that its decomposition might afford electronically excited states of the stilbenes. The possibility of examining the

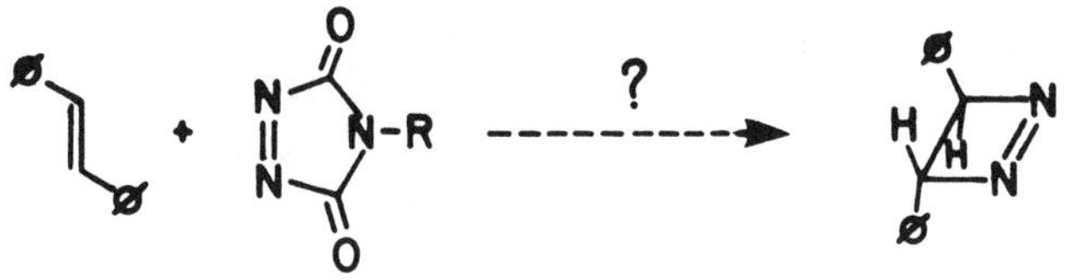

(1)

effect of substituents on the aryl rings was also of much interest to us. We wondered if we could make this system by a simple cycloaddition reaction, by analogy to the reaction of indene with phenyltriazolinedione to give the diazetidine derivative (eq. 2) reported by von Gustorf [2].

Reaction of trans-stilbene with triazolinedione (TAD) is rapid and affords two products, both of which are 2:1 adducts of TAD and the olefin (eq. 3). We see no evidence for a diazetidine in this

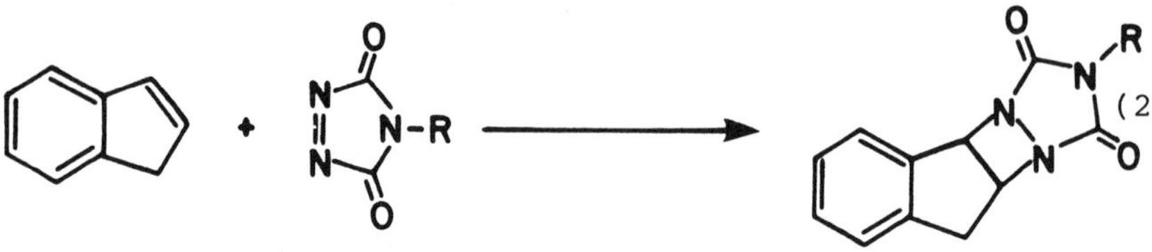

(2)

reaction. The 2:1 adducts are most simply accounted for in terms of a Diels-Alder reaction to afford A,

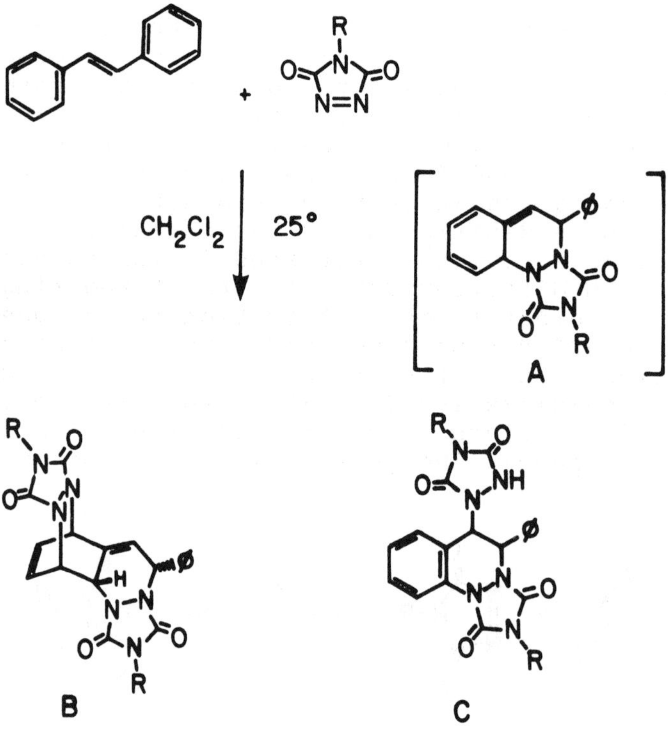

CH_2Cl_2 25°

A

(3)

B

C

followed by reaction of A with a second TAD in two different ways--Diels-Alder reaction across the 1,3-cyclohexadiene in A to afford B and ene reaction to afford C. One sees in the TAD-indene reaction and in the TAD-stilbene reaction that the triazolinedione is reacting in three ways--a [2+2] type reaction (with indene), a Diels-Alder reaction, and an ene reaction. The three types of reaction are shown in eq. 4.

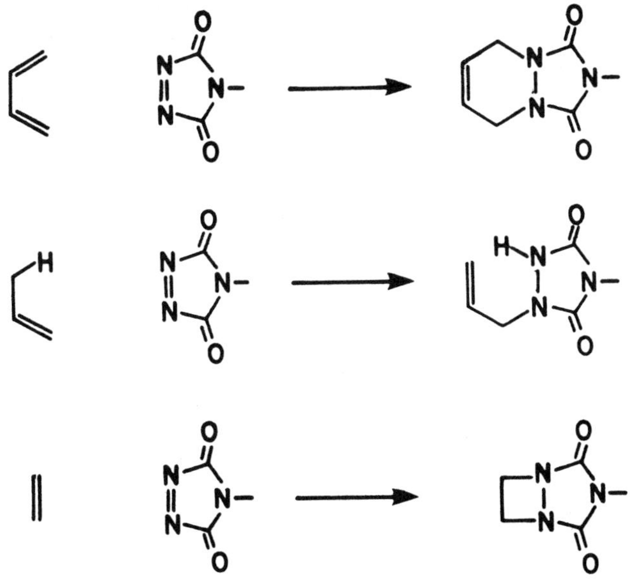

(4)

(Indeed, triazolinedione can also react in a fourth way--reaction with strained carbon-carbon single bonds [3].)

We were struck by the fact that these three types of reactions were also seen in reactions of singlet oxygen with unsaturated systems. Reaction

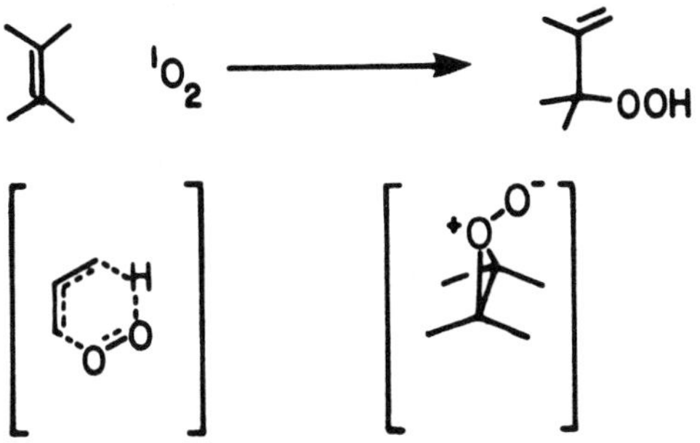

(5)

of singlet oxygen with an olefin possessing allylic hydrogen affords the hydroperoxide (eq. 5). For some time, this type of reaction was considered to proceed by a [4+2] path. Strong evidence against the [4+2] formulation was provided by Bartlett [4] and by Stephenson [5]. In the former study, an important finding not in accord with a [4+2] path to hydroperoxide was observation of an inverse isotope effect upon substitution of D for H at the β position of 4-methyldihydropyran (eq. 6) [4].

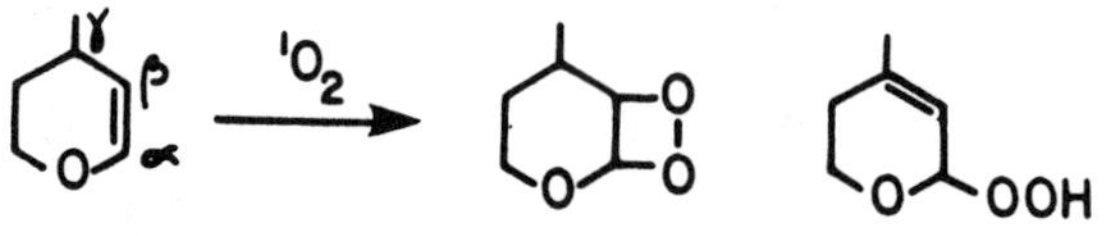

(6)

In the Stephenson study, cis- and trans-hexadeuteriotetramethylethylenes were allowed to react with singlet oxygen to form the hydroperoxide. In the cis-TME-d6, the isotope effect (an intramolecular one, determined by NMR analysis of the product) was approximately unity; in trans-TME-d the isotope effect was 1.4 (eq. 7) [5].

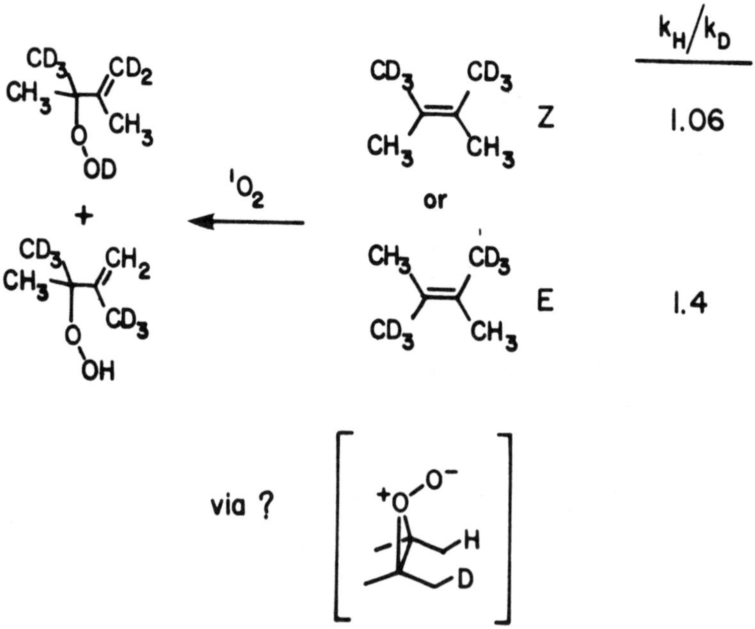

	k_H/k_D
Z	1.06
E	1.4

(7)

via ?

This difference in isotope effects of *cis*- and *trans*-TME-$\underline{d}_6$ was clearly inconsistent with a [4+2] formulation. The interpretation suggested by Stephenson was rate-determining formation of an intermediate--a perepoxide or a species with the structural characteristics of a perepoxide [5].

In view of the similarities between the types of reactions of triazolinedione and singlet oxygen with unsaturation, noted above, we wondered if intermediates were involved in the triazolinedione-olefin reactions. Consequently we subjected triazolinedione (both the *N*-phenyl and the *N*-methyl compounds) to the Stephenson isotope effect test (eq. 8). One sees very little isotope effect with *cis*-

(8)

TME	R = CH_3	R = C_6H_5
cis-$\underline{d}_6$	1.08	1.1
trans-$\underline{d}_6$	3.8	3.7

TME-$\underline{d}_6$ but a substantial isotope effect with the trans compound [6]. How is triazolinedione discriminating between hydrogen and deuterium in these systems? One way to account for this is via an "aziridinium imide." Triazolinedione and *cis*-TME-$\underline{d}$ could lead to D and E (eq. 9). In D the "imide" nitrogen only has access to hydrogen, in E access only to deuterium. The corresponding species from *trans*-TME-$\underline{d}_6$ is F in which both hydrogen and deuterium are accessible to the nitrogen, and isotopic discrimination is possible. The big difference in isotope effect between the *cis*- and *trans*-$\underline{d}_6$ compounds is difficult to reconcile in terms of anything like a [4+2] path (G eq. 9). At first sight, the aziridinium imide, H (eq. 10) may not look very good in that it places positive charge on a nitrogen atom adjacent to a carbonyl group. However, compounds related to H, such as I have been isolated [7]. Compound I obviously has a much better accommodation for positive charge than does

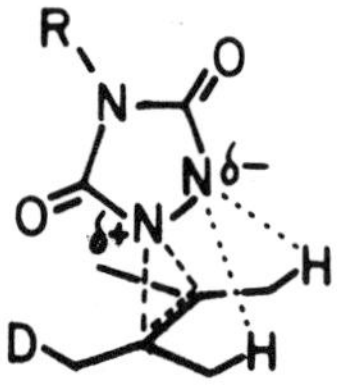

D 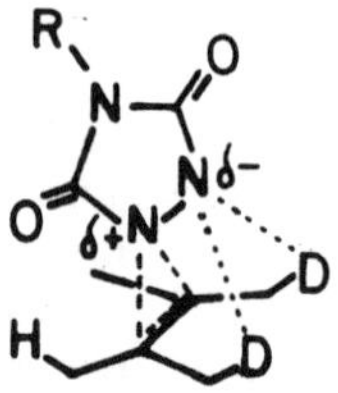 E

(9)

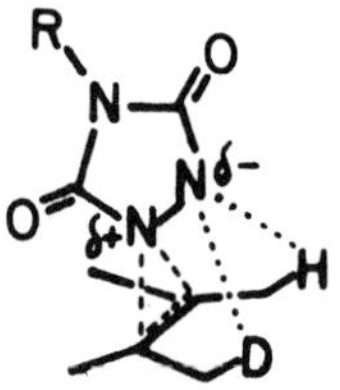

F 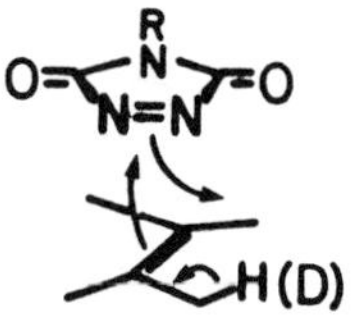 G

species H. However, the question is not whether H is isolable but whether it is a viable intermediate in the triazolinedion-olefin reaction.

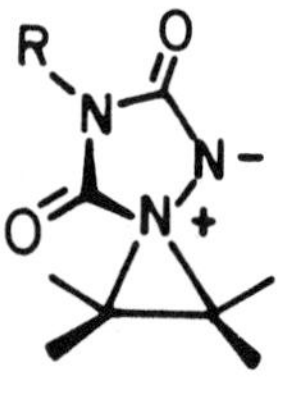

H 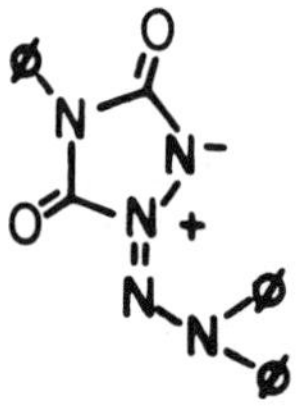 I

(10)

The large difference in isotope effects for cis- and trans-TME-d_6 strongly indicates that the reaction path involves rate-determining formation of an intermediate followed by subsequent (and faster) conversion of the intermediate to product. The lack of an isotope effect with the cis-d_6 compound carries a strong implication that the rate-determining step does not involve much C-H or C-D cleavage. Such cleaveage takes place in the product-determining step. (A classical test for an intermediate is to ascertain that the rate-determining step and the product-determining step are not the same.) The energy profile for the triazolinedione-olefin reaction is shown in Figure 1; we think an aziridinium imide (H in eq. 10) is the best candidate for the intermediate but the evidence to date (the isotope effects) is indirect.

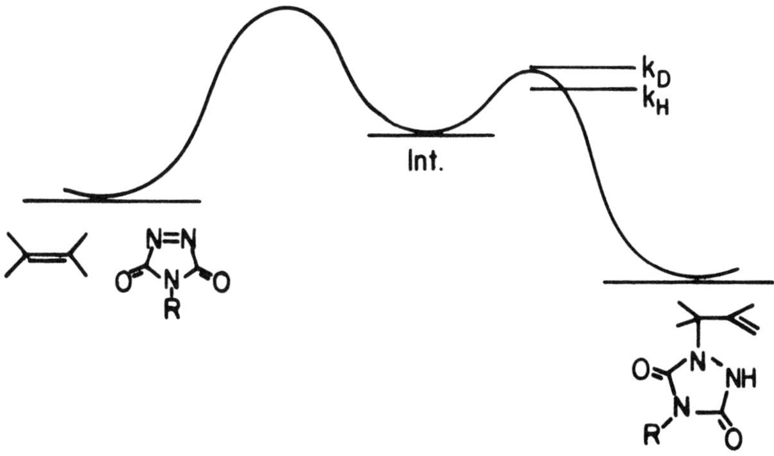

Figure 1. Energy profile for the triazolinedione-olefin reaction.

At this point we are interested in two questions: What can one learn about the nature of the intermediate that seems to be required in this case? What can one learn about the nature of the transition state leading to the intermediate? I will not try to pursue one of these questions to the exclusion of the other. They are obviously related. We will first consider some kinetics comparisons.

Reaction of triazolinedione with olefin follows second-order kinetics. Butler has published some work in this area [8], and we have extended the range of olefins. Increasing alkylation of olefin increases the rate of reaction, ie, electrophilic attack on an olefin in which electron-donating groups increase the reactivity [9]. One also notes

$$\text{Ph-TAD} + \text{olefin} \xrightarrow[CH_2Cl_2]{k_2} \text{ene product} \tag{11}$$

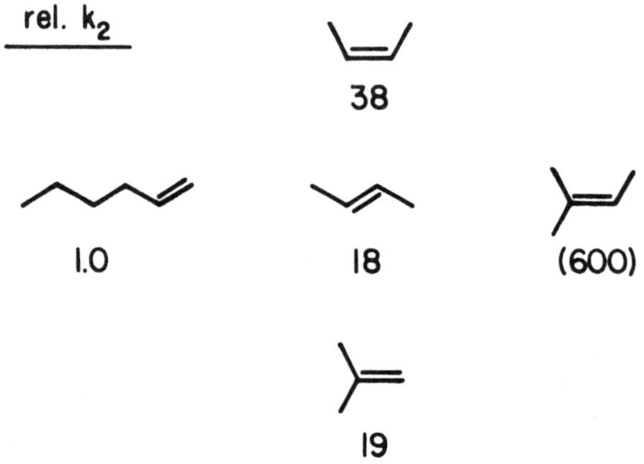

rel. k_2

38

1.0 18 (600)

19

that approximately the same increase in rate is produced by placement of the two alkyl groups on the same end or on opposite ends of the double bond. This pattern is observed in reactions of various other electrophiles, such as bromine, with alkyl-substituted olefins [9]. The implication of this pattern is that the transition state involves a more or less symmetrical addition of the electrophile into the π system of the olefin.

The comparison of rates of reaction of triazolinedione and of bromine [10] with olefins is rather striking (Figure 2). Note that the reactions involved are the ene reaction of triazolinedione with these olefins versus the reaction of bromine with the olefins to form 1,2-dibromides. The two reactions show a similar sensitivity to substituent effect--both with simple alkyl olefins and with less simple cyclic ones. Of course, in the bromination reactions for a long time now the implication has been that the rate-determining step involves the formation of something like a bromonium ion, a three-center species. The similarities reflected in Figure 2 raise the question of whether the transition state for the triazolinedione-olefin reaction is three-centered. Is it three-centered, four-centered, or just what is it?

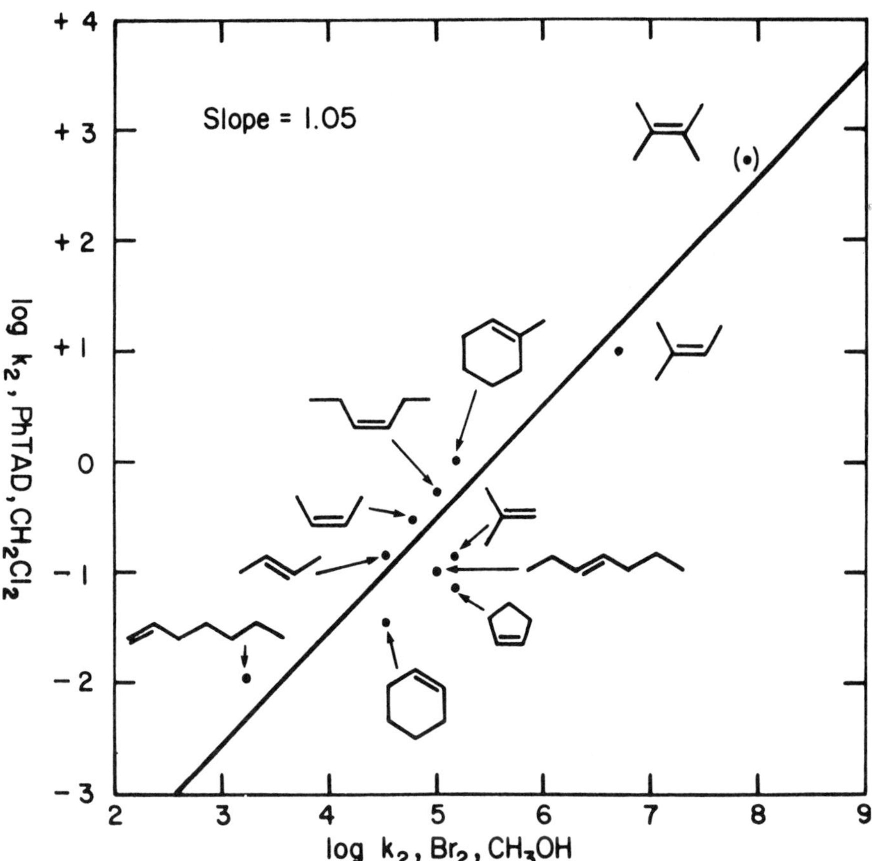

Figure 2. A comparison of the rates of reaction of triazolinedione and of bromine.

Three possible ways in which triazolinedione might be interacting with the π system are shown in eq. (12). Representation a corresponds to a [2s-2a] type of interaction in what I would call a four-center transition state. Alternatively, the interaction might be as in b with π* at nitrogen (LUMO) interacting with the π bond of olefin (HOMO). Another three-center alternative is c. Offhand, c probably looks much worse to you than a or b in that the nitrogen lone-pair orbital has two electrons in it. In representation c a filled orbital of the electrophile seems to be jammed into the olefin π bond. That looks bad; I will come back to this matter near the end of the talk.

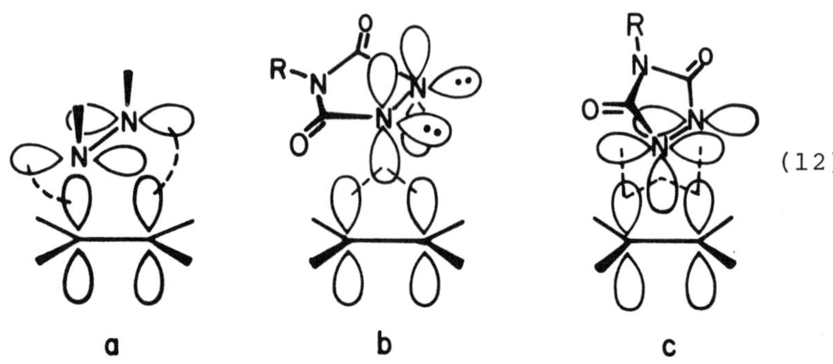

(12)

We were interested in devising experiments to tell us something about the geometry of approach of the triazolinedione to the olefin. Comparisons between cyclopentene, cyclohexene and norbornylene are shown in eq. 13.

$$\text{Ph-TAD} + \text{olefin} \xrightarrow[CH_2Cl_2]{k_2}$$

(13)

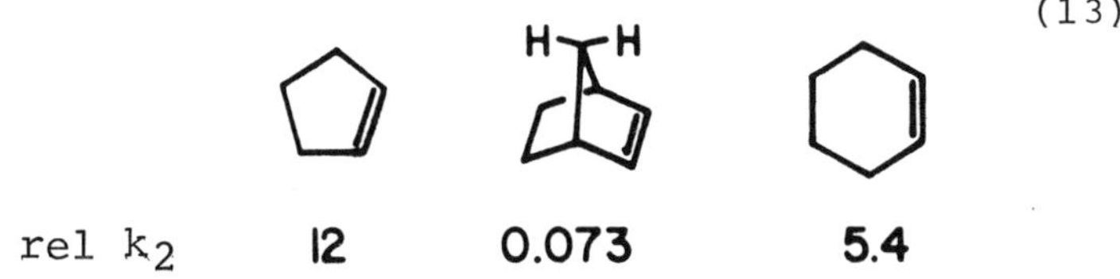

rel k_2 12 0.073 5.4

Typically, norbornylene is far more reactive than cyclohexene in 1,3-dipolar cycloadditions [9]. In three-center processes, norbornylene is often more reactive than cyclohexene [9]. In reaction with phenyltriazolinedione, norbornylene is substantially slower than cyclohexene [11]. By reference to a and c (eq. 12) one can see that interference between the triazolinedione moiety and the methylene bridge might be a reason for the slower rate of TAD with norbonylene compared to cyclohexene. A transition state such as b (eq. 12) would not appear to account for the rate differences.

An olefin of rather different geometry is adamantylidenadamantane (Ad=Ad). It is sterically difficult to bring the triazolinedione up to this olefin in parallel planes (eg, b in eq. 12) unless the transition state occurs at a long distance of separation of the two reactants. We find that triazolinedione does react quite readily with Ad=Ad to form a 1:1 adduct (eq. 14)[6]. At first consideration, the infrared bands at 1730 and 1675 cm^{-1}

(14)

did not seem to be in accord with a diazetidine-urazole system. Such systems usually show frequencies of approximately 1780 and 1730 cm^{-1}. However, Dr. Blount of Hoffmann-La Roche has carried out the X-ray analysis on this adduct, and the structure is, indeed, a diazetidine (Figure 3). In the adduct the diazetidine ring is skewed. The reason for this is primarily associated with non-bonded H..H interactions of the four methylenes opposite the diazetidine ring. The same situation is also found in the dioxetane formed from singlet oxygen and Ad=Ad [12].

Upon heating to 70°, the triazolinedione-adamantylidenadamantane adduct reverts to the reactants. In the presence of tetramethylethylene, the triazolinedione thus liberated is irreversibly converted to the ene product, and adamantylidenadamantane may be recovered in high yield.

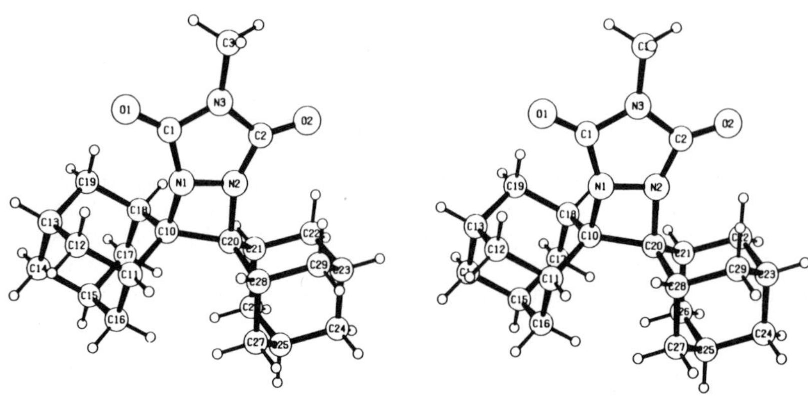

Figure 3. Stereo drawing of diazetidine-urazole adduct.

Is an intermediate, specifically an aziridinium imide (ie, compound H in eq. 10) involved in the triazolinedione-adamantylidenadamantane reaction? To study this reaction in the forward direction is difficult in that triazolinedione itself is reactive toward many species, but to study the reaction in the reverse direction offers advantages in that various kinds of trapping agents (nucleophile, acids, etc) are compatible with the 1:1 adduct. By the principle of microscopic reversibility, if we form the adduct via an aziridinium imide, then we must also decompose the adduct via this same aziridinium imide. We have tried various trapping agents (eg, ones that might protonate an aziridinium imide and ultimately afford some aziridine-derived compounds) but with no success. Thus at present we have no evidence that requires an intermediate in the triazolinedione adamantylidenadamantane reaction, although we favor that interpretation. For triazolinedione to react with adamantylidenadamantane, the easiest path would seem to involve an orientation of reactants in a perpendicular array such as a and c (eq. 12). which could further collapse to an aziridinium imide and/or shunt on over to the diazetidene product.

There are various other cases in which we and others have made efforts to trap intermediates in reactions of triazolinediones with olefins. The work of von Gustorf describes some experiments [2]. Also, Butler and his co-workers examined the reaction of triazolinediones with enol ethers in the presence of acetone [13]. Tetrahydrooxadiazines were obtained; these were considered to arise from the trapping by acetone of a 1,4-dipole intermediate formed by reaction of triazolinedione with enol ether. We have reexamined that type of reaction using 4,4-dimethyldihydropyran (a in eq. 15).

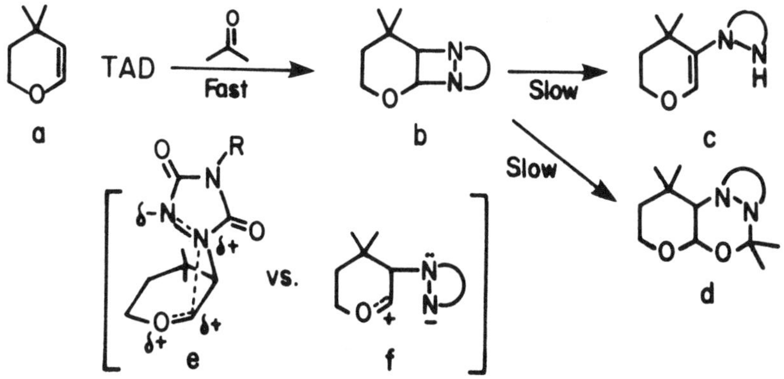

TAD Fast Slow Slow a b c d e f vs.

In a nonpolar solvent. the reaction affords the diazetidine (b), an isolable species. In acetone, the first-formed product is again the diazetidine (observed in the NMR) which is subsequently converted to vinyl urazole (c) and to a tetrahydrooxadiazine (d). These findings illustrate the difficulty of obtaining evidence on intermediates involved in the initial step--conversion of reactants to the diazetidine (b)--in that this product is quite labile in polar media, readily undergoing carbon-nitrogen bond cleavage to afford species (eg, an aziridinium imide, e or a 1,4-dipole, f) convertible to the vinyl urazole by intramolecular prototropic shift and to the tetrahydrooxadiazine (d) by reaction with acetone. The aziridinium imide or 1,4-dipole may also be involved in the initial reaction sequence by which the diazetidine is formed from the triazolinedione and the dihydropyran, but here again the available data do not require it.

There are some other kinds of trapping experiments in this area. Evidence for an intermediate was described in the reaction of triazolinedione with vinyl acetates in which the product was an aldehyde urazole acyl derivative (eq. 14) [14]. The intermediate was formulated as a 1,4-dipolar species; the results could also be accommodated in terms of an aziridinium imide (eq. 16).

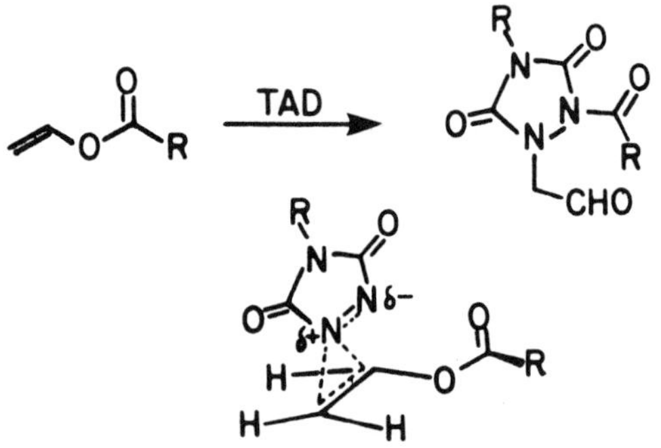

TAD (16)

These and other observations raise questions of how much charge is present at the "olefinic" carbons of an intermediate and about the degree of bonding between these carbons and the nitrogen. Some experiments that relate to these questions involve further comparisons with singlet oxygen. One of the rather surprising things to emerge from singlet oxygen investigations, both by Stephenson [15] and by Schulte-Elte [16] is the "preference for syn ene addition" (PSEA), ie, the preference for hydrogen abstraction from the more substituted <u>side</u> (the "syn" side) of the olefin in the reaction of singlet oxygen with olefin to afford allylic hydroperoxide. In this ene reaction, one also observes little preference for hydrogen abstraction at the more substituted <u>end</u> versus the less substituted end of the double bond. These findings are illustrated by the reaction of singlet oxygen with E-3-methyl-2-pentene and with methylcyclopentene in eq. 17 [16]. The preference for syn ene addition has been interpreted by Stephenson and others in terms of species such as a perepoxide with some stabilizing interaction between the oxygen and the hydrogens, as shown in eq. 17 [17]. (Professor Houk also has a paper related to this matter and he may wish to comment further about that in the discussion period [18].)

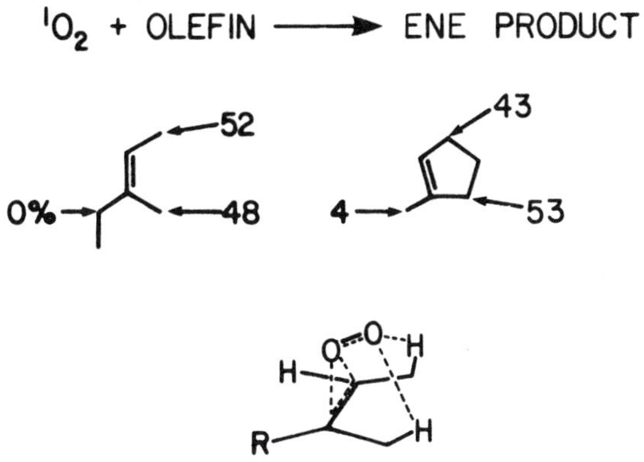

$$^1O_2 + \text{OLEFIN} \longrightarrow \text{ENE PRODUCT} \quad (17)$$

0% 52 48 4 43 53

H O O H R H

Reaction of triazolinedione with these same two olefins affords quite different results (eq. 18).

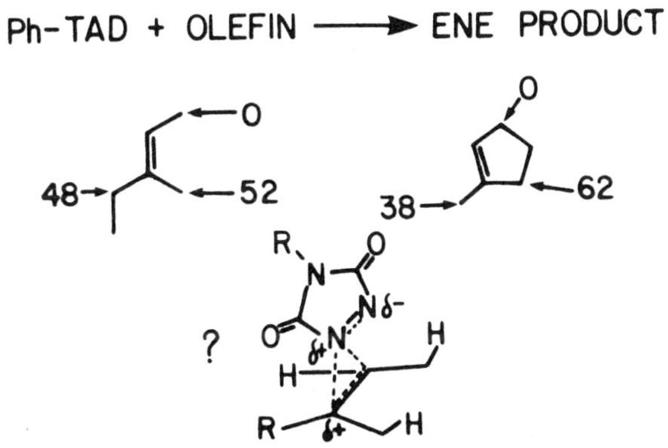

$$\text{Ph-TAD} + \text{OLEFIN} \longrightarrow \text{ENE PRODUCT} \quad (18)$$

48 0 52 38 0 62

R O N N δ- ? O δ+ N H H R δ+ H

One sees approximately equal reaction on both sides of the olefins, ie, no preference for syn ene addition. We do see a large preference for hydrogen abstraction from the more substituted ends of the olefins, also in marked contrast to the singlet oxygen results (compare the findings in eqs. 17 and 18). A contributing factor in these differences between singlet oxygen and triazolinedione may lie in the much lower activation energies for the singlet-olefin reactions than for the triazolinedione-olefin reactions.

Other comparisons bearing on charge distribution in an intermediate (eg, an aziridinium imide) are available. In the reaction of 1-phenylpropene with triazolinedione, there is no ene product from attack at the methyl. The products observed are similar to those described earlier with the stilbenes, involving the formation of a 1:1 Diels-Alder adduct followed by reaction of this adduct with another triazolinedione in two ways--to give a second Diels-Alder or an ene product (eq. 3). The lack of ene products involving C-H abstraction from the methyl group is consistent with an unsymmetrical intermediate with positive charge at the benzylic position (eq. 19); it is also consistent with Diels-Alder addition into the benzene ring.

TAD → D-A [adduct] → ✓

? [intermediate: N−, Nδ+, H, Ar, δ+, CH$_3$, H] ENE [product: N-NH] X (19)

These possibilities are seen more clearly in the reaction of triazolinedione with indene. As reported originally by von Gustorf [2], the major product is the [2+2] adduct, the diazetidine. One does not see products of ene reaction or of Diels-Alder addition. These findings are consistent with reaction via the unsymmetrical intermediate shown in eq. 20. The lack of Diels-Alder addition, in contrast with its importance with other arylethylenes, may be due to the added strain associated with a [2.2.1] bicyclic system (eq. 20). Why the lack of ene reaction in this case? Perhaps because that process would involve breaking the more fully formed carbon-nitrogen bond in the unsymmetrical intermediate. The finding of the [2+2] reaction to form diazetidine here is in accord with an unsymmetrical bonding between the nitrogen and the olefin carbons, in which the simplest way out may be collapse of the "imide" nitrogen with the partially positive benzylic carbon.

TAD

D-A ? NO
ENE ? NO
"2+2" YES

(20)

X ?

The indications of unsymmetrical character to the two carbon-nitrogen bonds in the intermediate

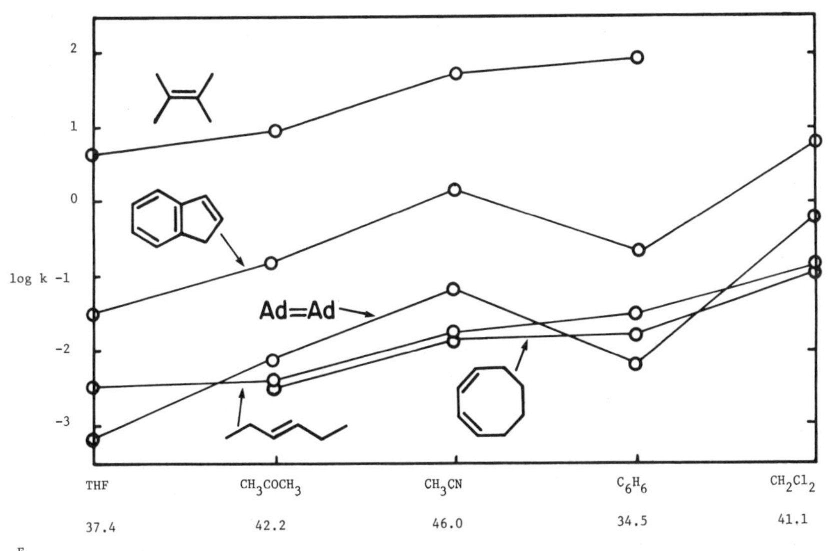

Figure 4. Solvent effect data for triazolinedione-olefin reactions.

raises, of course, the question of how polar is the transition state leading to this intermediate? In Figure 4 are summarized some solvent effect data [19] on triazolinedione-olefin reactions: tetramethylethylene and trans-3-hexene [19a] are ene reactions; indene and adamantylidenadamantane involve the formation of diazetidines (overall, 2+2); 1,3-cyclooctadiene and triazolinedione give a Diels-Alder adduct [19b]. All of these reactions in Figure 4 show a rather similar response to solvent, a response showing little sensitivity to solvent polarity. Adamantylidenadamantane and indene do show stronger response to solvent than do the other olefins. Perhaps this implies some greater degree of charge separation at the transition states for those cases than for the others. Note also that benzene has a rather atypical effect on rate. It is the least polar of these solvents and yet seems unusually effective in several of the reactions.

We have also made a limited comparison of activation parameters for two of these cases in benzene at 25°. Both the ene reaction of phenyltriazolinedione with trans-3-hexene and the [2+2] (overall) addition to adamantylidenadamantane show comparable energies and entropies of activation.

	Reaction	ΔH^* (kcal/mol)	ΔS^* (e.u.)
trans-3-hexene	ene	9.3	-34
adamantylidenadamantane	[2+2]	10.6	-33

The similarity in entropies of activation implies a comparable degree of rigidity of the triazolinedione with respect to the two double bonds, ie, no unusual steric problem is revealed for the rather encumbered double bond of adamantylidenadamantane. This finding is consistent with formation of a transition state in which the triazolinedione is perpendicular to the π bond--in a [2s+2a] or in a three-center type of formulation (as in a or c of eq. 12).

Thus we come to the view that the triazolinedione-olefin reaction proceeds through a transition state that looks like an aziridinium imide or like a [2s+2a] interaction (eq. 21). From the transition state we proceed to an intermediate for which we suggest the aziridinium imide species. If an allylic hydrogen is available, the "imide" nitrogen abstracts it, affording the ene product. If one is not available, the intermediate collapses to a diazetidine (eq. 21).

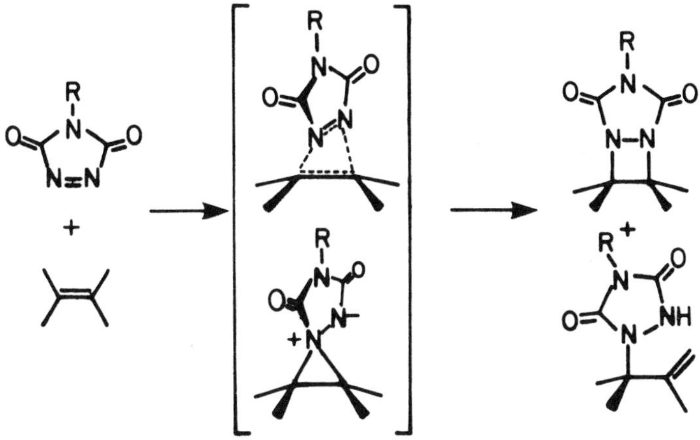

(21)

To what extent do these notions carry over to other reactions? We have mentioned that in reactions with olefins singlet oxygen and triazolinedione have much in common. Are these patterns observed elsewhere (eg, eq. 22)?

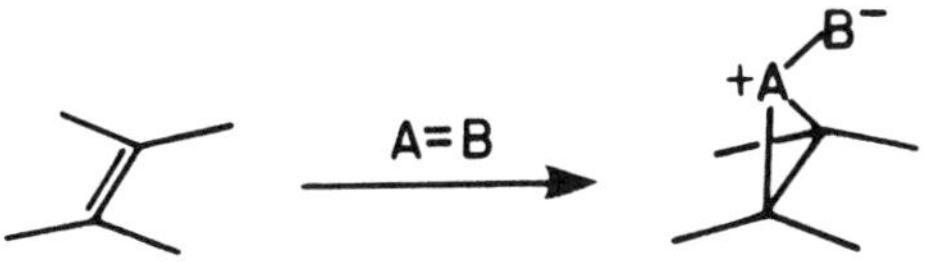

A=B (22)

O=O

R-N=N-R

R-N=O

We have looked at a few other systems, for example, nitroso compounds. Nitroso compounds do react with olefins. The results are rather varied

[20]. ESR signals are often seen, and products may be complex. In simple cases, the first-formed products may be hydroxylamines. Hydroxylamines may interact with nitroso compounds to give various sorts of electron transfer-hydrogen transfer products, giving rise to ESR signals. In some instances, a rather clean reaction is observed [21]. For example, reaction of pentafluoronitrosobenzene with tetramethylethylene affords the hydroxylamine (eq. 23) [22]. *A priori*, numerous mechanisms are possible, some of which are shown in eq.(23).

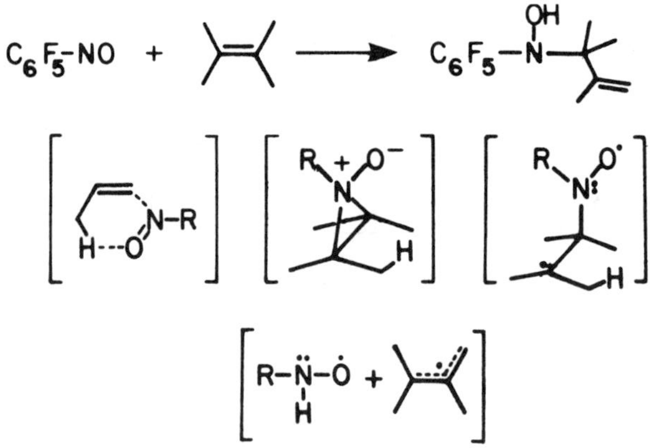

C_6F_5-NO + TME → C_6F_5-N(OH)-R (23)

We subjected this case to the Stephenson isotope effect test [5]. Here again we find a low isotope effect with *cis*-TME-*d*$_6$ and a large isotope effect with the *trans*-*d*$_6$ compound (eq. 24). As with the triazolinedione-TME case, these findings imply that the reaction of the nitroso compound with the olefin proceeds in two steps--a rate-determining step with a small (or negligible) isotope effect and a product-determining step with a large isotope effect when hydrogen and the deuterium are on the same side of the double bond. The results are accommodated by reaction via an aziridinium *N*-oxide intermediate (eq. 24). Indeed, species of this type may be intermediate in the ozonolysis of aziridines [23].

We looked at another case of this sort, reaction of bis-tosyl sulfur diimide with tetramethylethylene. Overall, the reaction effects replacement of allylic hydrogen by allylic nitrogen, a reaction

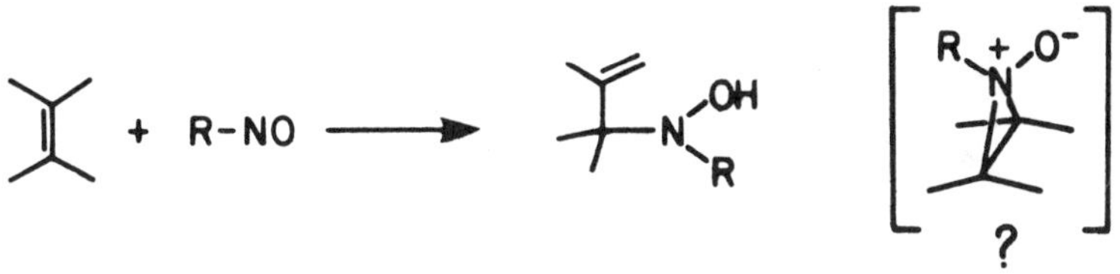

(24)

TME	k_H/k_D
cis-d_6	1.2
trans-d_6	3.0

investigated by Sharpless [24] and also by Kresze [23]. The reaction presumably proceeds by an ene reaction, followed by a 2,3-sigmatropic rearrangement (eq. 25). Now in this particular case, if

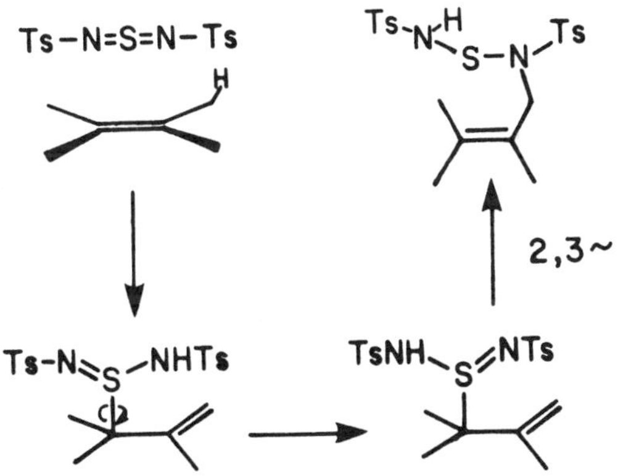

(25)

the ene reaction proceeds via a three-center species involving the sulfur atom and the carbon-carbon double bond, the resulting intermediate is symmetrical. Because of its symmetry, both cis- and trans-TME-d_6 would show similar isotope effects. The results are shown in eq. (26). The finding of significant and generally comparable, isotope effects with the different d_6-olefins is, of course, also consistent with many other mechanisms--with and without intermediates. Evidence of an intermediate in this reaction might be found in a comparison of intramolecular and intermolecular

(26)

TME	k_H/k_D
cis-d_6	3.9 ± 0.3
trans-d_6	3.4 ± 0.4
gem-d_6	4.9 ± 0.2

isotope effects. The intermolecular isotope effect should be accessible by means of a competition experiment with tetramethylethylene (TME-d_6) and perdeuteriotetramethylethylene (TME-d_{12}). A significant difference in inter- and intramolecular isotope effects would imply a two-stage process involving some type of intermediate. We hope to report on this aspect at a later date. The intramolecular isotope effects for the three doubly-bonded (A=B) electrophiles are summarized in eq. (27).

(27)

A=B	cis-d_6	trans-d_6	gem-d_6
Ph-TAD	1.1	3.7	5.6
C_6F_5NO	1.2	3.0	4.5
TsNSNTs	3.9 ± 0.3	3.4 ± 0.4	4.9

I might also call attention to the greater isotope effects for gem-TME-d_6 compared with trans-TME-d_6, a pattern seen with all three of these A=B electrophiles. The larger isotope effect observed with the gem-d_6 probably is due to β-deuterium isotope effects on the developing double bonds in the transition states for H (and D) transfer. In the case of trans-TME-d_6, k_H involves formation of $CH_2=C(CD_3)-$ and k_D involves formation of $CD_2=C-(CH_3)-$. With gem-TME-d the corresponding comparisons are k'_H, $CH_2=C(CH_3)-$, and k'_D, $CD_2=C(CD_3)-$. Thus, in the product-determining transition states to these olefins it is reasonable to find k'_H/k'_D (the observed isotope effect for gem-TME-d_6) greater than k_H/k_D (the isotope effect for trans-TME-d_6). A steric deuterium isotope effect may also be involved in that the transition state for the step in which the carbon-hydrogen bond is broken may involve some twisting around the R-N-O bond (eq. 28). In the gem case such a twisting

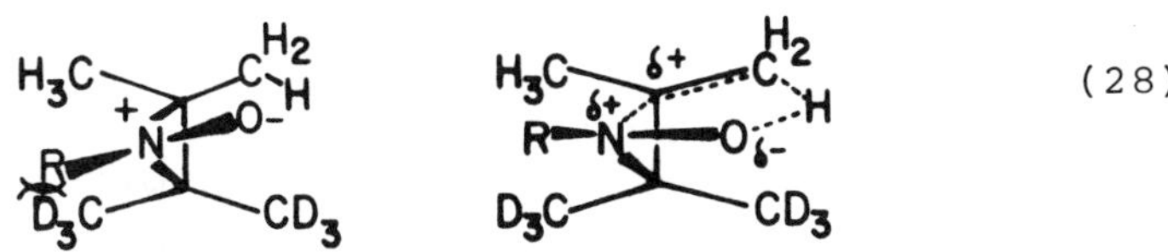

(28)

would bring the R group of the electrophile closer to CD_3, a sterically smaller group than CH_3. These considerations may also pertain to the preferential abstraction of allylic hydrogen from the more substituted end of the double bond systems shown in eq. 18.

The isotope effects with the double-bonded A=B electrophiles, triazolinedione and the nitrosoarene, imply rate-determining formation of an intermediate followed by conversion to products in a subsequent fast step (Figure 5). What can one learn through these types of experiments about reactions of singly bonded A-B electrophiles with olefins? Compelling evidence for intermediates in these cases is known [9]. Are there orientational characteristics related to the type of three-center species shown in Figure 5 but now necessarily involving only partial bonding between A and B? We have examined this point briefly in three cases, tert-butyl hypochlorite, N-chlorosuccinimide, and phthaloyl peroxide. Reaction of tert-butyl hypochlorite with olefins in general proceeds by a free radical chain reaction, affording mixtures of the possible allylic chlorides.

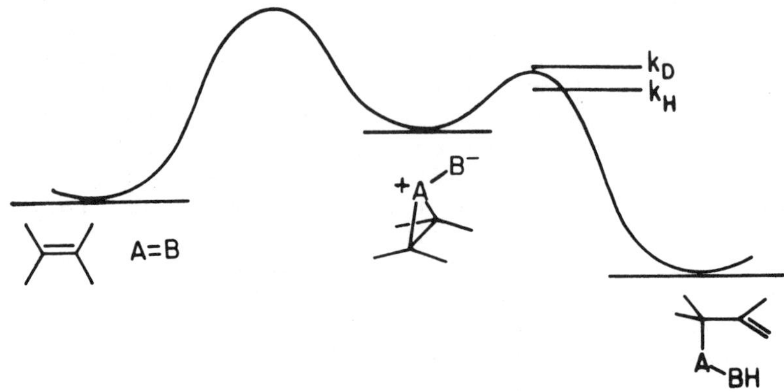

Figure 5. Profile for a reaction with a rate-determining formation of an intermediate followed by conversion to products in a subsequent fast step.

With tetramethylethylene, tert-butyl hypochlorite reacts rapidly in the dark affording only the tertiary chloride (eq. 29). This reaction has been described recently by Wynberg and co-workers [26]; it does not involve a free radical chain reaction.

O-Cl + $\xrightarrow[25^\circ]{CDCl_3}$ Cl ✓ Cl NONE (29)

? δ^+Cl···Oδ^- H H

To what extent, if any, does this reaction involve an oriented chloronium tert-butoxide species, ie, a species in which alkoxide oxygen is oriented on one side of the olefin, for example by a stabilizing interaction with neighboring hydrogens (eq. 29)? We have used the gem-TME-d_6 to screen some electrophile-olefin reactions. The gem-olefin is more easily made than the trans-TME-d_6 and, as noted above, gives larger isotope effects in ene reactions. So, if there is isotopic discrimination we should see it in the gem case. t-Butyl hypochlorite and N-chlorosuccinimide were examined (eq. 30). In both cases, the isotope effect is

$$(CH_3)_2C{=}C(CD_3)_2 + A{-}B \longrightarrow CH_2{=}C(CH_3){-}CA(CD_3)_2 + (CH_3)_2CA{-}C(CD_3){=}CD_2 \quad (30)$$

AB	k_H/k_D
t-butyl hypochlorite	1.26
N-chlorosuccinimide	1.27
methyltriazolinedione	5.7

rather small, particularly in comparison with the value of 5.7 found in the reaction of triazolinedione with gem-TME-d_6. The smallness of the isotope effects does not provide support for the type of oriented intermediate shown in eq. (29), and consequently we did not examine these reactions with cis- and trans-TME-d_6. We did look at one other reaction of a singly bonded A-B electrophile--one with a greater likelihood for an oriented intermediate than the two cases just considered. This is the reaction of phthaloyl peroxide with olefins. In some work carried out many years ago, phthaloyl peroxide was shown to react with olefins to afford mixtures of products including cyclic phthalates, lactonic ortho esters, anhydride, epoxide and also some ene product (eq. 31) [27]. In more recent times, Schuster and co-workers have examined reactions of phthaloyl peroxide with a variety of substrates, including simple olefins [28]. They concluded that the reactions, at least in the cases with good electron donors, proceeded by single electron transfer rather than by some type of concerted reaction. We were interested in possible involvement of intermediates in the reactions, and specifically, in whether an oriented intermediate was involved in the ene reaction. Potentially, this was another case--one concerning a singly bonded A-B electrophile--in which the Stephenson isotope effect test [5] might indicate the presence of an oriented intermediate. If the phthaloyl peroxide interaction with tetramethylethylene proceeds via an electron-transfer process, the first-formed species would be the phthaloyl peroxide radical anion and the olefin radical cation.

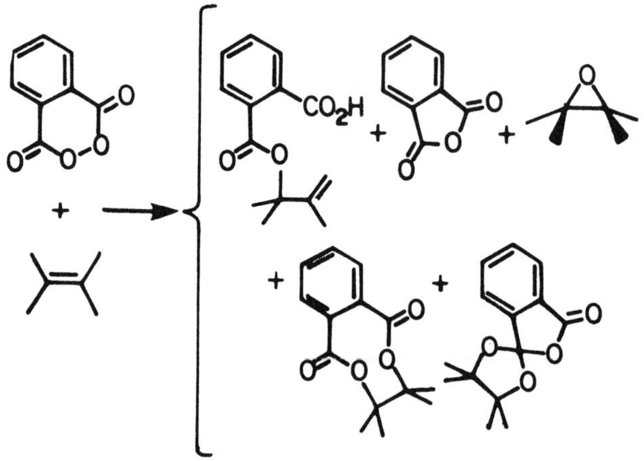

(31)

Schuster suggested that the peroxide radical anion would open rapidly and irreversibly to the carboxy radical-carboxylate anion of phthalic acid; recombination with the olefin radical cation would then lead on to products [28]. On the other hand, if phthaloyl peroxide and olefin react to form some kind of three-center species the intermediate might show some isotopic discrimination starting with trans-TME-d_6, and no isotopic discrimination starting with cis-TME-d_6 (eq. 32).

TME + phthaloyl peroxide ⟶ ene product (half ester acid)

TME	k_H/k_D
trans-d_6	1.35
cis-d_6	1.2
gem-d_6	1.2

However, we find no evidence for this kind of isotopic discrimination. We see only small, and comparable, isotope effects for trans-, cis-, and gem-TME-d_6.

R-D-S ? ? R-D-S

(32)

? (PhP)$^{\dot{-}}$(OLEFIN)$^{\dot{+}}$

PRODUCTS

What does this mean? The diversity of products in the phthaloyl peroxide-olefin reaction is suggestive that an intermediate is involved in the reaction. But the smallness of the isotope effect, and more particularly the similarity in isotope effect for cis- and gem-TME-d_6, indicate that such an intermediate (if involved at all) does not proceed directly to ene product. In summary, large isotope effects have only been observed with doubly bonded (A=B) electrophiles, consistent with three-center intermediates of the general type shown in Figure 5.

Most of the reactions I have been discussing involve interactions of electrophiles with π systems, many of which contain allylic C-H bonds (eq. 38). With doubly bonded (A=B) electrophiles, reaction with olefins ultimately affords products with four-membered rings or products involving cleavage of an allylic C-H bond. Previously, these reactions have been formulated as proceeding via [4+2] reactions or via 1,4-dipolar species. Reduced to its simplest terms, our suggestion is that some of these may be better explained in terms of three-center species. Those of you who remember some of the early work of Winstein will see that these considerations bear a close resemblance to his considerations on neighboring group participation [29]. In systems of the types shown in eq. 33 there may be many arrays in which the energy balance between extremes (open forms: 1,4-dipoles, 1,4-di-

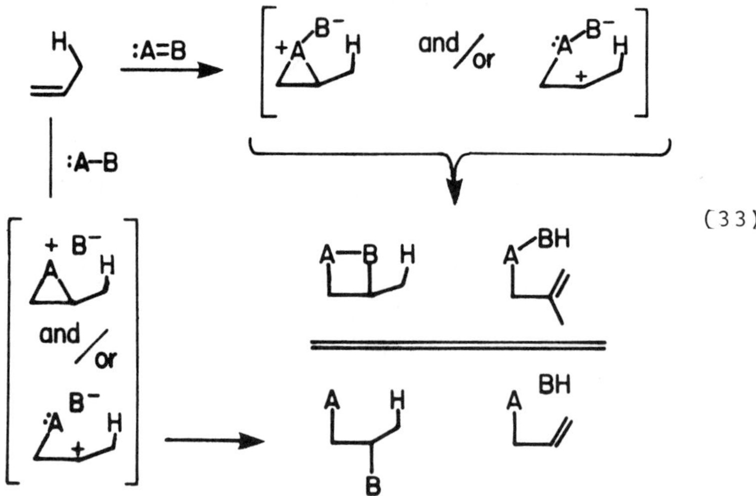

(33)

radicals; closed form, three-center species) is easily shifted one way or the other by variations in the electrophile or olefin. If one starts with singly bonded (A–B) electrophiles, evidence for the involvement of three-center species is certainly strong, eg, as seen in the formation of trans-dibromide by bromide ion attack on cyclic bromonium ion. Depending on the specific nature of A and B, the anion may remove an allylic hydrogen. The tert-butyl hypochlorite reaction with tetramethylethylene may be an example of this. Here also the balance between addition or ene reaction by A–B may be a sensitive function of A, B, and the olefin.

One thing that has intrigued me about these reactions, particularly with regard to triazolinedione, is the following. In going from reactants to products we start with an electrophile having a lone pair on atom A (eg, on azo nitrogen of triazolinedione). In the product we again have a lone pair on that atom A. To what extent, if any, is the lone pair of A involved at the transition state or in an intermediate? If the reaction goes via an aziridinium imide then the lone pair certainly is required to form part of the bonds in the aziridine ring. Is that lone pair of A of any help in the transition state for the interaction of this A=B system with olefins? If the olefin and the N=N system of the triazolinedione come together in a [2s+2a] fashion (a of Figure 6), then two electrons in the olefin π bond (the highest occupied molecular orbital) would interact with the lowest unoccupied π* orbital of the triazolinedione, an "allowed" interaction. If the transition state looks like a of Figure 6, then the lone pairs on nitrogen are just along for the ride; and the reason that triazolinedione reacts with olefins faster, for example, than does maleic anhydride, might simply be the fact that the two nitrogens and their lone pairs of electrons are smaller than the corresponding two C-H groups of maleic anhydride.

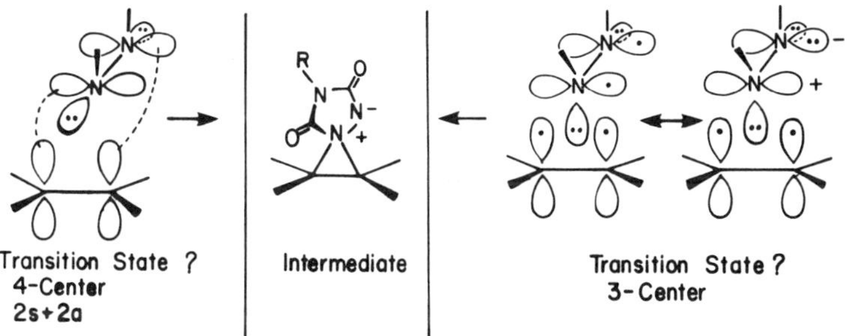

Figure 6. Possible transition states for the olefin-triazolinedione reactions.

These latter might pose a big steric problem to [2s+2a] cycloaddition. (Of course, they also affect the energy levels of the HOMO-LUMOs of the electrophile-olefin pair.) On the other hand, is it possible to bring triazolinedione into the olefin in a three-center transition state (c of Figure 6)? I raised that as a possibility much earlier in this talk, and now I come back to it. If I bring the triazolinedione and olefin together as in c, the lone pair points straight at the π electrons of the olefin--ostensibly a bad interaction. The orbitals are of the right symmetry but each orbital has two electrons! (Actually, the lone pair on one nitrogen interacts with the lone pair on the adjacent nitrogen leading to two new orbitals, $N_1 + N_2$ and $N_1 - N_2$; this does not alter the aspect we are considering.) An important distinction to keep in mind is whether one is trying to describe interaction between electrophile and olefin at an early point in the reaction or at the transition state. By the time one reaches the latter, numerous changes and electron reorganization may have taken place. If I think in terms of moving a π electron from N-1 to N-2, that leaves an empty p orbital at N-1 (see c of Figure 6). One sees that this formulation is reminiscent of the Walsh representation of a cyclopropane and also reminiscent of the type of interaction involved at some point in the addition of a carbene to an olefin. Another way to view the role of the lone pair of electrons on the nitrogen is shown in Figure 7. Interaction of the n

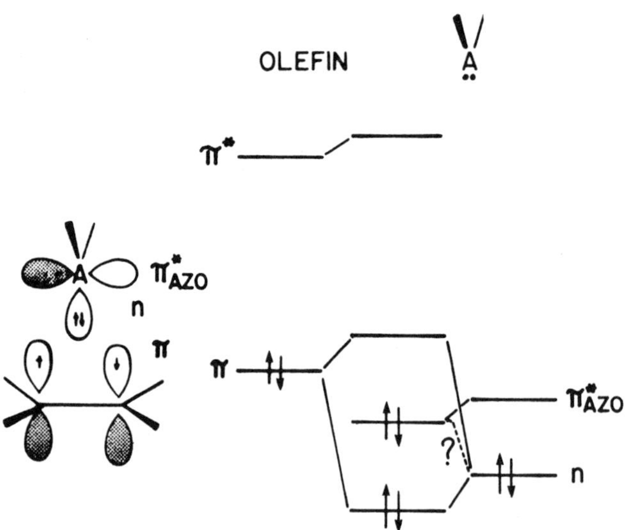

Figure 7. Possible electron reorganization if HOMO and LUMO of the electrophile lie lower than the HOMO of the olefin.

orbital of electrophile with the π orbital of the olefin obviously results in a net destabilization if both n and π are filled (as they are in the reactants). An additional feature in this case is the location of π* of the electrophile relatively close to the n level. (I call your attention to the fact that triazolinedione and nitroso compounds are colored; both have relatively small gaps between highest occupied and lowest unoccupied orbitals.) If both the HOMO and the LUMO of the electrophile lie lower than the HOMO of olefin, the opportunity exists for the type of electron reorganization shown in Figure 7; ie, viewed overall, the cost of promotion of electrons from n to π* of the electrophile is paid for by the new bonding possible between the resultant electron-deficient n orbital and the π orbital of the olefin.

Are these considerations also applicable to the singly bonded (A-B), colored, electrophile bromine in its reactions with olefins? The feature of a low-lying LUMO on the electrophile is again present but the orbital interaction diagram is quite different (Figure 8) from that for triazolinedione with olefin. Now the LUMO of bromine (σ*) is of the same symmetry, and approximately the same energy as the HOMO (π) of olefin. (The I.P. of Br_2 is 10.55 eV [30], the n-π* for Br_2 is ≃2 eV; the IP for tetramethylethylene is 8.3 eV [30]). This favorable interaction provides a powerful driving force for the reaction, and in combination with the interaction of HOMO of electrophile (p) with LUMO of olefin (π*), could lead straight to bromonium bromide. Thus the olefin-bromine reaction may proceed like an SN2 reaction with the π electrons of the olefin effecting a nucleophilic displacement on Br_2. (There is also evidence for complexes between some electrophiles and olefins [9]; in general, these may be weak, reversible interactions, describing the reaction coordinate at an early point in the electrophile-olefin interaction.)

In the bromine-olefin case, it is easy to see a simple transformation of reactants into a three-center transition state (Figure 8). In the triazolinedione (and nitroso)-olefin cases transformations of reactants to a three-center intermediate is much less obvious since the initial interactions (n with π) are clearly unfavorable. However, as pointed out above, the availability of a LUMO of electrophile lying close to HOMO of electrophile (and near or below the HOMO of olefin) may have some bearing on the possibility of three-center transition states for these cases.

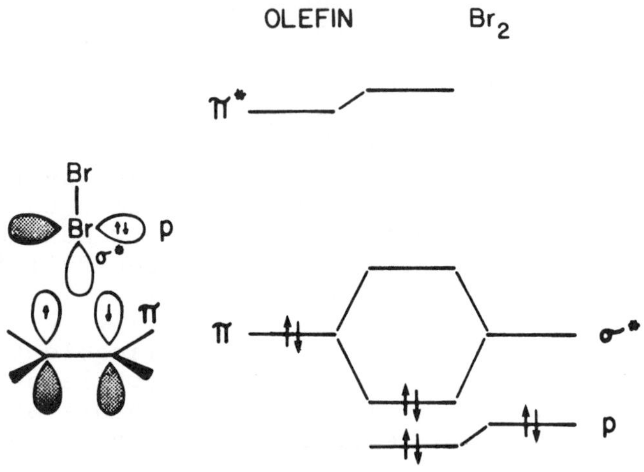

Figure 8. Orbital diagram for Br interaction with an olefin.

In summary, these studies have raised additional questions on a fundamental class of reactions; electrophiles with carbon-carbon multiple bonds. These studies, and those by other groups, have shown that several reactions, particularly ene reactions, which were formerly considered to proceed by a [4+2] route definitely do not proceed in that way. Rather, they proceed from reactants to products via intermediates. In some cases, these intermediates appear to be most simply described in terms of three-center species. A more difficult question concerns the nature of the transition states betweeen reactants and these intermediates: are they better described as [2s+2a] interactions, as three-center interactions, or something else? On this matter, we have called attention to some factors for consideration of three-center transition states.

Acknowledgements. I wish to call special attention to those who carried out the work reported here: Catherine A. Seymour and Chen-Chih Cheng, both graduate students, and Michael A. Petti, an undergraduate. Also, we wish to thank the National Science Foundation for support of this work (on NSF Grant No. CHE-8022783).

Discussion

Traylor: You do not get any nonstereospecific removal of the ene hydrogen?

Greene: No. Always specific. It can also depend on other steric factors.

Traylor: The reason I asked the stereochemistry question is because some years ago we were looking at the TCNE reactions with allyl metals. In that case I believe the electrophile, TCNE, comes in the side opposite the metal. So that is a system in which you could set up this reaction to lose a metal anti and that would be different. It would mean that it definitely would be in two steps because the loss, which corresponds to your hydrogen, would be on the opposite side of the molecule. The other comment is that this is very reminiscent of some oxymercuration reactions. In the case of cis additions to norbornene or other strained olefins you get a lot of acetate with mercuric acetate and methanol. So it looks as if the ligand is simply folding over the way you are writing it. If it goes trans there is no acetate or very little acetate also using methanol. This is very consistent with the picture you are writing of the ligand in the case of 4-center closure simply folding over in a sort of front side displacement. At least that is how we explain that you get all of this acetate with mercuric acetate and norbornene or other rigid olefins but not those which add trans. I was wondering if there was any dichotomy of reactions in which you can get something happening at the other side, maybe by neighboring group or by anything which would differentiate these two steps?

Greene: I think there would be a lot of possible cases. We are in the process of trying to look at some which would involve other kinds of neighboring group effects in which triazolinedione comes into the olefin on one side and some internal nucleophile interacts with the other side of the olefin such as iodolactonization reactions in the addition of iodine to unsaturated acids. Indeed, there may be cases already in the literature specifically on this point. There is a vast amount of work on triazolinedione reactions and my inability to come up immediately with some example which bears specifically upon your question may be one of just forgetting something in the literature.

Foote: I must say the more I hear about these reactions, the less I understand. I had pretty well convinced myself that the singlet oxygen [2+2] and ene reactions were distinct mechanisms. You have convinced me that I may have to rethink that one. You have drawn all three reactions coming from the same intermediate. Do you want to comment on that?

Greene: There are many papers on triazolinedione and on singlet oxygen. One thing that unsettles me about the singlet oxygen, from a theoretical standpoint, is the substantial difference between various theories put forward to rationalize the same experiments, for example, even with respect to the Stephenson experiments on the addition of singlet oxygen to cis and trans d tetramethylethylene. A recent paper by Jefford [Helvetica Chimica Acta 64, 2534 (1981)] provides some results on singlet oxygen-olefin reactions which he thinks are better explained by means of open ion-biradical species or "anion-cation"-like species. He also has extended those interpretations to explain the tetramethylethylene isotope effect of Stephenson in terms of "open" species. I find his extension to these Stephenson cases very unconvincing. I do not know about your own reaction to that particular paper, but I cite it here as an example of the possibility of rationalizing these types of reactions by a wide variety of paths.

Gleiter: You mentioned a low-lying π^* orbital of triazolinedione. How do you exclude any biradical intermediate, because there might be electron transfer from olefin to the π^*?

Greene: I do not think one could exclude something that was "biradical"-like. I think the triazoline reaction is much less sensitive to solvent than is something like the bromine reaction. Bromine is a good electron acceptor--directly or via a charge transfer complex. In the case of the triazolinedione all we can say is that the available solvent effect data do not indicate much charge difference in the transition state of TAD-olefin reactions compared to the ground state. Beyond the rate-determining transition state, whether things are better formulated as biradical-like or as ion pairs, or to what degree these species are better represented by partial charges, I do not feel strongly about. I should mention one problem with the triazolinediones. A common impurity is a paramagnetic species, presumably a urazolyl radical. This may prove to be of some consequence in some reactions of TAD.

Vogel: In the beginning of your lecture you said it is somewhat difficult to admit intermediate structures with a positive charge adjacent to a carbonyl group. At the limit, a carbonyl group [see eg, X. Creary, J. Am. Chem. Soc. 103, 2463 (1981)] just like a carbonitrile group [see eg, P. G.

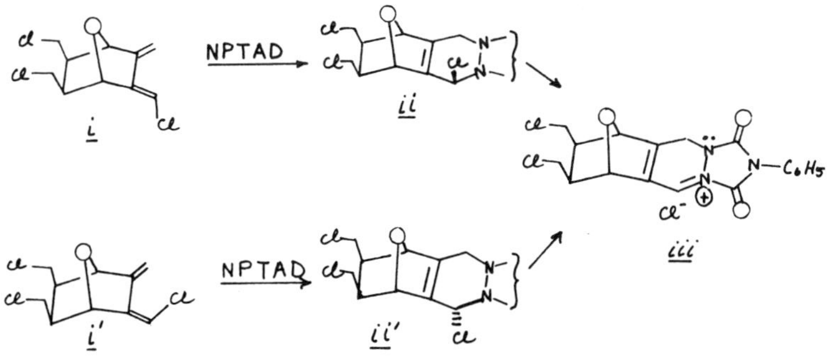

Gassman, K. Saito and J. J. Talley, J. Am. Chem. Soc. 102, 7613 (1980)] can act as an electron donating substituent because of its polarizability. We found recently [C. Mahaim, dissertation, Ecole Polytechnique Federale de Lausanne, 1982; C. Mahaim and P. Vogel, Helv. Chim. Acta 65, 866 (1982)] that the addition of N-phenyltriazoline dione (NPTAD) to 1-(E)- or 1-(Z)-chlorobutadienes i,i' gave the corresponding Diels-Alder adducts ii,ii' that ionized in a few minutes at 20° in benzene into the same unstable 2,3-dihydropyridazinium chloride iii. I think there is nothing wrong with writing a positive charge α to a carbonyl group.

Wagner: You compared the rates of the ene reaction with bromination rates. I guess this implies both ends of the double bond are serving equally as an electron donor. However, the product formation indicates one double bond is serving as a proton donor. If you had a charge separated species, that sort of implies you have two distinct transition states, one rate determining, and a second which is product determining. You did not comment specifically on this did you?

Greene: That was shown in the energy diagram in which I had the olefin and triazolinedione proceed over a high hill to an intermediate and then a smaller hill where k_H/k_D discrimination takes place.

Wagner: What I am getting at is if there is one intermediate, the transition state for formation implies it is a symmetrical intermediate, symmetrical with regard to the olefin. Yet the product it forms implies asymmetric charge distribution on the olefin. That is why I would think there are two distinct species. Perhaps, you get a charge transfer complex which is formed irreversibly and is then rearranged to a σ bond.

Greene: There certainly is agreement that two stages are needed to get from the reactant to the product. With regard to what the transition state for C-H cleavage looks like, I do not know. I think that transition state might have a fair amount of C-H cleavage, and a fair amount of C=C formation; and then the olefin that is formed should be the more substituted olefin. So whether one talks in terms of the second stage as something that is product-related or whether it is the stabilization of some incipient charge at these two unequal carbons of the original double bond, both views lead to the same product. I might say with regard to this step, the effect of unsymmetrical substitution on the olefin has a big effect, if one goes over to groups like aryl systems. For those cases the evidence is strongly on the side of a much less symmetrical intermediate. This type of pattern has been described by Dubeis and others in electrophilic addition to arylethylenes.

We have not carried out the kind of detailed comparisons to ascertain how unsymmetrical the triazolinedione-aryl substituted olefin intermediates are. Probably they will be quite different from the simply alkyl olefins.

Paquette: I am concerned about the first step of this mechanism. The issue here is those changes which may occur along the reaction channel. We have published some work on the directionality of eneophile additions to 7-isopropylidenebenzonorbornenes which also addresses a similar issue.

Utilizing the A=B notation, one sees that syn and anti products are possible. We have varied A=B considerably: singlet oxygen, *N*-methyltriazolinedione, *t*-butyl hypochlorite (not in your solvent but in ethyl formate where there is a greater guarantee of having the characteristics in which we are interested), and so on. Before we get into that, the reason for modifying X is because as you change X from an electron-withdrawing substituent, say tetrafluoro, to something much different electronically, say 1,4-dimethoxy, you get very different perturbation of the exocyclic π bond at C-7. In point of fact, electron-donating groups distort the π cloud to one side and electron-withdrawing groups distort it to the other. Our point of interest was the stereochemistry of addition to the double bond and the response of changing A=B to the stereochemical outcome of this reaction. We have found that singlet oxygen, N-methyltriazolinedione and t-butyl hypochlorite all give essentially identical stereochemical distributions as a function of X. That is, X will change the distribution, but within the range of the same substitution pattern on the aromatic ring, the same stereochemical distribution of syn and anti products is realized. That suggests to us that as you pass from one substituent to another to a third, that you do not really change too much the orbital character that is involved in the transition states as the reagents approach. Certainly, one does not want to vary too extensively from one set of conditions to very different ones. Nonetheless, irrespective of whether N-methyltriazolinedione is utilizing its filled or unfilled orbitals it must be kept in mind that one arrives at similar stereochemical consequences.

Greene: I think the comparison in your cases between the reactivity of bromine and triazolinedione also is stunning. Those syn:anti ratios are 80:20, exactly the same within experimental error.

Paquette: I do not know the significance of these observations either, except for the fact there has to be a great deal of transition state similarity. Otherwise, the alternative conclusion forced upon us is that the reagent is not particularly sensitive to which orbital construct exists in the overall system to begin with.

Greene: That is the other alternative of course.

Foote: I think my question is related. I wonder if you know of any evidence that excludes a charge transfer intermediate in these reactions. It would explain the fact the rates do not depend much on N-substitution, yet products sometimes do, although not with singlet oxygen.

Greene: I think my comments on that would be pretty diffuse. One thing that troubles me when I think about electron transfer goes back to an analogy with acids. When one has a strong acid you greatly affect the character depending upon what kind of solvent it is in. If you take toluenesulfonic acid in a solvent like benzene, it is a very strong acid. When you put it in some solvent that can coordinate with it, it is a very different system. It is a weak acid. I do not have a very clear feeling as to what might be involved when we have species which might be capable of electron-donation and their interaction with solvent. Obviously, there are going to be different degrees of solvation of these species as we change solvent. So as I make a solvent more polar I probably also affect some special stabilization, say of the triazolinedione component, so that the mere fact that one does not see some big response of rate to solvent is clouded by the fact that at the moment I do not have data pertaining to what the energies of triazolinedione solvation might be. I remind people of the observation by Haberfield some years ago in a Diels-Alder reaction. He measured the heats of solution of TCNE in o-xylene and in chlorobenzene and they were different by 5 kcal/mol. If one simply looked at the overall rates you would conclude quite a different thing than if you looked at the fact that the solvents were exerting 5 kcal more stabilization of TCNE in o-xylene. Because of those facts, I find it difficult to answer your question without knowledge of the solvation data pertaining to these reactions. Another thing which pertains to that is the question, should things that involve single electron transfer be very sensitive to solvent? Intuitively, I would have thought yes, particularly in the kinds of things Gary Schuster is suggesting with phthaloyl peroxide-olefin reactons and things of that sort.

Foote: You do not really have to transfer charge completely. You get a photochemical exciplex formation. This can form at a diffusion controlled rate even in non-polar solvent, although the energy is solvent dependent.

Noland: You mentioned that the bisadamantylideneadamantane adduct of triazolenedione is reversed at 70°. I wonder if you have looked at the behavior of any of your triazolinedione adducts in a mass spectrometer. It seems to me that if they dissociate at that low a temperature you may be getting a base peak which is due to dissociation on the probe before it ever gets hit by an electron.

Greene: I think we have looked at some of those. I think field desorption might get those up into the gas phase without dissociation, but I do not remember specifically what the abundance of the ions is that has both pieces together. I would comment that several other of these adducts are reversible. The one between dimethyl-2,2-dihydropyran is reversing in polar solvents, at least in part. Not coming all the way apart, it cleaves at C-H and then abstracts an adjacent hydrogen to give vinylurazole. In the case of the indene adduct it is reversible but at a temperature higher than the adamantylideneadamantane adduct.

Noland: Do you picture utilizing the unshared pair on nitrogen to accommodate the decreased reactivity of azodicarboxylate esters similar to these?

Greene: In the azodicarboxylate ester the orbitals will be different because they now have the lone pairs away from each other. The energy levels of the azodicarboxylates are quite different from the triazolinediones. Triazolinediones are very colored compounds while azodicarboxylates are not. The other point is there would be other kinds of steric effects if I bring the azodicarboxylate into the same kind of aziridinium imide-like geometry. But I am interested in that question, and we are trying to make some azodicarboxylate systems where I can wrap the azo group around intramolecularly and sit it over a double bond. Starting with something like the hydrazo precursor of the azo, oxidize it, and then see if the resulting azodicarboxylate grabs into the olefin in that kind of perpendicular geometry (i.e. to form an aziridinium imide).

Houk: I would like to comment on the confusion that I share with Chris and everyone else. Although it would be difficult to calculate, I think there is no doubt that the preferred approach of the pyrazolinedione to the alkene would be like that of a carbene. The considerations you are making about the incorporation of another configuration are the same as you would make for a carbene. That is, the direct ramming of a carbene in a least-motion fashion into an alkene is forbidden, although the product ultimately is stabilized. The way you avoid that is to bring it in sideways and then flip it over later. I expect the same would happen here. The situation is more complicated with singlet oxygen and probably with pyrazolinedione because, in the case of a carbene and these other species, there probably is a preliminary charge transfer complex, but in the case of the carbene the next thing you get is product. In the case of singlet oxygen and in your case, you may have yet another intermediate species which is kinetically insignificant and which is a face-to-face complex. On the other hand, what you are grapplng with could be that charge-transfer species, although it is difficult to imagine a 9-10 kcal/mol barrier to the formation of face-to-face complexes, at around 2.35 Å separation. I suspect the formation of such a complex would have a very low barrier, and it seems more likely that the activation energy is to convert this complex to product. But I still think the approach is going to be nonleast-motion and the way you get the intermediate will not be by bringing it in straight, but in by face-to-face first, with a later change.

Greene: One comment that I would make to that, and I think it would be different from what you are saying: in the case of the carbene, at long distances the interaction would be (as I understand it and as Hoffmann described it some years ago) be-bond of olefin. When the carbene comes in closer to it, it tips over. It will ultimately be formed in a three-membered ring. In the case of triazolinedione, I sense what you are saying is that at long distances the first and best thing is coming in something like that, with π^* of TAD and π of olefin in parallel planes. Why do you exclude the [2s+2a]? Why is it worse?

Houk: Because you get less overlap. An electrophilic reaction, generally, is the place where the application of frontier orbital consideration is most valid. If you bring up a strong electrophile to an alkene, it just goes to the site of the biggest HOMO density. Although it is not the same as your case, that is what happens initially. So I would just bring it up in the π-fashion to maximum overlap.

Greene: If one brings it in that way, one thing that pertains to this question is the adamantylideneadamantane adduct. This (the approach in parallel planes) is not good unless you are saying this thing is operating at a distance of many Angstroms. By the time I am forming a three-center bond, I have to be much closer; this suggests to me that the transition state of triazolinedione is already "down the chute" (ie, close to the three-center intermediate). I think I understand what you are saying. In answer to the question, what evidence is there that requires them (the electrophile and the olefin) to be close? Maybe the norbornene case is a case in point. In terms of this kind (the parallel plane approach) of overlap why is norbornene not off to the races? If the transition state like these two is farther apart, norbornene should have reacted without problems (ie, at a rate comparable to or greater than cyclohexene). If the transition state is already looking more like three-center, I can understand why norbornene is slow and adamantylideneadamantane is fast.

Herndon: The rate is fast or it just gives a better yield of product?

Greene: The rate is much faster. The rate constant for triazolinedione at 25° in dichloromethylene is 0.0007 $M^{-1}S^{-1}$ while adamantylideneadamantane is 0.7. These are second order rate constants.

Traylor: You cannot compare tetrasubstituted with disubstituted olefins. You lose a factor of 100 in going from tetramethylethylene to adamantylideneadamantane and in the disubstituted case you also lose a factor of 100 in going from cyclohexene to norbornene. What are the steric factors? Are they looking similar? Is that wrong?

Greene: I do not think they are altogether wrong; however, I think there are other ways of looking at that comparison. In terms of most of the electrophiles that have been added to norbornene and cyclohexene, norbornene is much faster. Whereas things like adamantylideneadamantane obviously are hindered severely on both sides, norbornene is quite open on one side. However, your point is well taken and the comparison between those systems if you look only at substitution is not a profound difference.

References

[1] D.K. White and F.D. Greene, J. Am. Chem. Soc. 100, 6760 (1978).
[2] E.K. von Gustorf, D.V. White, B. Kim, D. Hess, and J. Leitich, J. Org. Chem. 35, 1155 (1970).
[3] (a) R.L. Amey and B.E. Smart, J. Org. Chem. 46, 4090 (1981). (b) M.H. Chang and D.A. Dougherty, J. Org. Chem. 46, 4092 (1981).
[4] A.A. Frimer, P.D. Bartlett, A.F. Boschung, and J.G. Jewett, J. Am. Chem. Soc. 99, 7977 (1977).
[5] (a) Sr. B. Grdina, M. Orfanopoulos, and L.M. Stephenson, J. Am. Chem. Soc. 101, 3111 (1979) (b) L.M. Stephenson, Tetrahedron Lett., 21, 1005 (1980).
[6] C.A. Seymour and F.D. Greene, J. Am. Chem. Soc. 102, 6384 (1980).
[7] J.E. Weidenborner, E. Fahr, M.J. Richter, and K.H. Koch, Angew. Chem., Int. Ed. Engl. 12, 236 (1973).
[8] (a) S. Ohashi, K. Leong, K. Matyjaszewski, and G.B. Butler, J. Org. Chem. 45, 3467 (1980). (b) S. Ohashi and G.B. Butler, ibid. 45, 3472 (1980).
[9] (a) G. Freeman, Chem. Rev. 75, 439 (1975). (b) G.H. Schmid and D.G. Garratt in "The Chemistry of Double-Bonded Functional Groups," Part 2, S. Patai, Ed.; John Wiley and Sons, New York 1977, Ch. 9.
[10] (a) J.E. Dubois and G. Mouvier, Bull. Soc. Chim. France, 1426 (1968). (b) See also L.A. Paquette et al, J. Am. Chem. Soc. 103, 7106 (1981) for a study showing marked similarity in stereoselectivity by a variety of weak electrophiles including triazolinedione.
[11] W. Adam, O. DeLucchi, and I. Erden, J. Am. Chem. Soc. 102, 4806 (1980).
[12] J. Hess and A. Vos, Acta cryst. B33, 3527 (1977).

[13] S.R. Turner, L.J. Guilbault, and G.B. Butler, J. Org. Chem. 36, 2838 (1971).
[14] K.B. Wagner, S.R. Turner, and G.B. Butler, J. Org. Chem. 37, 1454 (1972).
[15] M. Orfanopoulos, M.B. Grdina, and L.M. Stephenson, J. Am. Chem. Soc. 101, 275 (1979).
[16] K.H. Schulte-Elte and V. Rautenstrauch, J. Am. Chem. Soc. 102, 1738 (1980).
[17] L.M. Stephenson, M.J. Grdina, and M. Orfanopoulos, Acc. Chem. Res. 13, 419 (1980).
[18] K.N. Houk, J.C. Williams, Jr., P.A. Mitchell, and K. Yamaguchi, J. Am. Chem. Soc. 103, 949 (1981).
[19] (a) Some of these results are from ref. 8b. (b) H. Isaksen and J.P. Snyder, Tetrahedron Lett. 889 (1977).
[20] C. Chatgilialoglu and K.U. Ingold, J. Am. Chem. Soc. 103, 4833 (1981).
[21] M.G. Barlow, R.N. Haszeldine, and K.W. Murray, J. Chem. Soc., Perkin Trans. I, 1960 (1980).
[22] R.A. Abramovitch, S.R. Challand, and Y. Yamada, J. Org. Chem. 40, 1541 (1975).
[23] (a) J.E. Baldwin, A.K. Bhatnagar, S.C. Choi, and T.J. Shortridge, J. Am. Chem. Soc. 93, 4082 (1971). (b) See also Y. Hata and M. Watanabe, J. Am. Chem. Soc. 101, 1323 (1979).
[24] K.B. Sharpless and T. Hori, J. Org. Chem. 41, 176 (1976).
[25] R. Bussas and G. Kresze, Liebigs Ann. Chem., 627 (1980) and references cited therein.
[26] (a) E.W. Meijer, R.M. Kellogg, and H. Wynberg, J. Org. Chem. 47, 2005 (1982). (b) See also ref. 10b.
[27] F.D. Greene and W.W. Rees, J. Am. Chem. Soc. 80, 3432 (1958).
[28] J.J. Zupancic, K.A. Horn and G.B. Schuster, J. Am. Chem. Soc. 102, 5279 (1980).
[29] S. Winstein and E. Grunwald, J. Am. Chem. Soc. 70, 828 (1948) and references cited therein.
[30] K. Kimura, S. Katsumata, Y. Achiba, T. Yamazaki, and S. Iwata, "Handbook of HeI Photoelectron Spectra," Japan Scientific Societies Press, Tokyo, 1981.

7. THE DI-Π-METHANE REARRANGEMENT REVISITED

Waldemar Adam, Néstor Carballeira, Ottorino De Lucchi, and Karlheinz Hill

Institut für Organische Chemie, Universität Würzburg, Am Hubland, D-8700 Würzburg (FRG) and Departamento de Química, Universidad de Puerto Rico, Rio Piedras, Puerto Rico

The rich menu of π dishes that is being offered at this symposium would hardly be complete if one of the most abundant dishes, namely the di-π-methane rearrangement, were not included. As newcomers, as we are, it requires no little courage to "revisit" such an exciting and challenging area of photomechanistic chemistry. Despite the numerous publications on this topic [1], we felt that the various diradical species that have been postulated as reaction intermediates required more rigorous confirmation through independent synthesis. For this purpose we chose appropriate five-membered ring azoalkanes, which on thermal or photochemical denitrogenation were expected to provide a convenient entry into the diradical manifolds of such photochemical rearrangements. What follows recollects our experiences on this problem during the past couple of years.

The accepted mechanism of the di-π-methane rearrangement, recently renamed the Zimmerman rearrangement [1b], is shown in eq. (1). The di-π-methane substrate <u>1</u> is converted on excitation into the vinylcyclopropane, the di- -methane product <u>2</u>. Presumably at first the 1,4-diradical <u>3</u> is produced via vinyl-vinyl bonding, which then cleaves one of the lateral bonds of the cyclopropane ring to afford the 1,3-diradical <u>4</u>. Cyclization at the radical sites affords the product <u>2</u>.

A specific example is shown in eq. (2), in which 6,7-benzobicyclo[3.2.1]octa-2,6-diene (<u>5</u>) is photochemically rearranged into 3,4-benzotricyclo-

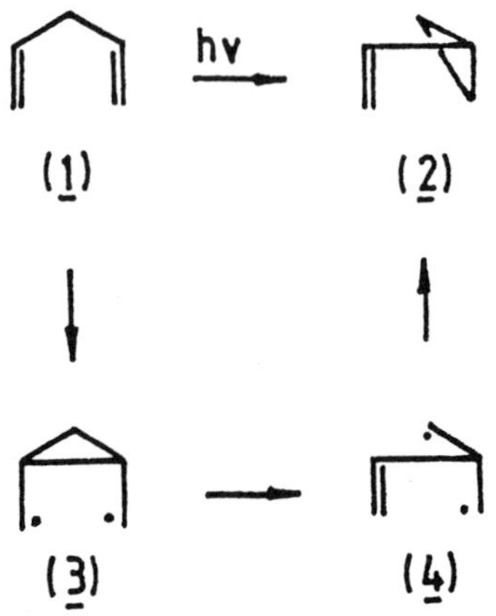

(1) hv (2)

(3) (4) (1)

[3.2.1.0^{2,7}]oct-3-ene (6) [2]. This example will be particularly important, as we will see, because it constitutes one of the very first substrates that we have investigated. The di-π-methane moiety in the substrate 5 and the vinylcyclopropane grouping in the product 6 (eq. 2) are marked darkly for emphasis. Through vinyl-benzo bonding across

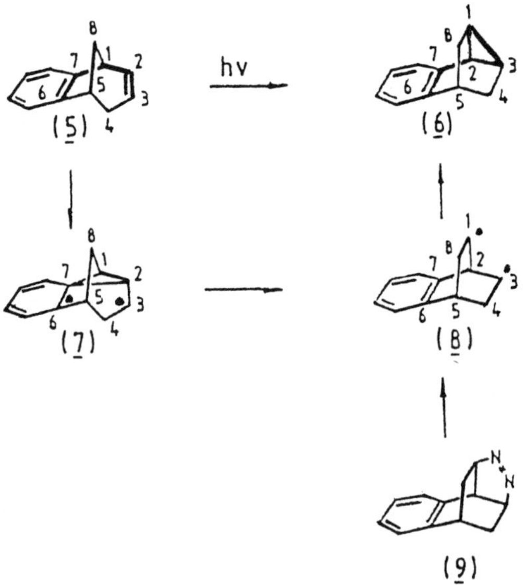

(5) hv (6)

(7) (8) (2)

(9)

C-2 and C-7 the 1,4-diradical 7 is first generated, which on breaking of the C-1-C-7 bond leads under rearomatization to the 1,3-diradical 8. Final cyclization at the C-1 and C-3 sites affords 6.

A logical entry into the diradical manifold postulated in eq. (2) would be independent generation of the 1,3-diradical 8 via denitrogenation of the azoalkane 9 [3]. If azoalkane 9 could be conveniently prepared, then we would have a unique opportunity of exploring the chemical fate of diradical 8. For example, besides closure at C-1 and C-3 to give the di-π-methane product 6, diradical 8 might revert via 7 to the di-π-methane substrate 5. The sequence 8 → 7 → 5 would then constitute a *retro*-di-π-methane process. Other points of mechanistic interest concerned how different spin states, ie, singlet versus triplet diradical 8, would influence the product distribution between cyclization and the *retro*-process.

The key question, of course, is how could one synthesize such a convenient diradical precursor as azoalkane 9? Those of us working on azo compounds quickly sense that via classical routes the synthesis of azoalkane 9 is not a trivial matter. Fortunately, a few years ago we stumbled onto a most unusual reaction of *N*-phenyl-1,2,4-triazoline-3,5-dione (PTAD) with strained bicyclic olefins, which makes such azo compounds efficiently and

(10) + (PTAD) → (11) → (12) (3)

conveniently available [4]. The generalized sequence is given in eq. (3), in which bicycloalkene 10 affords on cycloaddition with PTAD the rearranged urazole 11. Subsequent hydrolysis and oxidation converts urazole 11 into azoalkane 12.

With respect to the required azoalkane 9, application of this synthetic sequence to olefin 5 gives the rearranged urazole 13 (eq. 4). Of special interest, which will become apparent toward the end, is the fact that the same olefin 5, which on one

$$\text{(5)} \xrightarrow{\text{PTAD}} \text{(13)} \longrightarrow \text{(9)} \qquad (4)$$

hand serves as substrate in the di-π-methane rearrangement, constitutes also the starting point for the preparation of azoalkane 9, which in turn is to provide an entry into the diradical manifold postulated for olefin 5.

$$\text{(5)} \xrightarrow{\text{PTAD}} \text{(14)} \longrightarrow \text{(15)} \longrightarrow \text{(16)} \longrightarrow \text{(13)}; \quad \text{(14)} \longrightarrow \text{(16)} \qquad (5)$$

The mechanism of this novel reaction mode of PTAD is presently under active investigation in our group. However, our preliminary results suggest initial electrophilic attack by PTAD on olefin 5 to give in a slow step the aziridinium ion 14, postulated by Greene (eq. 5) [5]. Formation of the rearranged urazole 13 can then be readily rationalized in terms of Wagner-Meerwein shifts either directly from aziridinium ion 14 or via the 1,4-dipole 15 to give the dipolar intermediate 16. Cyclization at the dipole sites affords urazole 13.

With this general and powerful synthetic tool on hand, the essential azoalkanes for the generation of diradical intermediates, which are postulated in di-π-methane rearrangements, can in principle be made and thereby we can elucidate the mechanistic features of such complex photorearrangements. Several such mechanistic applications will now be illustrated, beginning with azoalkane 9 and its connection with the di-π-methane rearrangement of diene 5 (eq. 2). Thus, our *modus operandi* is to generate diradical 8 via photolysis of diene 5 and via denitrogenation of azoalkane 9 and compare the

(5) $\xrightarrow{h\nu}$ (8) $\xleftarrow[(-N_2)]{\Delta, h\nu}$ (9) (6)

products that are formed (eq. 6), in the hope that a mechanistic link between these two processes can be established. We describe the investigation of this system in particular detail, because it shall serve as a model study for the other azoalkanes to be presented here.

The denitrogenation results [6] are summarized in Table 1. While in solution the azoalkane 9 is stable toward heating up to 250° for long periods of time (negligible amount of denitrogenation after several days), on vacuum flash pyrolysis (VFP) by sending the volatilized 9 through a hot tube at 350-400° and 0.2 Torr, efficient denitrogenation occurs, affording 82% of the di-π-methane product 6 and 18% of the *retro*-product 5 (first entry in Table 1). A control experiment showed that the *retro*-product 5 is stable toward the VFP conditions. Consequently, the *retro*-product 5 is a primary product of the thermal denitrogenation 8 → 5 + 6. Clearly, the thermally generated 1,3-diradical 8 mainly cyclizes to give the expected di-π-methane product 6, but a significant fraction suffers rearrangement into the *retro*di-π-methane product 5 via the 1,4-radical 7.

Table 1. Product Composition of the Denitrogenation of Azoalkane 9 (b)

Denitrogenation conditions	Conversion (%)(b)	Yields 5	6
VFP(350-400°C/0.2Torr)	100	18±1	82±2
hν/Ph CO(300-330 nm; n-pentane)(c)	100	(d)	100
TMD(80-85°C/2h, toluene)(e)	40	(d)	100
h (350 nm/Pyrex; pentane)(f)	100	18±1	82±2
h (254 nm/Quartz; pentane)(f)	5	9±3	91±3(g)

(a) Determined by quantitative VPC (flame ionization) using a 6 ft x 1/8 in. stainless steel column packed with 100% SE-30 on Chromosorb P and operated at injection and detector temperatures of 250°C and a column temperature program of 16°/min, starting at 60°C and finishing at 220°C and a carrier gas flow of 20 mL/min.

(b) Normalized to 100% conversion; olefin 5 is stable toward the thermolysis conditions and the photolyses conditions, except at 254-nm irradiation.

(c) [9] = 0.01M and [Ph_2CO] = 0.1 M; Rayonet Photoreactor, using 300 nm lamps.

(d) Below GC detection limit (0.01%).

(e) TMD stands for tetramethyl-1,2-dioxetane; [TMD] = 0.1 M and [9] = 0.02 M.

(f) [9] = 0.041 M; Rayonet Photoreactor.

(g) Yields corrected for the direct photochemical conversion of 5→6.

In the benzophenone-sensitized denitrogenation of azoalkane 9 at 300-330 nm (second entry in Table 1), using a dichromate filter to avoid n,π* excitation of the azo chromophore [3a], only the di-π-methane product 6 is produced. Not even traces of the retrodi-π-methane product 5 could be detected by GC. This seemed very unusual to us, because for a triplet state reaction the propensity for direct bonding of the radical sites is low in view of spin forbiddenness, unless a triplet-state cyclopropane product is formed. The latter process is energetically not feasible. Consequently, the lifetime of the triplet-state diradical 8 should be appreciably increased and thereby competitive reactions such as reversion via the 1,4-diradical 7 into retro-product 5 expected. We felt obligated to provide additional proof for the exclusive formation of 6, particularly since the diene 5 on triplet sensitization affords the di-π-methane product 6 [2], and thus might have been formed but suffered in situ di-π-methane rearrangement. A control experiment showed that the di-π-methane rearrangement of diene 5 is too inefficient under the benzophenone-sensitized photolysis of azoalkane 9. Yet, to be completely sure and thus mechanistically meaningful, it was essential to avoid such problems.

For this purpose we employed triplet sensitization by means of triplet acetone, chemically generated via the thermolysis of tetramethyl-1,2-dioxetane (TMD). It is known [7] that TMD is a clean source of triplet acetone (eq. 7). Therefore, on heating of a solution of TMD and azoalkane 9 in benzene at 80-85° until total decomposition of TMD (ca. 2 h), ca. 40% of azoalkane had been converted exclusively into the di-π-methane product 6 (third entry in Table 1). Thereby it is confirmed that triplet sensitization leads only to cyclization.

$$\underset{\text{(TMD)}}{\text{TMD}} \xrightarrow[C_6H_6]{\Delta} \underset{\sim 30\%}{{}^{T}(CH_3)_2C{=}O^{*}} + \underset{\sim 0.01\%}{{}^{S}(CH_3)_2C{=}O^{*}} \qquad (7)$$

In this context it should be mentioned that attempted photochemical triplet sensitization with acetone (even as solvent) gave significant amounts of retrodi-π-methane product 5. With acetone as sensitizer it is in practical terms difficult to avoid direct n,π* excitation of the azo chromophore. On the other hand, singlet sensitization by acetone also might play a role. Whatever the reason, the advantage of dioxetanes in such photomechanistic problems is clearly demonstrated with this example.

We come now to the direct photolyses of azoalkane 9, which we conducted at 350 nm (n, π* excitation of the azo chromophore) and 254 nm (π,π* excitation of the benzene chromophore). The direct photolysis at 350 nm (fourth entry in Table 1) gave the same ratio of products 5 and 6 as the thermal denitrogenation, ie, 18% retro-product 5 and 82% di-π-methane product 6. However, on direct photolysis at 254 nm (fifth entry in Table 1) the product ratio shifts still further in favor of cyclization, ie, 92% di-π-methane product 6 and only 8% retro-product 5. Since diene 5 is efficiently di-π-methane active at 254 nm yielding 6, the above yields are run at low converson, ie, less than 5% denitrogenation of azoalkane 5 and extrapolated to zero conversion.

The qualitative energy diagram in Figure 1 attempts to connect the mechanistic features of the di-π-methane rearrangement of diene 5 (left-hand side) with those of the denitrogenation of azoalkane 9 (right-hand side). Clearly, since the triplet-state process (benzophenone sensitization as well as TMD chemienergization) of azoalkane 9 gives only the cyclized product 6, the retro-product 5 is presumably a singlet-state product. The fact that the same ratio of products 5 and 6 are formed in the thermal and photochemical (350 nm) denitrogenation of azoalkane 9 suggests that identical diradical intermediates are engaged. Since the thermal denitrogenation must, because of spin conservation, initially generate a singlet-state 1,3-diradical 8, either such a singlet diradical 8 efficiently cyclizes into the di-π-methane product 6 or suffers efficient spin flip into the triplet diradical 8, which subsequently cyclizes into 6. On the basis of the present results we cannot distinguish between these two alternatives, but it would be truly unusual if a singlet diradical 8 would not be capable of cyclization. In this respect, mechanistically puzzling is the finding that on 254 nm photolysis the azoalkane 9 leads to more cyclization product 6.

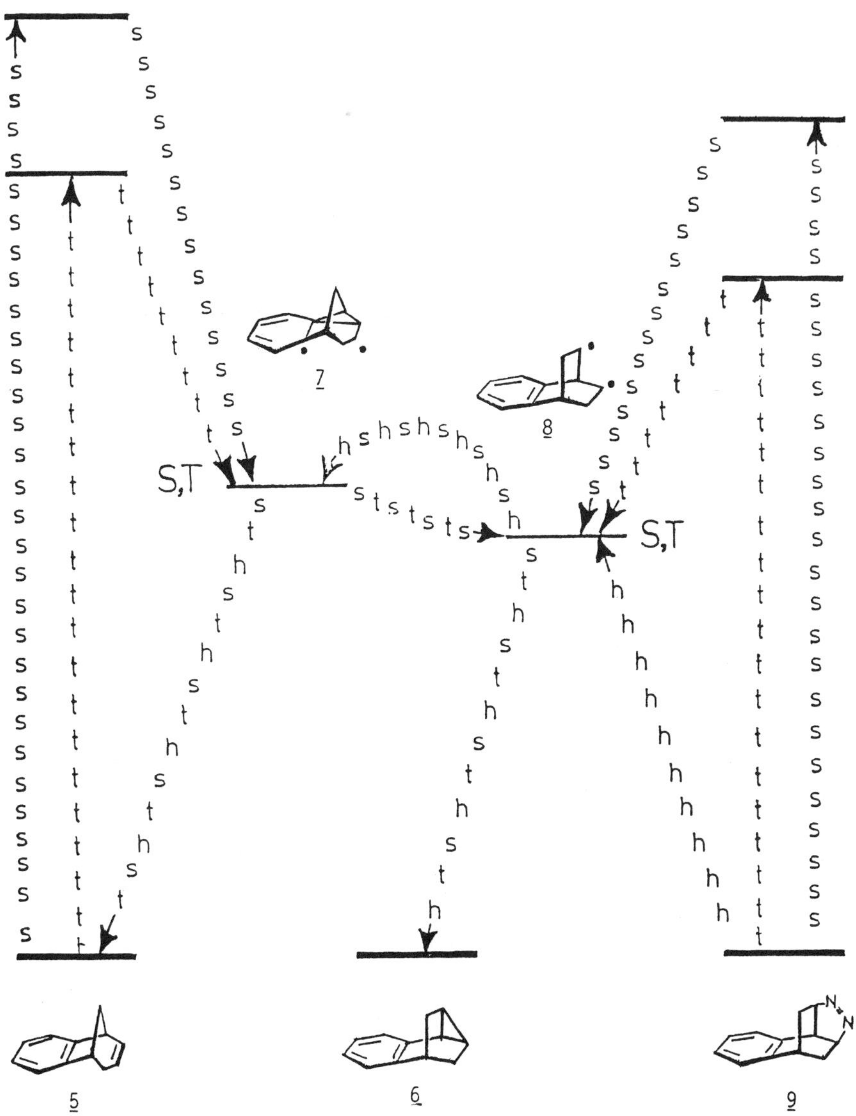

Figure 1. Qualitative energy diagram interconnecting the thermal (hhh), the singlet (sss), and the triplet denitrogenations of azoalkane 9 with singlet (sss) and triplet (ttt) excited-state di-π-methane rearrangements of bicycloalkadiene 5 into tricycloalkene 6 via the 1,4-diradical 7 and the 1,3-diradical 8.

The most valuable mechanistic finding of this study is the demonstration that the 1,3-diradical 8 that is postulated for the di-π-methane rearrangement of diene 5, but independently generated via denitrogenation of azoalkane 9, reverts into the 1,4-diradical 7. This conclusion is based on the fact that significant amounts (ca. one-fifth) of diene 5 are formed in the thermolysis as well as direct photolysis (350 nm) of azoalkane 9. This

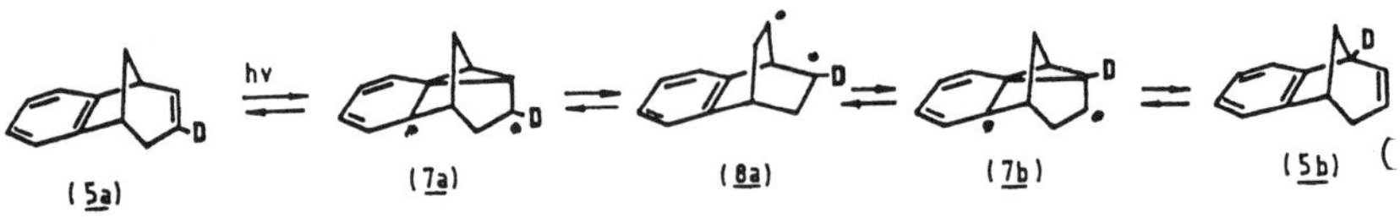

(5a) (7a) (8a) (7b) (5b) (8)

implies, although it needs to be proved, that in the di-π-methane rearrangement of diene 5 the diradicals 7 and 8 also interconvert. An experiment to establish the interconversion of diradicals 7 and 8 directly in the di-π-methane rearrangement of diene 5 is in progress using 3-deutero-6,7-benzobicyclo[3.2.1]octa-2,6-diene (5a). As shown in eq. 8, except for a possible secondary isotope effect, the 1,3-diradical 9a has the options of regenerating diene 5a via the 1,4-diradical 7a or producing diene 5b via 7b.

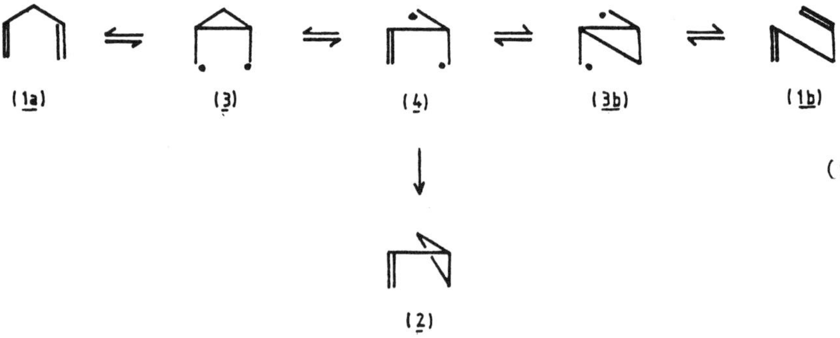

(1a) (3) (4) (3b) (1b)

(9)

(2)

If indeed the 1,4- and 1,3-diradicals of the di-π-methane rearrangement interconvert, the mechanistic scheme for the general process of eq. 1 should be extended as shown in eq. (9), in analogy to the more complete picture for diene 5 (eq. 8). Of course, in eq. 9 the generalized di-π-methane substrates 1a and 1b are structurally identical and thus not differentiable, unless an isotopic label is introduced, as in the case of dienes 5a and 5b (eq. 8).

In principle, however, the possibility exists that a di-π-methane substrate could be chosen for which interconversions in the diradical manifold of eq. (9) would generate distinct dienic products 1a and 1b. Such a case is shown in eq. (10), in which the two distinct dienes 17a and 17b interconvert into one another and afford the di-π-methane product 20 via the distinct 1,4-diradicals 18a and 18b and the common 1,3-diradical 19.

Indeed, the photochemistry of 5-methylenebicyclo[2.2.1]hept-2-ene (17a) has been investigated [8], affording the di-π-methane product 20 in low quantum efficiency. No mention is made whether the isomeric 7-methylenebicyclo[3.2.0]hept-2-ene (17b) is formed as well, nor apparently has the photochemistry of diene 17b been investigated since it is an unknown compound. It appeared to us, therefore, worthwhile to test the generality of the extended mechanistic scheme of eq. (10), by independently generating the 1,3-diradical 19 via denitrogenation of the azoalkane 21. However,

(21)

(17a) ⇌ (18a) ⇌ (19) ⇌ (18b) ⇌ (17b) (10)

(20)

the problem arose now of how to prepare this essential azoalkane. For example, attempted PTAD cycloaddition with diene 17a gave only ene product 22 (eq. 11), a severe limitation of the applicability of PTAD in the preparation of azoalkanes from the rearranged urazoles (eq. 3).

(17a) (22) (11)

Fortunately, 5-norbornenone (23) provided the solution to this synthetic problem (eq. 12). The results of the denitrogenation of azoalkane 21 are given in Table 2, in which only the relative yields of the dienes 17a, 17b, and the di-π-methane product 20 (normalized to 100%) are listed. For example, in the VFP at 360°C the major product is pyrazole 26, formed via *retro*-Diels-Alder (eq. 13),

(23) PTAD (24) Ph_3PCH_2 (25) (21) (12)

analogous to the azoalkane derived from norbornene [9]. Clearly, the interconversion between the various diradicals 18a, 18b and 19 is best illustrated in the 350-nm photolysis of azoalkane 21, since besides the di-π-methane product 20 also the *retro*-products 17a and 17b are produced. The major

(21) VFP 350°C (26) (13)

route for the 1,3-diradical 19 is cyclization to give 20, but significant amounts (ca. 15%) of 19

Table 2. Product Composition of the Denitrogenation of Azoalkane 21 (a)

Denitrogenation conditions	Conversion (%)	Yields(%)(b) 17a	17b	20	26
21, VFP(360°C/22 Torr(c)	100	13±2		35±2	52±2
21, hν(350 nm, C_6H_6)(d)	13	8±2	7±2	82±2	
21, hν/Ph_2CO(350 nm, C_6H_6)(e)	4	96±2		4±2	
17a (185 nm, n-C_5H_{12})(f)	20		12±2	84±2	
17b (185 nm, n-C_5H_{12})(f)	20	45±2		46±2	
17a (450W medium pressure Hg-lamp(g)	50			16(h)	

(a) Determined by capillary GC; 50-m OV-101 capillary column operated at injector, column, and detector temperatures of 200, 80 and 200°C, respectively.

(b) Normalized to 100% conversion; olefins 17a to 20 are stable towards photolysis conditions.

(c) Under these conditions ca. 15% of authentic 17b is transformed into an unknown product.

(d) [21] = 0.013 M; Rayonet photoreactor supplied with 350 nm lamps.

(e) Same conditions as in footnote d, except in the presence of 0.130 M benzophenone.

(f) [17a] = [17b] = 0.010 M; the lamp output is ca. 20% 185 nm and 80% 254 nm light.

(g) Taken from Ref. [8a].

(h) Plus another six, not identified minor products.

rearrange via the 1,4-diradicals 18a and 18b into the respective dienes 17a and 17b in approximately equal proportion. Control experiments confirm that the di-π-methane product 20 and the dienes 17a and 17b are photostable under the 350-nm photolysis conditions of azoalkane 21. Therefore, interconversion must take place in the diradical manifold. However, it is of interest to point out that preliminary experiments show that both dienes 17a and 17b interconvert and give the di-π-methane product 20 on 185-nm irradiation. Di-π-methane chemistry in 185-nm photolyses has been reported recently [12] and we are presently extending our work into this novel area of organic photochemistry.

While in the thermolysis (VFP at 360°C) the 1,3-diradical 19 undergoes mainly cyclization to give the di-π-methane product 20, of the two possible retro-processes it chooses exclusively the route 19 → 18a → 16a. Again, control experiments confirm that 17b and 20 are stable under the VFP conditions of azoalkane 21. Such selectivity of the 1,3-diradical 19 is also exhibited in the benzophenone-sensitized photodenitrogenation of azoalkane 21, in that only the retro-product 17a is formed. However, in contrast to the thermal denitrogenation of azoalkane 21, the retro route 19 → 18a → 17a predominates even over the cyclization route 19 → 20. Consequently, in this system the triplet state diradical 19 prefers retrodi-π-methane rearrangement into diene 17a, while the singlet state prefers cyclization into di-π-methane product 20.

This is to be contrasted with the previously discussed azoalkane 9 (Table 1), for which the opposite behavior obtains. Presumably the triplet state 1,4-diradical 18a provides an efficient spin inversion mechanism for the 1,3-diradical 19 via the "rotor" effect of the methylenic radical site. Such "rotor" effects have been invoked to rationalize the low quantum yields for triplet-sensitized di-π-methane rearrangements of substrates such as 17 [1,8]. We are pursuing this mechanistic query in detail.

(14)

(24) → → (27)

With the help of azoalkane 21 we were able to illustrate the potential generality of the anticipated interconversions in the di-π-methane diradical manifold of eq. (9). Since the rearranged urazole 24, which provided synthetic access to azoalkane 21 (eq. 12), can in principle be converted into the ketoazoalkane 27 (eq. 14), the opportunity presented itself to explore analogous diradical interconversions in oxadi-π-methane (ODPM) rearrangements.

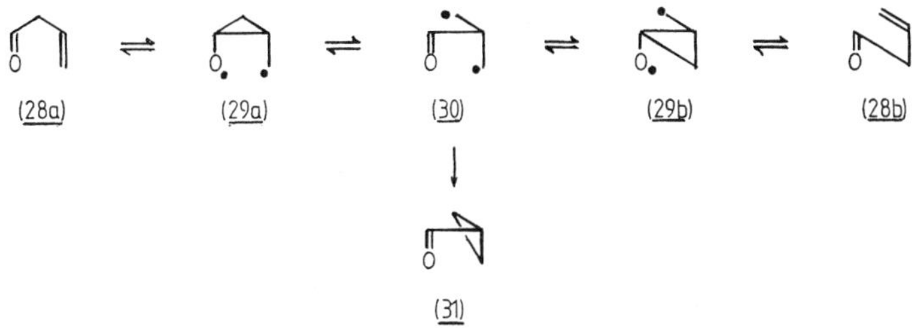

(28a) ⇌ (29a) ⇌ (30) ⇌ (29b) ⇌ (28b)

(30) → (31) (15)

This is illustrated in eq. (15) for the generalized ODPM process. Again, the enones 28a and 28b in eq. 15 are structurally identical and, therefore, not differentiable unless isotopic labeling is applied as in diene 5 (eq. 8). However, when applied to the azoalkane 27 (eq. 16), we note that enones 32a and 32b are distinct molecules, produced respectively via the sequences 34 → 33a → 32a and 34 → 33b → 32b. Indeed, the ODPM reactions of 32a and 32b affording 35 and the interconversions of

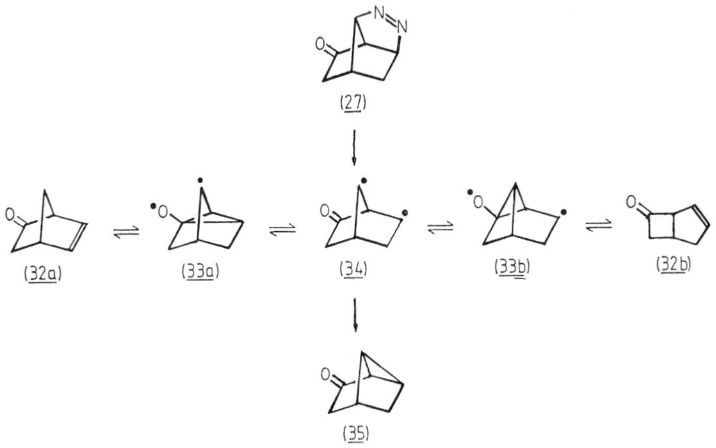

(27) → (34) (16)

(32a) ⇌ (33a) ⇌ (34) ⇌ (33b) ⇌ (32b)

(34) → (35)

32a and 32b have been the subject of active investigation [11]. Instead of the interconverting diradical scheme shown in eq. (16), the isomerization 32a → 32b has been interpreted mechanistically in terms of 1,3-acyl shifts, competing with the ODPM processes 32a → 35 and 32b → 35. It was, therefore, of interest and importance to probe the interconversion of the diradicals in eq. (16) through authentic generation of the 1,3-diradical 34 via denitrogenation of azoalkane 27.

Table 3. Product Composition of the Denitrogenation of Azoalkane 27 (a)

Denitrogenation conditions	Conversion (%)	32a	Yields(%)(b) 32b	35
27, VFP(300 °C/22 Torr)	100	87*	13	(c)
27, hν(350 nm, C_6H_6)(d)	40	50	42	7
27, hν(350 nm, 1,3-cyclohexadiene, C_6H_6)(e)	40	43	53	3.5
27, hν/Ph_2C=O (350 nm, C_6H_6)(f)	60	68	4	28
27 TMD(85 °C, C_6H_6)(f)	8	59	5	30
32a, hν/Ph_2C=O (364 nm, C_6H_6)	45		0.006 (h)	0.042 (h)

*Errors in % yields ±1 or ±2.

(a) Determined by CGC; 50-m OV-101 capillary column, operated at injector, column, and detector temperatures of 200, 90 and 200°C.

(b) Normalized to 100% conversion.

(c) Stable toward VFT conditions.

(d) Under these conditions ca. 7% of authentic 32a is transformed into 32b; the % yields have been appropriately corrected; [27] = 0.0096 M; Rayonet photoreactor with 350 nm lamps.

(e) Same conditions as in footnote (d), except in the presence of 0.102 M 1,3-cyclohexadiene.

(f) 15 mol benzophenone/mol of 27.

(g) 1 mol tetramethyl-1,2-dioxetane 32a transformed into 35; ca. 6% unidentified products.

(h) Quantum yields from ref. [11d].

The denitrogenation results are summarized in Table 3. Significant differences are noticeable in the behavior of the ketoazoalkane 27 and the methyleneazoalkane 21. For example, in the thermal denitrogenations only retro-products 32a and 32b are formed, although the di-π-methane product 35 is stable toward the VFP conditions of azoalkane 27. Of the two isomeric enones, the less strained isomer 32a is formed preferentially. Assuming that a thermally equilibrated singlet diradical 34 is produced in the VFP denitrogenation, then it appears that the ODPM product 35 is derived from a triplet-state diradical 34. Indeed, both in the benzophenone-sensitized and TMD-chemienergized denitrogenations, in which presumably the triplet state diradical 34 intervenes, significant amounts (ca. 30%) of the ODPM product 35 are formed. The major products also in the triplet-sensitized reactions are the retro-products 32a and 32b, respectively formed via the sequences 34 → 33a → 32a and 34 → 33b → 32b, of which the former process again predominates.

Interesting to mention is the fact that the ratio of quantum yields of ODPM product 35 and retro-product 32b in the benzophenone-sensitized reaction of enone 32a is within experimental error the same as the ratio of these products formed in the benzophenone-sensitized denitrogenation of azoalkane 27, namely 7:1 (Table 3). We feel that this fact can hardly be coincidence and propose that the same interconverting diradical manifold obtains, ie, 33a → 34 → 33b, in the ODPM processes of enones 32a, 32b, and in the denitrogenation of azoalkane 27.

Of mechanistic significance is also the direct photodenitrogenation of azoalkane 27 at 350 nm (Table 3). Control experiments show that at 350 nm the enones 32a and 32b are interconverted and the product yields have been appropriately corrected. Again, the retro-ODPM products, ie, the enones 32a and 32b, prevail over the ODPM product 35; however, 32a and 32b are formed in essentially equal proportions. Presumably a "hot" 1,3-diradical 34 is produced here, which is sufficiently energetic to afford high yields of the more strained enone 32b. When this 350-nm photolysis is carried out in the presence of a triplet quencher such as 1,3-cyclohexadiene, the ODPM product 35 is reduced and the sum of the retro-ODPM products 32a and 32b correspondingly increased. This experiment corroborates our previous mechanistic conclusion that the ODPM product 35 is predominantly triplet-state derived.

Although not all the features of the photodenitrogenations of the azoalkanes 21 and 27 are mechanistically understood, it is evident that with the help of these azoalkanes we have been able to demonstrate that interconverting diradical manifolds are involved in di-π-methane (eq. 9) and ODPM (eq. 15) rearrangements. Whether such schemes are general must be established through additional work. For the two specific examples investigated here, the 1,3-diradical 19 represents a branching point in the product distribution of the di-π-methane rearrangement of dienes 17a and 17b, while the 1,3-diradical 34 represents such a branching point for the ODPM process of enones 32a and 32b.

The mechanistic question to be raised now is whether also the 1,4-diradical 3 in the generalized di-π-methane rearrangement in eq. (9) can function as branching point, leading to distinct di-π-methane products? This possibility is illustrated again in general terms in eq. (17). As we can observe, a common 1,3-diradical 3 serves as product branching

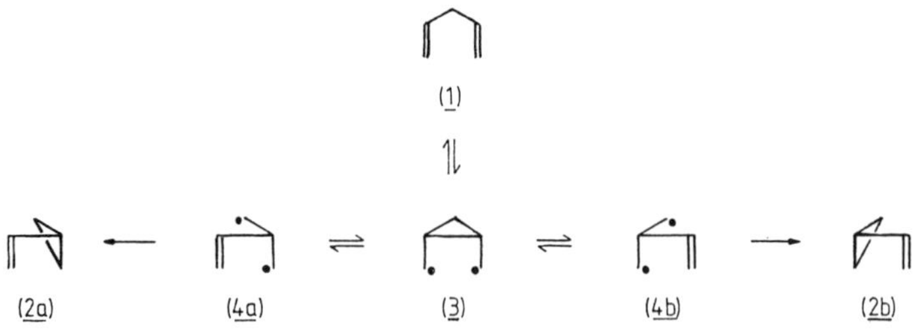

(1)

(17)

(2a) (4a) (3) (4b) (2b)

point, affording the retrodi-π-methane product 1 by fragmentation of the central bond of the cyclopropane ring, or the two di- -methane products 2a and 2b by fragmentation of the lateral bonds via the diradicals 4a and 4b, respectively. Again, except for being mirror images, the di-π-methane products 2a and 2b are structurally identical and thus indistinguishable. However, a sufficiently complex bicyclic substrate can in principle be conceived for which the vinylcyclopropenes 2a and 2b are distinct products.

Such an example constitutes the di-π-methane rearrangement of bicyclo[3.2.1]octa-2,6-diene (36) [13], which gives rise to the two different vinylcyclopropanes 39a and 39b via the respective 1,3-diradicals 38a and 38b that are produced from the common 1,4-diradical 37 (eq. 18). Entry into this interconverting diradical manifold could be achieved via denitrogenation of the azoalkanes 40 - 42, which serve as precursors to the diradical 37 and 38a and 38b, respectively. The denitrogenation of this set of azoalkanes comprises the final problem to be presented here.

(41) (36) (42)

(38a) (37) (38b) (18)

(39a) (40) (39b)

Fortunately, PTAD cycloaddition with the diene 36 leads to all three urazoles 43 - 45, for which the respective azoalkanes 40 - 42 can be prepared via oxidative hydrolysis [4e]. The product distributions of the denitrogenations of azoalkanes 41 and 42 are given in Table 4. The results for the denitrogenation of the azoalkane 40 are not listed, because only the diene 36 is produced. Presumably the energetically favorable concerted homo-Diels-Alder retrocyclization (eq. 19) is engaged, thereby circumventing formation of the 1,4-diradical 37.

Table 4. Product Composition of the Denitrogenation of Azoalkanes 41 and 42 (a)

Denitrogenation conditions	Conversions (%)	Yields(%)(b) 36	39a	39b
41, VFP(∿280°C/15 Torr)	22	14.5*	83.5	2.0
41, hν(350 nm; pentane)(c)	80	5	81	13
41, hν/Ph_2CO (300-330nm; pentane)(d)	40	13	24	62
42, VFP(250°C/14 Torr)	90	18	traces	82
42, hν (350 nm, pentane)(c)	100	24	traces	72
42, hν/Ph_2CO (300-330 nm; pentane)(c)	30	6		94

*Errors in % yields range from ±0.5 to ±2.

(a) Determined by capillary GC; 50-m OV-101 capillary column, operated at injector, column and detector temperatures of 150, 90 and 150°C, respectively.

(b) Normalized to 100% conversion; in all cases 100% product balance was obtained when the unidentified minor products were included; the products were stable towards the thermolysis and photolysis conditions of the azoalkanes.

(c) [41] = [42] = 0.01 M; Rayonet Photoreactor supplied with 350 nm lamps.

(d) [41] = [42] = 0.037 M and [Ph_2CO] = 0.3 M; Rayonet Photoreactor supplied with 300 nm lamps, using Pyrex vessels and a dichromate filter (Ref. [3a]).

This is, of course, disappointing because generation of the common 1,4-diradical 37 via denitrogenation of azoalkane 40 would provide the most convincing test whether 37 serves as branching point in the product distribution. But as the product data in Table 4 reveal, for both azoalkanes 41 and 42, respectively the precursors to the 1,3-diradicals 38a and 38b, the interconversion of these diradicals via the common 1,4-diradical 37 is demonstrated. The greater degree of interconversion is experienced when entering the diradical manifold via 38a by denitrogenation of azoalkane

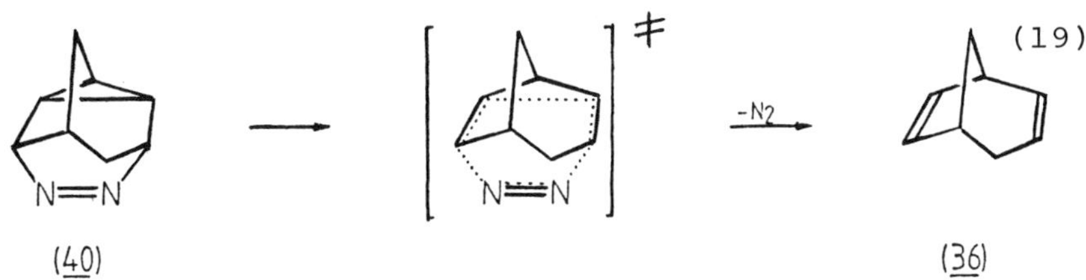

(19)

41, since both expected di-π-methane products 39a and 39b and the *retro*-product 36 are formed. Thus, in the thermolysis and direct photolysis at 350 nm, presumably singlet-state reactions, the major product is the vinylcyclopropane 39a. On the other hand, in the benzophenone-sensitized and TMD-chemienergized denitrogenations of azoalkane 41, presumably triplet-state reactions, the major product is the vinylcyclopropane 39b. However, irrespective of the mode of denitrogenation and thus also the nature of the spin state of the diradical intermediates, azoalkane 42 affords the vinylcyclopropane 39b as major product (Table 4). Only traces of the vinylcyclopropane 39a are produced, but analogously to azoalkane 41 the *retro*-product 36 is also here the minor product.

While the mechanistic details of these product distributions in the denitrogenation of azoalkanes 41 and 42 and their spin-state dependence are far from being understood at this point, our investigation clearly brings out that interconverting diradicals are involved in the diradical manifold of the di-π-methane rearrangement of diene 36. Its extended mechanistic scheme is given in eq. (20), in which the product branching points of the distinct 1,4-diradicals 37a, 37b, and 37c as well as those of the distinct 1,3-diradicals 38a and 38b are included. The dienes 36a and 36b are enantiomers and could be distinguished through deuterium labeling experiments quite similarly as its

(42) (39b)

(38b) ⇌ (37b) ⇌ (36b)

(36a) ⇌ (37a) (20)

(38a) ⇌ (37c) ⇌ (36c)

(41) (39a)

benzo derivative in eq. (8). However, diene 36c is a distinct retro product. Although an authentic sample of diene 36c was prepared, not even traces of it could be detected by means of capillary GC in the denitrogenations of azoalkanes 41 and 42. Apparently the interconvertibility of such 1,3- and 1,4-diradical species has its limitations.

In conclusion, the most extended mechanistic scheme for the generalized di-π-methane rearrangement is displayed in eq. (21). By means of systematically disconnecting and reconnecting distinct bonds in the manifold of interconverting 1,3- and 1,4-diradicals, a convenient mechanistic vehicle is on hand to recognize the complete set of vinylcyclopropane and diene products in di-π-methane rearrangements. Such an extended scheme permits one readily to conceive what azoalkanes should be useful as precursors to the various 1,3- and 1,4-diradical intermediates in the elucidation of the

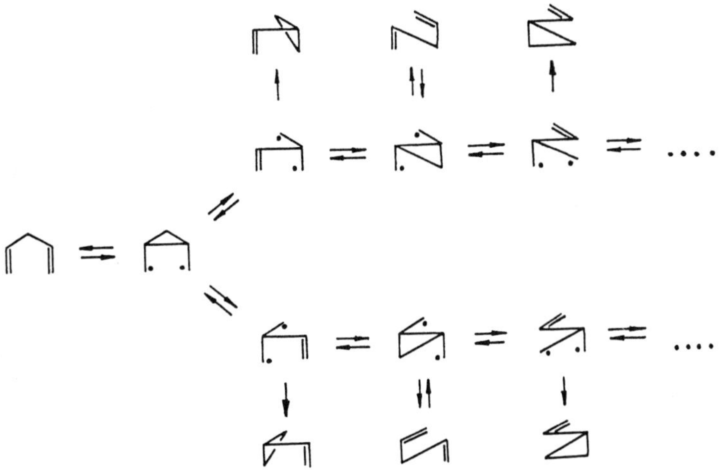

(21)

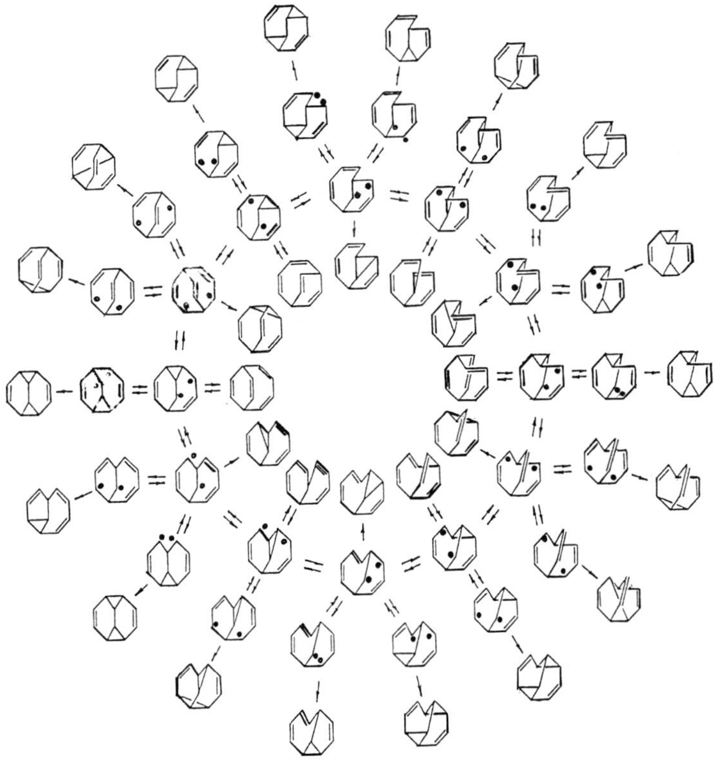

(22)

mechanistic details of complex photorearrangements. An impressive illustration of this point is given in eq. (22), in which the interconverting diradical species of the bullvalene photochemistry are portrayed. A first step [14] in the elucidation of this complex mechanistic scheme has been made by investigating the denitrogenation of the azoalkanes 46 and 47 affording the hydrocarbons 48 - 50 (eq. 23).

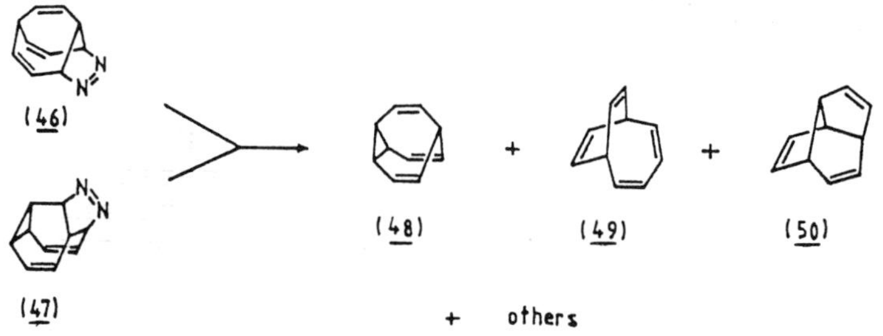

(46) (47) → (48) + (49) + (50) + others (23)

The limiting step, however, in such mechanistic studies is the availability of the essential diradical precursors, namely the appropriate azoalkanes. In our investigations, the cycloaddition of PTAD to bicyclic olefins leading to rearranged urazoles provided the synthetic access to the desired azoalkanes (eq. 3). In this context it must be

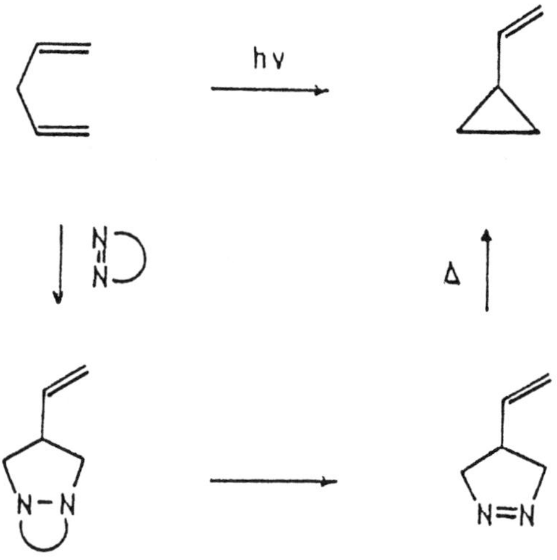

hv Δ N=N (24)

pointed out that the very dienes that serve as substrates in the photochemical di-π-methane rearrangement, also serve as a starting point in the preparation of the azoalkanes required for the generation of the diradical intermediates postulated in the di-π-methane rearrangement. In other words, in the lower loop of eq. (24), triazolinedione functions so to speak as a "photon equivalent" for the di-π-methane rearrangement. This novel concept should be a useful and powerful mechanistic tool for the elucidation of complex photorearrangements.

Acknowledgements. Thanks are extended to the Deutsche Forschungsgemeinschaft, the Fonds der Chemischen Industrie, the National Institutes of Health, the National Science Foundation, and the Petroleum Research Fund, administered by the American Chemical Society for generous financial support of our work in this area. N. Carballeira thanks the F. Thyssen Stiffung for a grant.

Discussion

Traylor: I have a question about the singlet versus triplet products. Did you do those as a function of temperature, that is photolyze at 300°C?

Adam: We intend to do that. In the sensitized and direct photolysis of azoalkanes 41 and 42 (eq. 18), the system that shows the most branching; over a temperature span of 80° we do see some changes in the product distribution.

Traylor: You are saying that if you photolyze at thermal temperatures such as 350°, you would get those products?

Adam: We did not go to those temperatures. You would have to do that in the gas phase. That would be interesting, but we have not done that. We have gone from room temperatue down to -60°C and did not see anything really astoundingly different.

Herndon: In the thermal reactions there are some possible biradical intermediates which would lead to triene products with one ring. I was wondering if you see any of these other products.

Adam: Would you please write on the board what you mean?

Herndon: For example, if you break this bond you

break

get the biradical. This can now open into a diene and there are many others like that.

Adam: Under the conditions at which we do our thermolysis control experiments we see that a variety of additional products are formed, but they are not related to those derived from the thermolysis of the azoalkanes. It is not simple to separate these complex product mixtures and identify the individual components since most products are formed in minor amounts. Capillary GC is absolutely necessary. Once we have our capillary GC/MS combination, we will be able to do such analytical work more routinely. I would not be surprised that the triene products are also formed.

Bartlett: In the oxa-di-π-methane rearrangement do you ever get any epoxide type of products corresponding to forming the three-membered ring of the oxygen on the carbonyl side?

Adam: No. Certainly not as major products since these denitrogenations are really clean reactions.

Bartlett: What would the thermodynamics of that be? Is it a thermodynamically unfavorable product compared to cyclopropane?

Adam: Well, I really have not considered that possibility. I would think that the epoxides would be energetically uphill from the cyclopropanes because bridgehead-centered epoxides such as structures A and B should result.

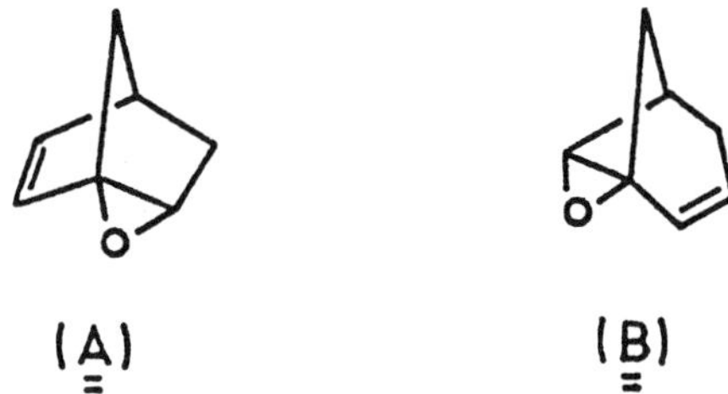

Vogel: About the different products that can be generated, don't you get cyclobutane derivatives? They could reasonably be formed from a 1,4-diradical intermediate.

Ring closure could occur here.

Adam: We looked for such products particularly in the low temperature photolysis of azoalkane 40 (eq. 18), hoping to isolate the yet unknown structure C; but apparently the 1,4-diradical 37 does not undergo such cyclization.

Vogel: Don't you think evidence for the intermediates could be found by labeling experiments?

Adam: Oh, if you mean whether the observed products, eg, the diene 36, could be derived from such labile intermediates as precursors, of course that is possible. We are trying something along these lines to elucidate the interconversion of dienes 36a and 36b (eq. 20) by means of deuterium labeling. With respect to the possibility of forming structure C, as already stated above, we have carried out the photochemical denitrogenation of azoalkane 40 down to dry ice temperature and could

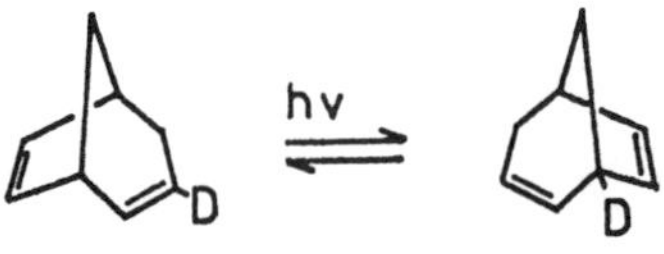

(36a) hν ⇌ (36b)

not detect C by NMR. One should try a matrix isolation experiment at liquid nitrogen temperature.

Foote: It seems to me that the azo compound which could either give norbornadiene or quadricyclane gives quite a significant amount of quadricyclane.

Adam: That is right (Ref. [5b,c]). It is one of the very few examples. We are looking at that system, by the way, as a synthetic entry into the di-π-methane diradical manifold of norbornadiene.

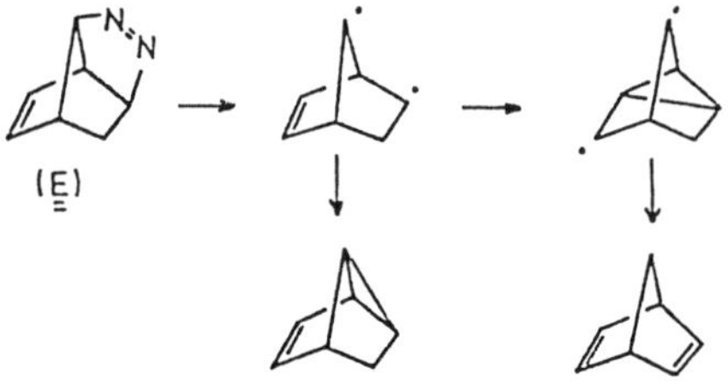

(E)

(D)

As you know, except for Prinzbach's example [Chem. Ber. 104, 2489, 1971] of a highly substituted derivative, the photolysis of norbornadienes affords quadricyclanes rather than the di-π-methane product D. By means of the azoalkane E, we could enter via denitrogenation into the di-π-methane chemistry of norbornadienes. Unfortunately, we have not yet succeeded in making azoalkane E.

Houk: In the case of the oxa-di-π-methane rearrangement in competition with the 1,3-shift, the usual interpretation of this is the 1,3-shift is a singlet n,π* process involving α-cleavage which, perhaps not to give a free diradical, but gives a

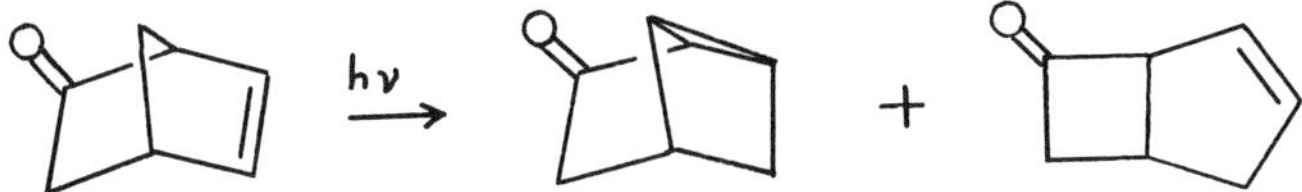

different set of diradicals than the ones you found. In the triplet sensitized reaction of norbornenone, I believe the oxa-di-π-methane occurs to the exclusion of the 1,3-shift.

Adam: It depends on the system.

Houk: With norbornenone?

Adam: You get all three possible ketones 32a, 32b and 35 (eq. 16).

Houk: Not in the sensitized reaction, you don't get the 1,3-shift in the sensitized norbornenone reaction, do you?

Adam: In the photochemical reaction of norbornenone (sensitized process) you get all three ketones, irrespective from which enone you start. This is Engel's work (Ref. [11d]). If the norbornenone case is general, it needs to be established; but I must say in reading the existing literature on the photochemistry of β,γ-enones (Ref. [12]), the ODPM rearrangement and the 1,3-acyl shift are considered as separate photochemical events. But this situation appears to be due to the fact that the substrate undergoes one or the other process, but not both. Thus, the norbornenone case is a neat example since the ODPM product (ketone 35) and the 1,3-acyl shift product (enone 32b) are both produced. More significantly, also enone 32b gives the ODPM product 35 and norbornenone (32a). Since all three ketones, ie, 32a, 32b and 35, are also produced in the denitrogenation of azoalkane 27 (eq. 16), we postulate that the 1,3-diradical 34 is a common intermediate for these products. Whether the 1,4-diradicals 33a and 33b are *bona fide* intermediates is open to debate and our data provides no evidence; but these intermediates are also not necessary. For example, a 1,2-acyl shift in the 1,3-diradical to the 2-position affords norbornenone (32a), while a 1,2-acyl shift to the 7-position yields the [3.2.0]-enone 32b.

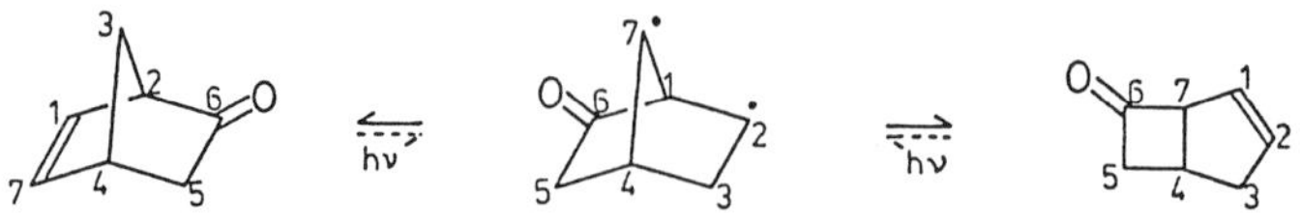

(32a) (34) (32b)

Similarly, such reverse 1,2-acyl shifts in the enones 32a and 32b would generate the 1,3-diradical 34, the precursor to the ODPM product 35. I want to stress here that we have no direct evidence for the 1,4-diradicals 33a and 33b. They are included in eq. (16) in view of the established ODPM mechanism. We are trying to generate such 1,4-diradicals via denitrogenation of the corresponding azoalkanes, but as was seen for azoalkane 40 (eq. 19), denitrogenation leads directly to the diene 36 via retro-Diels-Alder.

Wagner: As regards the di-π-methane, Wald's evidence is fairly significant. Engel's report is the only one in which you see sensitized 1,3-shifts and it has been used as evidence that they come from n,π* triplet. Now Wald is saying maybe we are looking at one biradical going in two different ways. So that throws what already is a confusing subject in photochemistry into a little more confusion.

Adam: I apologize for propagating the confusion!

Paquette: The azo compounds you showed today are unsymmetrical so one has the option of breaking one carbon-nitrogen bond ahead of the other. Given the particular relationship of those σ bonds with the remaining double bonds, the preference for kinetic cleavage of one is faster than the other with the result that the nitrogen may be extruded from the molecule by a C-C σ bond forming process. I have not had time to analyze your data at length, but I would like to hear your comments on the possibility of interpreting your data in terms of assistance of nitrogen departure.

Adam: I was afraid you would do that. Certainly, that is throwing the wrench into the works. You could explain many of our results by considering unsymmetrical cleavage of the azoalkane, affording a diazeno radical, as you pointed out. Therefore, unless one sees directly a diradical species, either spectroscopically or chemically, and shows that the product distribution is related to the species one sees, one always suffers the nightmare that unsymmetrical cleavage leading to a diazeno radical might be involved. In the thermal process I grant you that your suggestion is very reasonable. In the photochemical process, however, I find it intuitively a bit difficult to accept diazeno radicals as precursors to the products. Of course, if it should turn out that unsymmetrical denitrogenation pertains in the thermolyses and symmetrcal denitrogenation in the photolyses, then we have no right to discuss these two events under one common mechanism.

We are looking for evidence along these lines. Here is what we are trying to do. In collaboration with Schaffner (Mülheim), who has worked along these lines on ODPM and di-π-methane rearrangements, we plan to generate the postulated 1,3-diradicals by photodenitrogenation of the respective azoalkane and attempt to observe the matrix isolated 1,3-diradicals by ESR in the triplet sensitized reaction. So far these experiments have been unsuccessful. Perhaps it is a question of finding the right conditions? Such experiments are not easy to carry out.

(27) $\xrightarrow[h\nu]{Ph_2CO}$ (34) $\xrightarrow{^3O_2}$ (E)

Independently, we are trying to generate the 1,3-diradicals by benzophenone sensitization under a high pressure of oxygen, analogous to the Wilson (Cincinnati) experiment for the synthesis of prostaglandin endoperoxides via trapping of the triplet 1,3-diradicals with molecular oxygen. The triplet diradical 34 would be trapped by oxygen in the form of the peroxide F. The product distribution should be accordingly altered. So far we have not made any progress along these lines.

Traylor: If you made the diradical in that system, assuming normal geometry, which p orbital is better aligned to overlap the π system?

Adam: Unquestionably overlap of the p orbitals is best for the formation of the 1,4-diradical 33a (eq. 16), according to Dreiding models. Furthermore, the 1,4-diradical 33b is the more strained species, so that on energetic grounds 33a would be preferred. This works out very nicely with the observed product data (Table 2). Thus, the norbornenone product (32a) predominates over the [3.2.0]-enone 32b and 32a, expected either from the strain of the 1,4-diradical or the orbital overlap.

Paquette: It is great to make the best radical center first. The carbonyl can push off nitrogen and then you should get to the [3.2.0] system. If you break the C-N bond which gives you the best radical center, then leave nitrogen still attached at the bridge and use the carbonyl to push off nitrogen, then you should go to the [3.2.0] system.

N=N• → [3.2.0]

Adam: It works out equally well and is an alternative explanation.

Traylor: You don't even have to have two steps to do that. It could assist simultaneous bond cleavage by the same mechanism.

Adam: While this discussion is all very interesting and one can rationalize the product data in this way, the question is whether such diazeno radicals hang around long enough to do all of the suggested transformations?

Perrin: Let me follow Traylor's question about temperature dependence. What disturbs me is that you got that same 82:18 yield regardless of whether it comes from the bicyclic [2.2.2] case and the same yield whether it was 350°C or room temperature. Yet you drew an energy diagram with an energy difference between the two intermediates. Is it legitimate? Are you really justified in identifying discrete intermediates? Is there really an energy barrier separating them that governs the product distribution?

Adam: Such diradical species certainly possess inherent energy differences. Now, whether such diradical species are sensible or not as *bona fide* discrete intermediates is debatable.

Traylor: My point is they should come together if these manifolds you showed were true. They should come together at high temperature. But if what Leo suggests is true, maybe not because they are entirely different mechanisms.

Adam: That is right. The experiment you mentioned at the beginning, ie, wide temperature variations, is probably the more amenable to carry out and we are doing it.

Caldwell: The other point is if you look at the wavelength dependence. If you essentially heat it up photochemically, ie, use shorter and shorter wavelengths of light so that the average energy of biradical at the time of reaction ought to be higher, then you go away from the thermal mixture. I have a feeling all of that turns out to be fortuitous.

Adam: Fortuitous in what sense?

Caldwell: In the sense there is either a temperature difference or maybe some kind of electronic state difference between thermally and photochemically generated biradicals. They are related but not identical. One should not really expect them to give identical product ratios. It is really an interesting question if you look at singlet versus triplet differences. In every case I managed to copy down, the triplet was more selective than the singlet.

Adam: Correct, except for the norbornenone case. In comparing the direct photolysis and the triplet sensitized photolysis, the triplet photolysis is the cleaner one in terms of less retro process, sometimes none at all.

Caldwell: One thing that particularly interested me is that you just write the dots there. Assuming you have diradicals at all, you obviously just can not write dots, you have to write arrows to indicate spin because there can be differences between them. What do you think about the prospect of photochemical processes proceding through a singlet excited biradical?

Adam: Correct! In fact, these singlet diradicals could be further differentiated in terms of electronically and vibrationally excited species. We certainly could expect such species in the direct photolysis, since sufficient energy is available to produce excited diradicals. One could interpret our results along these lines and we do so in the case of the direct photolysis of the azoalkane 9 (Table 1).

Caldwell: There is no evidence but it is nice because if you look at radical rearrangements they are not nearly as nice as ionic rearrangements, cationic rearrangements.

Adam: That is why we are trying to do these different modes of activation, ie, thermolysis and direct and sensitized photolyses to get some handle on the spin state character of these diradicals.

Herndon: In the norbornadiene system the activation energy that leads to the biradical manifold that finally gives cycloheptatriene is around 50 kcal/mol, which requires temperatures of 400-450°C. So are there two sets of manifolds of biradicals, one leading to those types of products and the others that you observe?

Adam: You are now in what particular chemistry?

Herndon: Thermal chemistry.

Adam: The one with the exocyclic methylene, ie, azoalkane 21 (eq. 10)? I guess that would be one way of looking at it. We have no evidence along these lines; however, our data on the azoalkane 21 are preliminary results and are intended here as qualitative trends concerning mechanistic interpretations.

Wagner: I have been sitting here trying to make some sense out of the spin selectivities you do not see. In one system the triplet almost completely retro-cleaves, in the next almost completely closes in the way it is formed. I cannot make any sense to tell you the truth.

Adam: I gave you the data, but also I am puzzled about the behavior of the triplet diradicals. What particular case bothers you especially?

Wagner: Almost all of them. For example, the first system where the triplet diradical mostly couples. That is the only system which shows 100% coupling. You said you were surprised that the triplet did not have more time to go back. I guess I would have thought just the opposite. Going back would have generated a triplet alkene. In other words the first thing formed in the di-π-methane reaction in the triplet manifold is normally considered to be a 1,2-diradical, an excited alkene system.

Adam: O.K. If you by-pass the 1,4-diradical and postulate that 1,2-shifts in the 1,3-diradicals lead to the observed products, that would be a good way to get out of the difficulty in that a triplet diradical precursor would generate a triplet alkene product and energetically this is not feasible.

Wagner: Yes. I would not expect the thing to go back up hill to the triplet alkene.

Adam: That would suggest that one should not write 1,4-diradical species in such photochemical rearrangements.

Traylor: I have a question about the partici-

pation again. I am stuck on that. The compound right there, the azo compound. Is that faster or slower than the saturated one?

Adam: We do not have azoalkane E as yet.

Traylor: Do you have any evidence for or against neighboring group participation, that is to say with and without a double bond in such a compound?

Adam: No, we have not done that yet.

Foote: I thought Crawford had done some of those systems.

Adam: No bicyclic ones, only monocyclic ones.

Foote: But he has bicyclic compounds with acyclic azo groups. In those cases he has evidence for π assistance.

Adam: If I remember correctly, in one of Crawford's last studies he investigated secondary isotope effects on the denitrogenation of simple azoalkanes and concluded that thermal denitrogenations proceed via stepwise, one-bond cleavage leading to diazenyl radicals.

Noland: What have you found to be the best conditions for the hydrolysis of the triazolinedione adducts?

Adam: None better than Paquette's, ie, wet DMSO and potassium t-butoxide. Each one who gets into the field makes his own favorite cocktail. The yields are usually modest, especially for labile systems. We are trying to build triazolinediones which will allow us to take off the urazoles in an oxidative way. We have not been successful yet, but this would allow us to get at azo compounds that are thermally labile.

References

[1] (a) S.S. Hixson, P.S. Mariano, and H.E. Zimmerman, Chem. Rev. 73, 531 (1973). (b) H.

E. Zimmerman in "Rearrangements in Ground and Excited States", P. de Mayo (Ed.). Academic Press: New York, 1980; Vol. 42, Part 3, Essay 16, pp. 131-164. (c) H.E. Zimmerman, Topics Curr. Chem. 100, 47 (1982).

[2] (a) R.C. Hahn and L.J. Rothman, J. Am. Chem. Soc. 91, 2409 (1969). (b) Z. Goldschmidt and U. Gutman, Tetrahedron 30, 3331 (1974). (c) R.C. Hahn and R.P. Johnson, J. Am. Chem. Soc. 99, 1508 (1977).

[3] (a) H.E. Zimmerman, R.J. Boetcher, N.E. Buehler, G.E. Keck and M.G. Steinmetz, J. Am. Chem. Soc. 98, 7680 (1976). (b) N.J. Turro, C.A. Renner, W.H. Waddell, and T.J. Katz, J. Am. Chem. Soc. 98, 4320 (1976). (c) N.J. Turro, W.R. Cherry, M.F. Mirbach, and M.J. Mirbach, J. Am. Chem. Soc. 99, 7388 (1977).

[4] (a) W. Adam, O. DeLucchi, and I. Erden, J. Am. Chem. Soc. 102, 4806 (1980). (b) W. Adam and O. De Lucchi, Tetrahedron Lett. 929 (1981). (c) W. Adam and O. de Lucchi, Tetrahedron Lett., 3501 (1981). (d) W. Adam, O. De Lucchi, and D. Scheutzow, J. Org. Chem. 46, 4130 (1981). (e) W. Adam, O. De Lucchi, K. Peters, E.-M. Peters, and H.G. von Schnering, J. Am. Chem. Soc. 104, 161 (1982). (f) W. Adam, O. De Lucchi, and K. Hill, Chem. Ber. 115, 1982 (1982). (g) W. Adam, L.A. Arias, and O. De Lucchi, Tetrahedron Lett., 399 (1982).

[5] C.A. Seymour and F.D. Greene, J. Am. Chem. Soc. 102, 2107 (1980).

[6] W. Adam, N. Carballeira, and O. De Lucchi, J. Am. Chem. Soc. 102, 2107 (1980).

[7] W. Adam, Pure Appl. Chem. 52, 2591 (1980).

[8] (a) R.G. Weiss and G.S. Hammond, J. Am. Chem. Soc. 100, 1172 (1978). (b) Z. Goldschmidt and M. Shefi, J. Org. Chem. 44, 1604 (1979).

[9] W. Adam, N. Carballeira, and O. De Lucchi, J. Am. Chem. Soc. 103, 6406 (1981).

[10] R. Srinivasan, L.S. White, A.R. Rossi, and G. A. Epling, J. Am. Chem. Soc. 103, 7299 (1981).

[11] (a) G.O. Schenck and R. Steinmetz, Chem. Ber. 96, 520 (1962). (b) D.I. Schuster, M. Axelrod, and J. Auerbach, Tetrahedron Lett. 1911 (1963). (c) J. Ipaktschi, Chem. Ber. 105, 1840 (1972). (d) M.A. Schexnayder and P. S. Engel, Tetrahedron Lett. 1153 (1975).

[12] D.I. Schuster in "Rearrangements in Ground and Excited States", P. de Mayo (ed.), Academic

Essay, pp. 167-279.

[13] R.S. Sauers and A. Shuzpik, J. Org. Chem. 33, 799 (1968).

[14] R. Josel and G. Schroder, Liebigs Ann. Chem. 1428 (1980).

8. ENONES WITH DISTORTED DOUBLE BONDS

Herbert O. House

Department of Chemistry, Georgia Institute of Technology, Atlanta, Georgia, 30332

In 1967 the isolation of the bridgehead olefin shown at the top left of Figure 1 both by Wiseman [1] and Marshall [2] required a revision in the statement of Bredt's rule. Wiseman suggested that the important feature is the size of the ring which contains the equivalent of a trans double bond. If the trans double bond is in an eight-membered or larger ring, the compound can exist. If the ring containing the equivalent of a trans double bond is smaller than eight-membered, one does not expect to isolate the compound. Thus, the bridgehead olefin at the top right of the figure has yet to be isolated.

Following this restatement of the possibilities for bridgehead olefins, a variety of bridgehead olefins has been prepared. Among the compounds prepared was a series in which the bridgehead double bond was conjugated with a second carbon-oxygen double bond. Throughout the early and mid 1970s a variety of these unsaturated carbonyl compounds was prepared as summarized in lower part of Figure 1 [3-7]. One unique feature of each of these materials is that the carbon-carbon double bond and carbon-oxygen double bond span the bridgehead position. Consequently, the approximate plane of the olefin and the approximate plane of the carbonyl group are not coplanar and there is not good conjugation. Furthermore, adding a nucleophile at the end of one of these conjugated systems not only would not relieve strain but would make matters worse. I believe that much of the success in preparing these compounds is due to the fact that they are very poor acceptors in Michael reactions and can be prepared in the presence of nucleophiles.

BRIDGEHEAD OLEFINS CAN BE ISOLATED IF THE
TRANS C=C IS IN AN 8-MEMBERED OR LARGER RING
ISOLATED
UNKNOWN

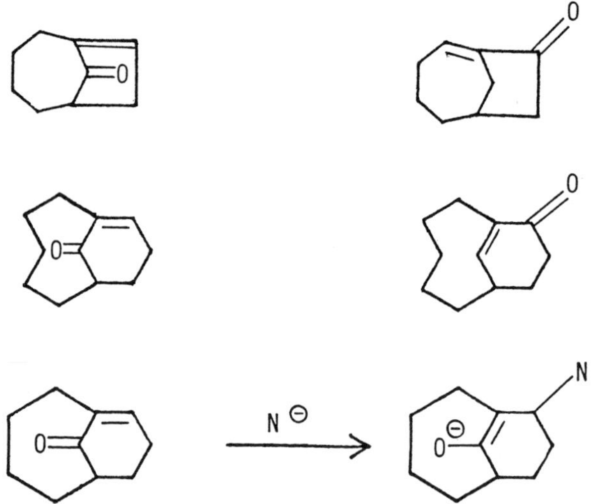
KNOWN BRIDGEHEAD ENONES WHERE C=C AND C=O ARE NOT COPLANAR
O
O
O
O
N
N
O

Figure 1. Examples of bridgehead olefins.

We became interested in a second series of bridgehead enones in which there could be effective conjugation between the carbon-carbon double bond and the carbonyl group. The compounds we envisioned are illustrated in eq. (1). We expected these materials to exhibit several properties which would make them rather different from the enones shown in Figure 1. The first property (which was to cause us some trouble throughout the preparation of these materials) was the expectation that all of these compounds would be good Michael acceptors.

(1)

GOOD MICHAEL ACCEPTORS

Nucl ⊖

Nucl

ADDITION OF NUCLEOPHILE

REDUCES STRAIN

Obviously, adding a nucleophile at the end of the conjugated system in these enones will relieve the

INCREASED C=C DEFORMATION SHOULD FACILITATE

POPULATION OF THE FIRST ANTI-BONDING ORBITAL

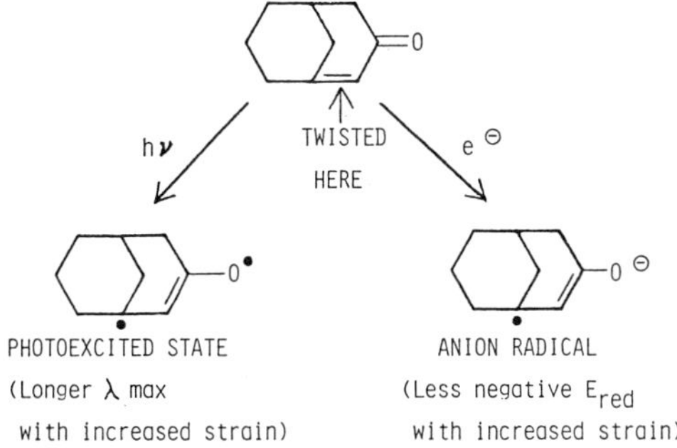

INCREASED C=C DEFORMATION MAY FACILITATE THE

THERMALLY INDUCED POPULATION OF A TRIPLET STATE

heat

TWISTED

HERE

Figure 2. Bridgehead olefins with conjugation.

strain associated with the distorted carbon-carbon double bond. This expectation, although interesting in offering a way to synthesize compounds with bridgehead substituents, was not the expectation that captured our interest in these materials. Rather, it was the geometry that we anticipated the compounds might have. Specifically, we expected that in these materials the portion of the molecule to the right of the vertical arrow in the structure in Figure 2 would be approximately planar with effective conjugation between the α carbon and the carbon of the carbonyl group. However, we expected a substantial twist approximately in the center of the carbon-carbon double bond in order to accommodate the bridgehead double bond. Distortion at this site corresponds to the geometry expected for the photoexcited state of the enones or for the anion radical of the enones. In both species there should be a node in the π system at the center of the carbon-carbon double bond. So we consequently expected that these enones to exhibit some interesting photochemical and electrochemical behavior. One would expect that as the strain increased the compound would absorb at longer wavelengths. Also, as the strain increased, the energy separating the anion radical from the ground state would become less and it would become progressively easier to reduce the compound. Another idea that intrigued us was the thought that if enough twist were placed on the carbon-carbon double bond, we would come to a point where the energy barrier between the ground state and a triplet diradical state would be small enough to allow thermal conversion to the diradical. Normally this change is extremely difficult to achieve thermally. This thermal population of a triplet state could lead to an interesting set of reactions that would be very different from those associated with a normal planar enone.

Since people planning a battle strategy should get to know their enemies, we wished to get estimates of the sorts of distortions likely to be present in our target molecules and to estimate just how much strain might arise in these molecules. To accomplish this, we made use of Allinger's molecular mechanics program.

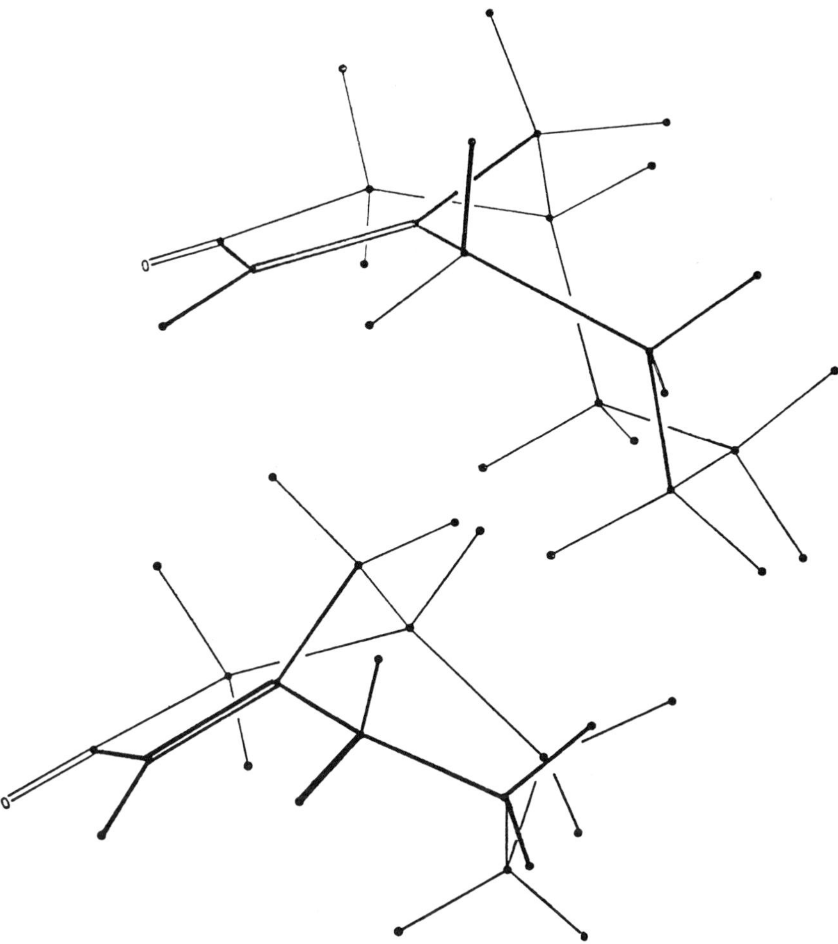

Figure 3. Molecular mechanics conformations for bicyclo[5.3.1]undecenone and bicyclo[4.3.1]decenone. C=C strain energies are 0.5 and 4.7 kcal/mol.

In Figures 3, 4, and 5 I show perspective views of energy-minimized conformations for various of these enones. In Figure 3 and the succeeding two figures are shown molecules with progressively fewer methylene groups in one of the bridges. I would like you to observe two things. As we make the methylene bridge in these molecules smaller and smaller, you will begin to see the carbon-carbon

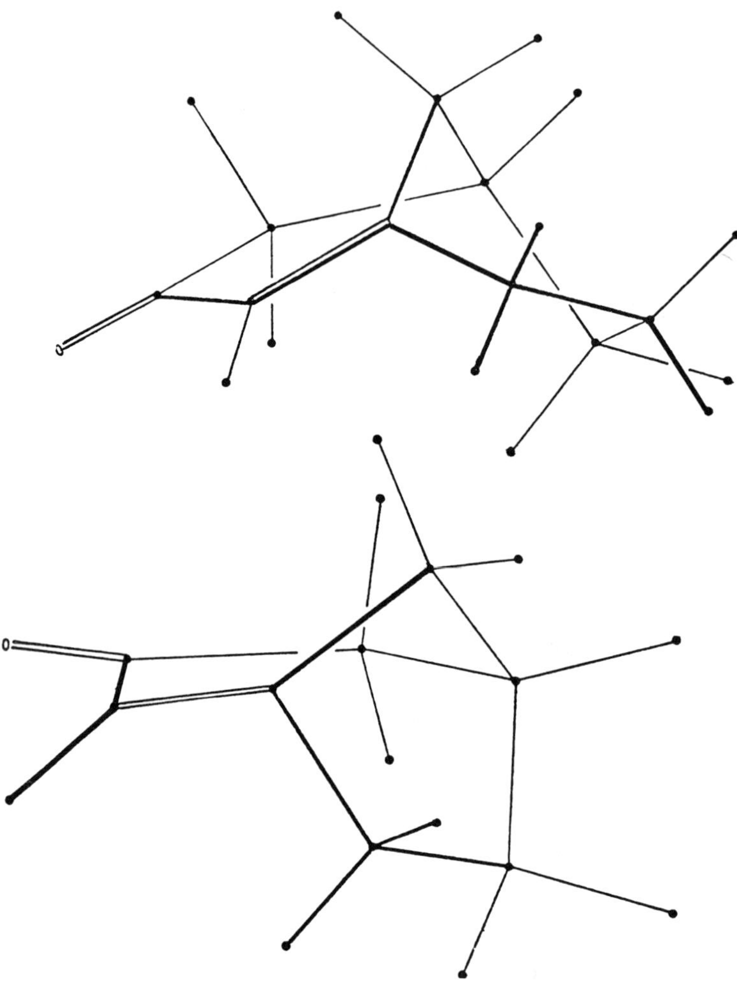

Figure 4. Molecular mechanics conformations for bicyclo[3.3.1]nonenone and bicyclo[3.2.1]octenone. C=C strain energies are 11.3 and 30.5 kcal/mol.

double bond twisting and you will also begin to see each end of the carbon-carbon double bond becoming more pyramidal. As one goes from the [5.3.1] to the [4.3.1] system there is a small amount of twist and not much pyramidalization. In the smaller [3.3.1] system, pyramidalization is more apparent. The pyramidal nature of the double bond is still more apparent in the [3.2.1] systems.

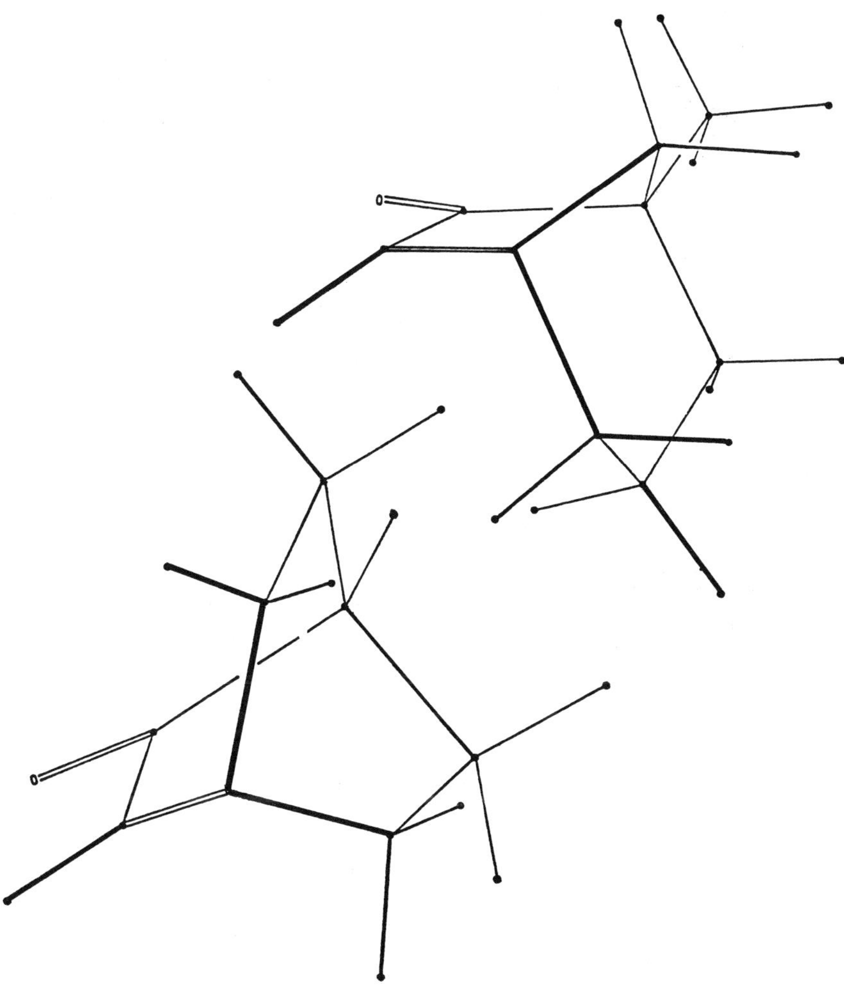

Figure 5. Molecular mechanics conformations for methyl bicyclo[3.2.1]octenone and bicyclo[2.2.2]-octenone. C=C strain energies are 32.8 and 59.2 kcal/mol, respectively.

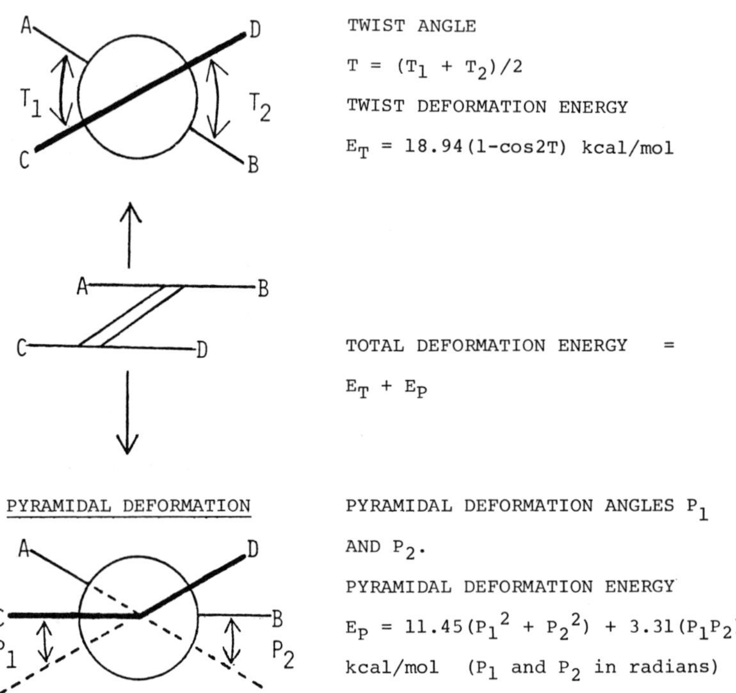

Figure 6. Estimation of C=C deformation energy.

From these estimates of geometry, we wanted to get some idea of what the relative degrees of strain might be in the various molcules. We were interested in two things: both the overall inherent strain in the molecule calculated by Allinger's program and the amount of strain that was concentrated in the carbon-carbon double bond. Because we anticipated that the carbon-carbon double bond would be the reactive spot in the molecule, we wanted an estimate of how much strain was compressed into this bond. To obtain this estimate we made use of the technique employed by Ermer to estimate the amount of strain within the double bond as a combination of two distortions. Both a pure twisting distortion (the amount by which the planes at each end of the carbon-carbon double bond are twisted) and also the extent to which each end of the double bond is distorted from a planar structure toward the pyramidal structure are used. The relationship shown in Figure 6 is that derived by Ermer to calculate the amount of excess strain or deformation in a carbon-carbon double bond.

If one tabulates these estimated geometries and strain energies for the series of homologs of related structure, it is seen that one starts with compounds having little or no strain, including monocyclic and large bicyclic compounds. With the [3.3.1] system we estimate about 10 kcal of strain

Table 1. Calculated C=C deformations

Compound	Average Twisting Deformation	Pyramidal Deformations	C=C Deformation (kcal/mol)	Inherent Strain (kcal/mol)
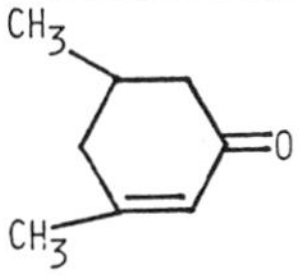 CH_3 / CH_3 =O	2.5°	2° and 4°	0.2 kcal	3.4 kcal
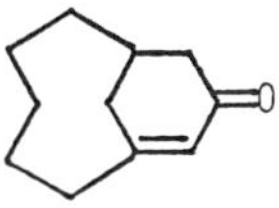 =O	4°	1° and 9°	0.5 kcal	16.9 kcal
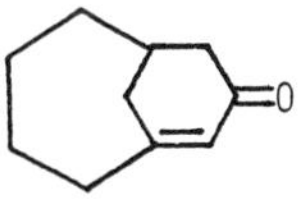 =O	14°	14° and 21°	4.7 kcal	17.7 kcal
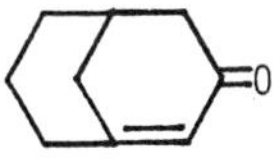 =O	21°	19° and 37°	11.3 kcal	21.3 kcal
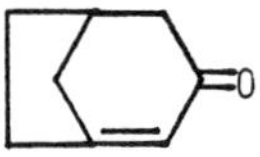 =O	36°	29° and 60°	30.5 kcal	32.3 kcal

Table 1 continued

Compound	Average Twisting Deformation	Pyramidal Deformations	C=C Deformation (kcal/mol)	Inherent Strain (kcal/mol)
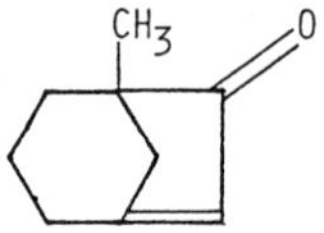	36^0	36^0 and 62^0	32.8 kcal	27.4 kcal
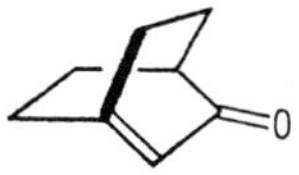	89^0	3^0 and 78^0	59.1 kcal	49.1 kcal

localized in the double bond and an inherent strain in the molecule of about 21 kcal. This molecule is beginning to distort significantly. Notice also an approximate 20^0 twist in the double bond and also a distortion at each of the double bonded carbons of the order of 20 to 35^0. In the [3.2.1] system the distortion is even greater and the energy is approximately doubled by removing just one additional methylene group.

In the continued section of Table 1 we compare the structures of two [3.2.1] systems. We were surprised that the calculated distortions were so similar in the two different [3.2.1] systems. As a result, we are led to expect that although both [3.2.1] systems will be extraordinarily reactive, there is no special reason to believe that one system will be more reactive than the other. Finally, we have the [2.2.2] system that can be expected to have a large amount of strain built into it. As will be shown in later sections, we have generated, at least transiently, every species discussed except for the [2.2.2] system.

This summarizes our efforts made in advance to know our enemy. What happened when we sought to prepare the compounds and put them in bottles? First of all what were the physical properties? The compounds that could be isolated readily and studied in a normal way are the first three enones shown in Table 2.

Table 2. A Comparison of Physical Properties

Compound	C=C Deformation Energy kcal/mol	^{13}C NMR (ppm)	λ_{max} (CH_3CN) $\pi \rightarrow \pi^*$	$n \rightarrow \pi^*$	$E_{1/2}$ (CH_3CN) V (vs SCE)
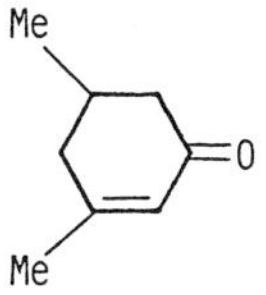 Me, Me, =O	0.2	198.8 161.1 125.6	232 (13,800)	323 (25)	-2.21
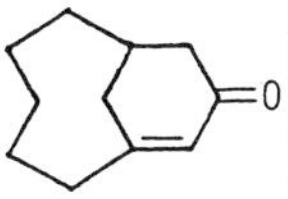 =O	0.5	198.9 165.0 126.4	240 (14,800)	335 (50)	-2.13
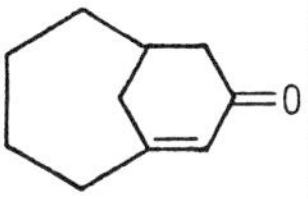 =O	4.7	198.8 164.1 122.7	250 (5,070)	345 (92)	-2.00
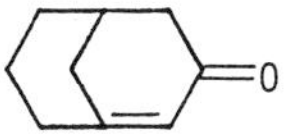 =O	11.3	199.5 173.2 123.8	?	?	?
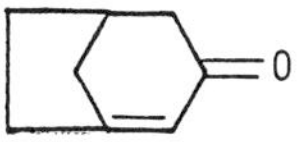 =O	30.5	?	?	?	?

These are compounds where the estimated strain in the double bond was of the order of 5 kcal or less. The trends we observed included an increase in the UV absorption wavelength with increasing distortion of the carbon-carbon double bond. Notice that this is true of both the π-π^* and the n-π^* bands. Also, we find that the electrochemical reduction potentials become less negative in the expected way as distortion of the carbon-carbon double bond increases. On examining the C-13 spectrum of these enones, some abnormality begins to show up in the C-13 signal for the β carbon of the [3.3.1] system.

Notice there is an approximate 10-ppm shift downfield associated in going from the [4.3.1] system to the [3.3.1] system. Whether that will be continued in more strained molecules we do not know.

What can we say about the chemistry of the compounds? Table 3 contrasts the chemical properties of the [5.3.1] and [4.3.1] systems with the [3.3.1] system.

Table 3. A Comparison of Chemical Properties

REAGENT	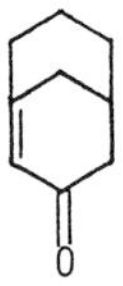 [3.3.1] enone (structure)	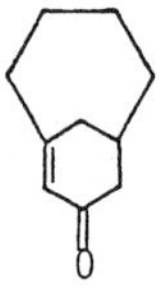 [4.3.1] enone (structure)	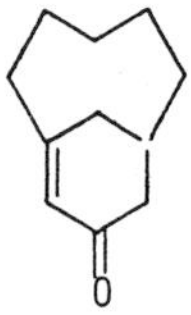 [5.3.1] enone (structure)
Na$^{\oplus}$ $^{\ominus}$OOH H_2O, CH_2Cl_2	FORMS HYDROPEROXIDE	FORMS EPOXIDE	FORMS EPOXIDE
Na$^{\oplus}$ $^{\ominus}$OCH$_3$ CH_3OH	FORMS ADDUCT	FORMS ADDUCT	NO ADDUCT
 (furan) REFLUX	FORMS ADDUCT	NO REACTION	NO REACTION
CH_2=CHCH=CH_2 100^{0}	--	FORMS ADDUCT	NO REACTION
HEAT	FORMS "2+2" ADDUCTS AT 25^{0}	NO REACTION AT 190^{0}	NO REACTION
UV LIGHT λ >290 nm	--	FORMS "2+2" ADDUCTS	FORMS "2+2" ADDUCTS

All of these compounds undergo certain conjugate addition reactions. The least-strained systems react normally with the sodium salt of hydrogen peroxide to form an epoxy ketone. The [3.3.1] enone has trouble in the second step of this epoxidation reaction and, depending upon ratios of reactants, either forms a hydroperoxide with excess of the reagent or forms a bisperoxide with an excess of the enone. The [4.3.1] and [3.3.1] enones spontaneously add methanol with no added catalyst. The reaction is even faster when a catalyst is present. The least strained [5.3.1] enone shows no tendency to add methanol. Instead, when one adds sodium methoxide an equilibrium is established between the conjugated and the unconjugated enones. That equilibrium position favors the conjugated isomer by a ratio of 83:17. The strained [3.3.1] enone is relatively reactive in Diels-Alder reactions or in other c cloaddition reactions. With refluxing furan, the enone ra idly forms an adduct whereas the two less strained [4.3.1] and [5.3.1] enones do not form an adduct with refluxing furan at atmospheric pressure. With butadiene the [4.3.1] system at 100° will form an adduct, although it will not do so at 25°. The [5.3.1] system at 100° did not form an adduct with butadiene. Perhaps the most dramatic difference between these com ounds lies in the fact that the [3.3.1] system forms a dimeric [2+2] adduct with iteself at 25° while the two less strained molecules are completely stable up to temperatures of 200°. This does not mean that the [2+2] adducts are unstable or strained, because both of these enones form [2+2] adducts when exposed to UV light filtered through pyrex.

From these summaries of chemical and physical properties it is apparent that the interesting chemical and physical properties are to be found with those molecules that are either [3.3.1] systems or more strained enones. So let us focus the rest of the talk on what we know about the preparation and reactions of the [3.3.1] and [3.2.1] systems.

First of all how do we prepare the [3.3.1] systems? What kinds of synthetic pathways are available? The simplest reaction path to generate the [3.3.1] system has proved to be the reaction of the bromoketone with triethylamine as shown in Figure 7. The reaction is clearly base catalyzed and is not a solvolysis. The base-catalyzed reaction resulting in the elimination of bromide ion generates the [3.3.1] enone that can be trapped with a

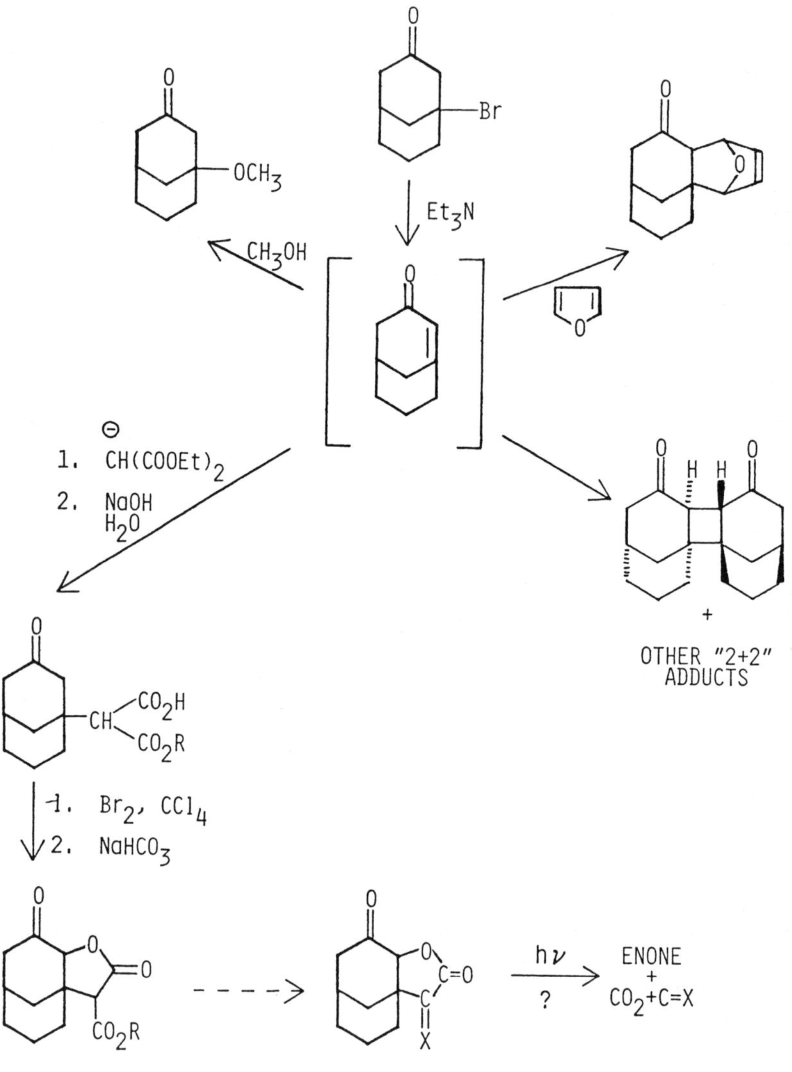

Figure 7. Synthesis of the bicyclo[3.3.1]nonane system from a bromoketone and triethylamine.

variety of nucleophiles such as methanol or the anion of diethyl malonate to form the corresponding ketone with a bridgehead substituent. In fact, this is an excellent way to form materials that have bridgehead substituents. If the enone is generated in the presence of refluxing furan, it reacts to form the furan adduct shown as a mixture of stereoisomers. If we remove both nucleophiles and dienes from the reaction mixture, the enone reacts with itself to form a set of [2+2] cycloadducts. There are at least three of these [2+2] cycloadducts formed. The one that I have drawn in the figure is the major stereoisomer. The second isomer is a head to tail dimer and the third minor isomer is another head to head dimer. One characteristic I should point out of all of these [2+2] cycloadducts is that they correspond to syn-syn additions across the double bonds. In other words, the isomers are not expected from a concerted thermal reaction. These reactions are clearly thermal; the same product is formed in the dark as in room light. We believe that these dimers are not being formed in concerted reactions but in a two-step process, presumably by way of a diradical that adds to a second enone in two steps. We have considered other methods for generation of this [3.3.1] enone which might allow us to form it at low temperatures and hence to be able to isolate the enone under conditions where it would not react with itself.

One possibility we have considered is the preparation of a substance that can absorb light in the visible range (see bottom of Figure 7). A suitable group "X", such as a phenyl-substituted nitrogen group. might be appropriate to give a compound that would absorb visible light and decompose to the enone, carbon dioxide, and a third product such as an isonitrile. Although we think this notion is still a viable one, our pursuit of it has only gone to the stage of demonstrating that we can efficiently convert either the monoester or the diacid efficiently to the lactone intermediates shown.

Three other methods we have explored for the generation of the [3.3.1] enone are of some interest. In the first method the β-phenyl selenide, generated from the enone, was oxidized to the selenoxide at about 5°. Then the methylene chloride solution was separated from excess oxidant and warmed to 25° to initiate decomposition of the selenoxide. Unfortunately, the product isolated was not the enone but the hydroxy selenide shown in Figure 8. We believe that the enone is being formed

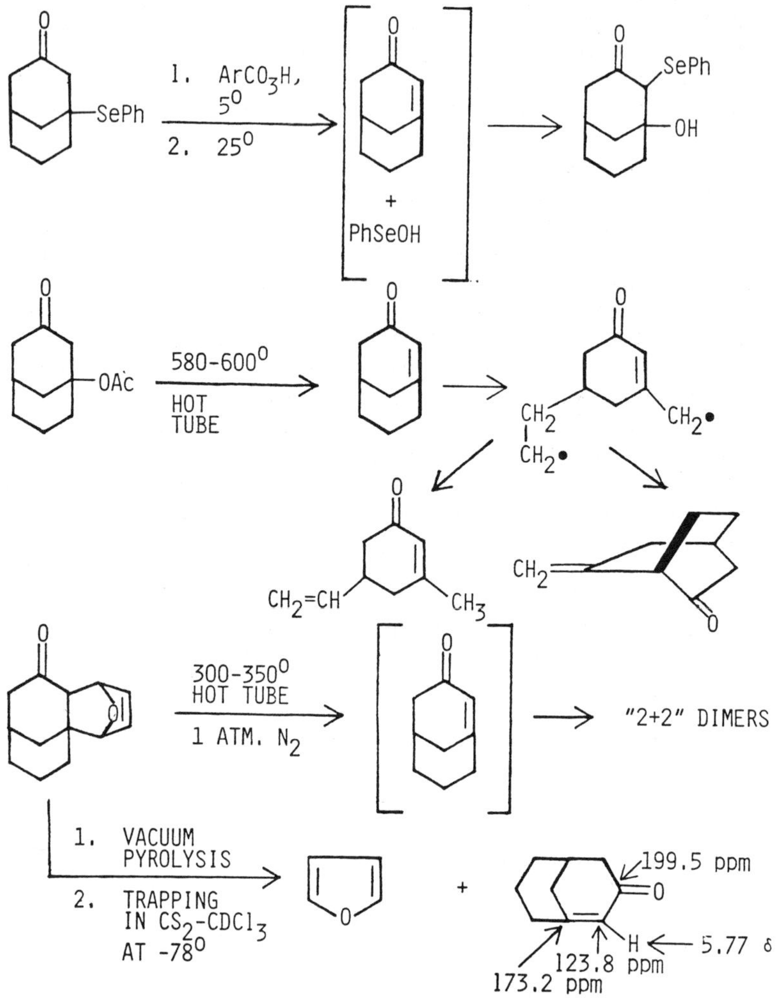

Figure 8. Additional synthetic routes to the bicyclo[3.3.1]nonane system.

but is a sufficiently good Michael acceptor to react with the phenylselenenic acid being generated. It seemed that a better method might be to generate the enone in the gas phase where it can be diluted with a large volume of an inert gas. The initial experiments involved pyrolysis of the β-acetoxyketone with a couple of seconds contact time in a hot tube at about 580-600°. Unfortunately, at this temperature the enone is sufficiently strained that further homolysis occurs and the reaction products turn out to be two isomers of the desired enone shown in Figure 8. The origin of the dienone and the isomeric bicyclic enone can most easily be seen if one writes a stepwise process in which a C-C bond is homolytically broken to form an allylic radical and a second radical. Hydrogen atom transfer can form the dienone or rebonding the diradical at the other end of the allylic system can form the isomeric enone. These processes may well be concerted, but one can most readily see what is happening by writing the discrete intermediates. In any case the problem with this reaction is not that we don't generate the enone, but rather that the enone won't survive temperatures in the range of 600°. We needed a precursor that could be pyrolyzed successfully at lower temperatures. We turned to the furan adduct described previously and found that this adduct could be pyrolyzed readily in the temperature range 300-350°. With short contact time and in the presence of a full atmosphere of nitrogen gas when the exit gasses were passed into a cold trap filled with methanol we could trap only 15-20% of the enone generated. The remaining enone dimerized. Our problem was caused by the difficulty in cooling a gas stream at 1 atm pressure from 350° to -70° in a short time.

After a discussion of this experiment last summer, Walt Trahanovsky noted that they were experimenting at Iowa State with a vacuum pyrolysis apparatus that might solve our problem. The difficulty was not that the pyrolysis was failing, but that we couldn't cool the gas stream fast enough to trap the enone before it dimerized. A vacuum pyrolysis system had the virtue that only a low pressure of gas needed to be cooled. So we sent samples of our furan adduct to Iowa State to be pyrolyzed in their vacuum pyrolysis system. The pyrolysis product was trapped in a cold trap, filled with a mixture of carbon disulfide and deuterochloroform and held at -78°. Under these circumstances they obtained a solution whose proton NMR and C-13 NMR spectra fit very well for a 1:1 mixture of furan and the corresponding [3.3.1] enone. I have listed in the figure a few of the salient C-13 and proton NMR values for this enone system. We are continuing to work with the [3.3.1] system generated in this way.

We also considered a second approach to the synthesis of [3.3.1] systems. Since our problem with the isolation of the parent enone generated at room temperature was the dimerization of the enone, we hoped to retard this cycloaddition process by putting a substituent at the enone α carbon to sterically impede cycloaddition. We were considering methyl, phenyl, or t-butyl substituents and wondered how those enones would compare with one another and with the parent system. We again made use of Allinger's molecular mechanics program to estimate the energies for these enones. Table 4 lists two sets of values. The first set of numbers

Table 4. Molecular Mechanics Calculations of C=C Deformation Energies (values in parentheses are for the saturated ring in a chair conformation).

Compound	Average Twisting Deformation	Pyramidal Deformations	C=C Deformation (kcal/mol)	Inherent Strain (kcal/mol)
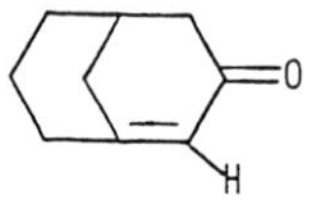 =O, H	21^0 (25^0)	19^0 and 37^0 (25^0 and 42^0)	11.3 kcal (16.5 kcal)	21.3 kcal (20.8 kcal)
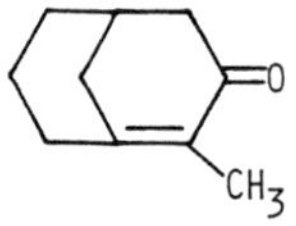 =O, CH_3	22^0 (27^0)	18^0 and 34^0 (23^0 and 40^0)	11.3 kcal (16.1 kcal)	20.7 kcal (20.3 kcal)
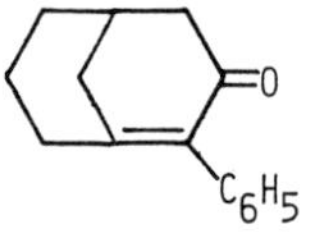 =O, C_6H_5	21^0 (27^0)	13^0 and 38^0 (17^0 and 44^0)	11.0 kcal (16.9 kcal)	27.4 kcal (25.6 kcal)
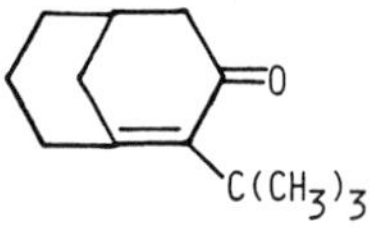 =O, $C(CH_3)_3$	22^0 (29^0)	13^0 and 32^0 (17^0 and 34^0)	9.6 kcal (14.7 kcal)	31.0 kcal (28.6 kcal)

represents the energies when the six-membered ring is in a twist-boat conformation. Although the overall calculated energies from the molecular mechanics program would suggest that the chair conformers are a little less stable, it is clear that the increased rigidity in the chair conformation localizes more strain in the carbon-carbon double bond. We are inclined to consider the flexible twist-boat conformers as being

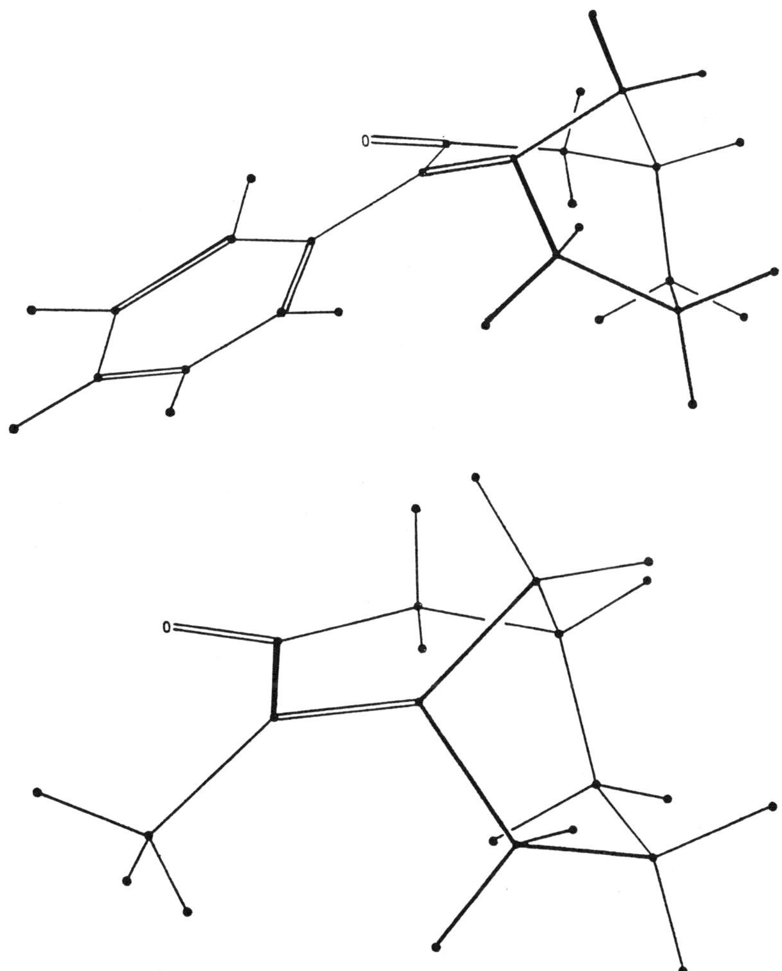

Figure 9. Molecular mechanics calculations for α-phenyl bicyclo[3.3.1]nonenone and α-methyl bicyclo-[3.3.1]nonenone. C=C strain energies are 11.0 and 11.3 kcal/mol, respectively.

the more probable in order to relieve distortion in the carbon-carbon double bonds. Notice throughout the series there is about a 5-kcal difference between twist-boat and chair conformers. In any case it is interesting that throughout the entire series of calculations the structure and degree of distortion of the parent enone are not markedly altered by placing substituents at the α position. Consequently, information that we might learn about compounds with α substituents can be expected also to be applicable to the parent compound. Perspective views of two of these compounds are shown in Figure 9.

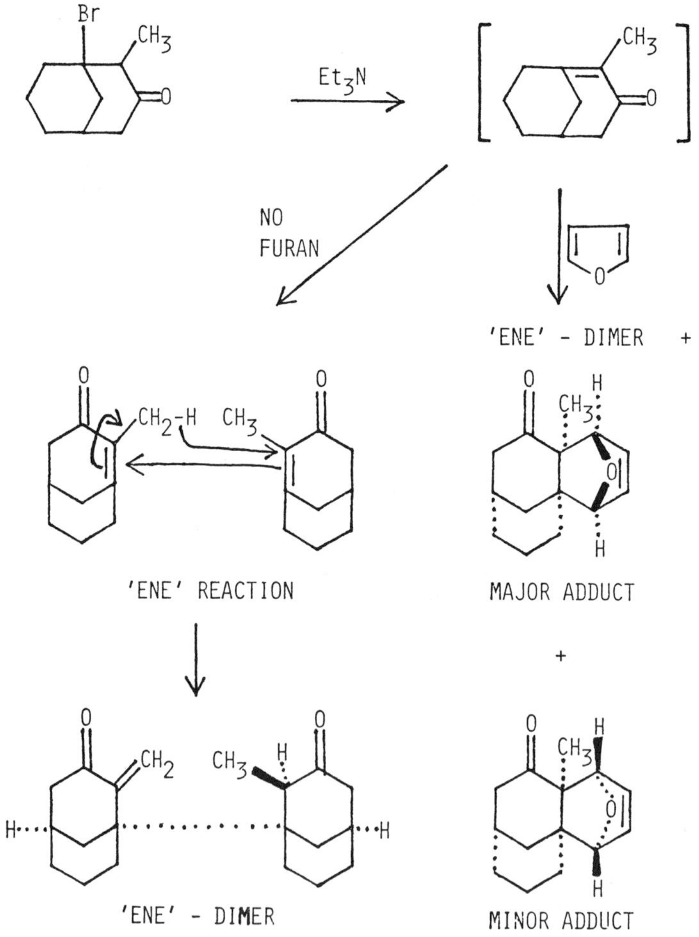

Figure 10. The α-methylbicyclo[3.3.1]nonane system.

Our conclusion from all of this was that we ought to prepare some compounds with α substituents. The compound examined first was the α-methyl compound which was generated from the bromo ketone as shown in Figure 10. This compound could be generated in refluxing triethylamine. Our expectation that this compound would not rapidly form a [2+2] adduct was correct. The compound in the presence of furan would form a furan adduct slowly. Unfortunately, the enone found an alternate way to escape our clutches by undergoing an ene reaction under remarkably mild conditions. This ene product formed rapidly enough to prevent us from isolating the enone. It became apparent that we needed to form enones with substituents which had no α hydrogen atoms. Substituents such as phenyl, *t*-butyl or adamantyl were the ones we considered seriously.

The phenyl-substituted compound proved to be the best choice for initial study. When the β-bromocarbonyl compound, shown in Figure 11, is dissolved in triethylamine at 25°, there is a slow reaction which requires approximately 8 h for completion. At that point the solution can be filtered in an anhydrous inert atmosphere to remove the triethylamine hydrobromide. Triethylamine can be removed from the yellow filtrate under reduced pressure at 25 to leave a pale yellow liquid. This pale yellow liquid has the spectroscopic properties listed in Figure 11. The yellow liquid is obviously a conjugated carbonyl compound. It has ultraviolet absorption suggestive of a conjugated system with an unusually long wavelength n-π* transition accounting for the yellow color. The proton NMR spectrum has the values shown and the C-13 NMR spectrum, very importantly, has only the requisite number of lines for the enone structure. In this C-13 NMR spectrum the lines are not broadened when measured at 0° or at 65°, rather than at 35°. This lack of broadening implies that there are no substantial concentrations of paramagnetic species in solutions of this compound in deuterochloroform in the range 0-65°. A reasonable model compound for the enone is shown at the bottom of Figure 11. The model obviously would be better had there been a methyl group at the β carbon atom. The greatest differences are observed in the UV spectra for the distorted and undistorted enones. In the C-13 NMR spectrum the position of the line for the β carbon

1. Et_3N, 25°
8 HR
2. FILTER UNDER
N_2 ATMOSPHERE
3. CONCENTRATE UNDER
N_2 ATMOSPHERE

PALE YELLOW
LIQUID

IR (CCl_4)	1680 cm^{-1}
UV MAX (CH_3CN)	242 nm (ε ∼ 4600) 350 nm (ε ∼ 300)
^1H NMR ($CDCl_3$)	δ 7.1-7.4 (5H, m) 1.6-2.8 (10H, m) 1.1-1.2 (1H, m)
^{13}C NMR ($CDCl_3$)	199.6 (s), 163.5 (s), 137.9 (s) 135.7 (s), 129.1 (d, 2C), 127.8 (d, 2C) 126.7 (d), 50.3 (t), 36.8 (t) 36.2 (t), 33.3 (d), 29.1 (t), 27.7 (t)

IR (CCl_4)	1685 cm^{-1}
UV MAX (CH_3CN)	217 nm (ε 10,900) 255 nm (ε 5,200) 328 nm (ε 53)
^{13}C NMR ($CDCl_3$)	197.4 (s), 147.7 (d), 140.2 (s) 136.5 (s), 128.4 (d, 2C), 127.8 (d, 2C) 127.3 (d), 38.9 (t), 26.4 (t) and 22.7 (t)

Figure 11. Properties of α-phenyl bicyclo[3.3.1]-nonane system and a model compound.

of the bridgehead enone (163 ppm) does not appear to be abnormal as was the signal at 173 ppm for the parent [3.3.1] enone system.

The reactivity patterns we see with this phenyl substituted enone are shown in Figure 12. Of course the enone adds methanol to form the β-methoxy

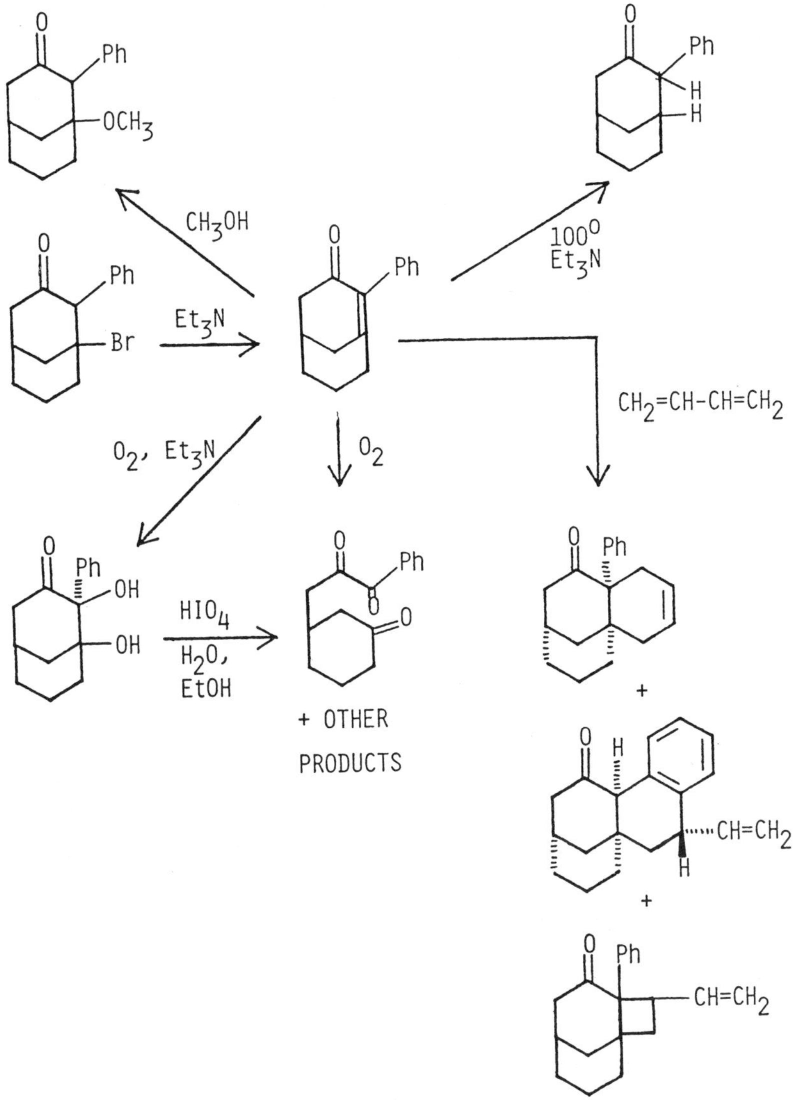

Ph
OCH_3
Ph
H
H
CH_3OH
Ph
100^{0}
Et_3N
Ph
Et_3N
Br
$CH_2=CH-CH=CH_2$
O_2, Et_3N
O_2
Ph
OH
OH
HIO_4
H_2O,
EtOH
Ph
Ph
\+ OTHER PRODUCTS
\+
H
$CH=CH_2$
H
\+
Ph
$CH=CH_2$

Figure 12. Reactions of α-phenyl enone.

derivative. It undergoes an interesting reaction if simply heated in triethylamine solution to 100^{0} for a period of time. Slowly and continuously over a period of 24 to 48 h or more the phenyl enone is converted to the corresponding reduced compound in which two hydrogen atoms have been added to the carbon-carbon double bond. Another reaction is the addition of butadiene to the phenyl enone. We have performed this reaction both in triethylamine solution and by isolating the enone and adding the butadiene subsequently. In both cases the mixture needs to be heated to 100^{0} and the time required for complete reacton is about 20 h. Under those circumstances we isolate the three products shown. The major product is the one that can be formulated as a [2+4] cycloadduct. The next most abundant is the one that can be formulated as a [2+2] cycloadduct, and the least abundant product is a material whose origin I will discuss in a moment. The enone reacts with oxygen in one of two interesting ways. If oxygen is passed into a solution of the enone in the triethylamine, oxygen is rapidly absorbed. The compound isolated is the diol shown with the stereochemistry established by X-ray crystallography. This diol can be cleaved to the structure shown to establish its structure. This triketone is also formed if the enone is exposed to oxygen in the absence of triethylamine and then warmed.

What do these observations mean? Although we have no direct evidence for thermally populating a triplet state of starting enone, it is a very convenient way to think about the observed reactions (see Figure 13). The reaction with butadiene is of course a reaction patterned after the classic experiments of Paul Bartlett with butadiene and the chlorofluoro olfins. The intermediacy of diradicals was demonstrated in those reactions. We used butadiene with the same thought that the initial reaction of transoid butadiene would form a diradical which, if it closed with rotation of the transoid allylic radical, would form the [2+4] cycloadduct. That portion of the diradical that closed without rotation of the allylic radical could form either the [2+2] cycloadduct or alternatively, via an alternate resonance structure of the benzylic radical, the second [2+4] cycloadduct shown. This species, after hydrogen atom transfer, could form the observed product. The stereochemistry of this product has been verified by crystallography. Thus, the reactions with butadiene are certainly consistent with the notion that there is a diradical intermediate. The slow formation of the reduced product from the enone and triethylamine also is readily explained by imagining that the diradi-

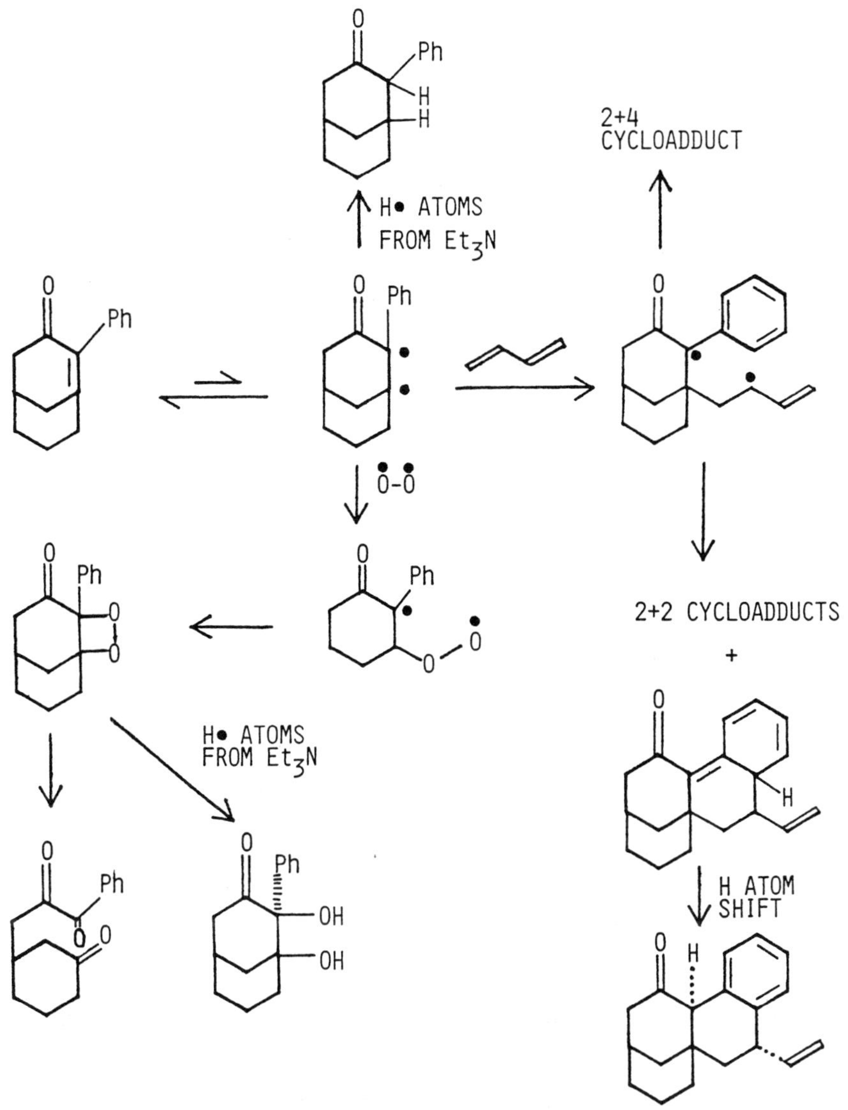

Figure 13. Reaction path for α-phenyl enone.

cal slowly abstracts hydrogen atoms from triethylamine, a good hydrogen atom donor. This process does not appear to be a chain reaction, but rather is a reaction that occurs very slowly over long periods of time. The reaction with oxygen again can

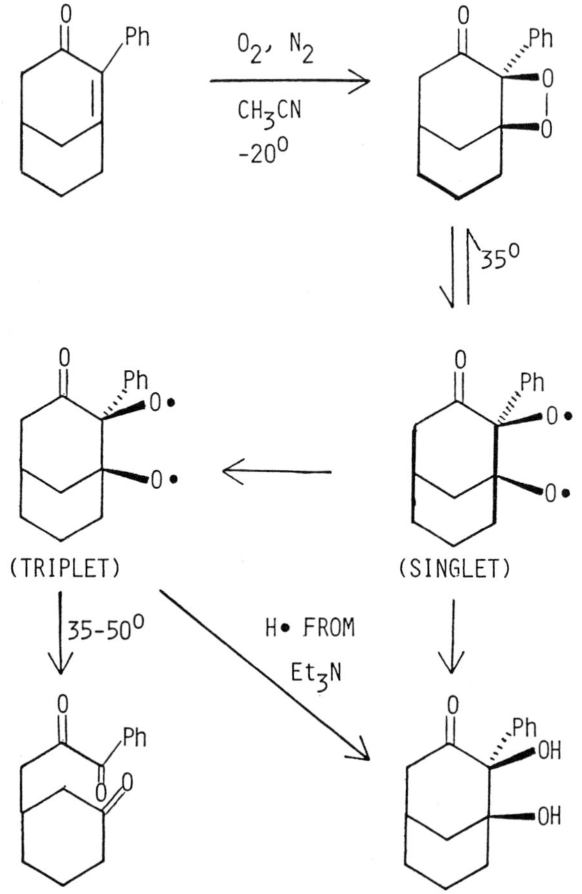

Figure 14. Reaction of α-phenyl enone with oxygen.

be envisioned as the reaction of the triplet form of the enone with triplet ground-state oxygen. I want to emphasize that this reaction, as far as we can tell, involves only ground-state oxygen; it occurs readily in the dark at ∿25°. One can formulate this reaction as forming an initial diradical which can close to a dioxetane or a related polymeric peroxide as shown in Figure 14. If the dioxetane or related peroxide breaks the oxygen-oxygen bond in the presence of a good hydrogen atom donor, such as triethylamine, the observed diol should be formed. Alternatively, if the dioxetane or related peroxide were to undergo isomerization, it would give the corresponding triketone, the observed product. There is a question of whether the intermediate is a dioxetane. The following experiments described our current information about this question.

If we dissolve the enone in acetonitrile, cool it to -20° and then pass in a stream of oxygen diluted with nitrogen, the oxygen is absorbed as rapidly as it is passed into the solution. A solid separates from the acetonitrile solution. This solid can be separated from the acetonitrile solution at temperatures in the range of -20 to 0° and then washed with additional cold acetonitrile and dried. The residual pale tan solid is at least transiently stable. We have tried taking a melting point of this material by placing it in a preheated bath; the crude product melts with decomposition, in the range $75\text{-}80^{\circ}$. Solutions of this crude product in deuterochloroform were relatively dilute, but nonetheless we were able to measure C-13 and proton NMR spectra. The NMR spectra either at 0° or at -20° exhibit lines that are compatible with the signals expected for the dioxetane or a related polymeric peroxide. That is, there are two C-13 signals at appropriate positions for saturated carbon atoms bound to oxygen. Also there are signals for various parts of the aromatic ring and the carbonyl group, as well as other unresolved signals. However, all of the signals associated with the dioxetane or related structure are broadened. This is not simply a problem of poor field or paramagnetic impurities in the sample because the lines for both deuterochloroform and residual amounts of acetonitrile in our sample remain sharp. There are a number of interpretations one could place upon these observed broad lines, including the presence of a mixture of polymeric peroxides or the occurrence of some reversible opening of the dioxetane ring to form some singlet diradical. The hyperfine splitting arising from the small amount of diradical present could cause broadening of all of the dioxetane lines. (In a subsequent study we have found that the amount of line broadening is the same at -20°, 0°, and $+20^{\circ}$ indicating that the broadening is caused by a mixture of polymeric peroxides.) When the deuterochloroform solutions were heated to 50° for approximately 30 min, the signals disappeared completely, and we obtained the NMR spectrum of the triketone shown at the bottom of Figure 14. When we observed the C-13 spectrum of the solution in deuterochloroform at 35°, in the half-hour accumulation time required approximately 20% of the dioxetane or related structure had decomposed to give a set of sharp C-13 lines corresponding to the triketone.

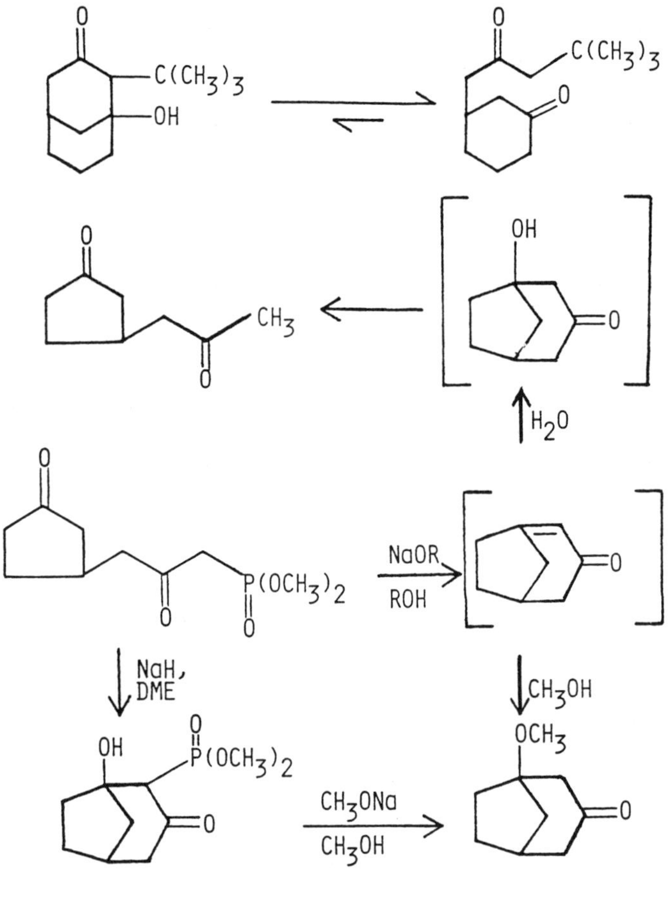

O
C(CH3)3
OH
O
C(CH3)3
O
OH
O
CH3
O
H2O
O
NaOR
ROH
P(OCH3)2
O
O
O
NaH, DME
CH3OH
OCH3
OH
O
P(OCH3)2
O
CH3ONa
CH3OH
O

Figure 15. The α-(t-butyl)[3.3.1]enone and the 3-keto[3.2.1]enone.

Thus, we can prepare at least one of the [3.3.1] enone systems and isolate it in the absence of water, oxygen, or other nucleophiles. The material reacts very rapidly with water or methanol to give adducts.

The phenyl group of course is not the ideal substituent to place at the α position. We have devoted some effort to trying to make compounds with *t*-butyl or adamantyl groups at this site. Our problem with the *t*-butyl compound has been our inability to effect the necessary aldol step shown in Figure 15 to give a precursor for the *t*-butyl compound. We have not been able to form enough aldol product at equilibrium to be able to trap it in some form that would be a precursor for the *t*-butyl enone. We have also worked with two [3.2.1] enones; our problem with the [3.2.1] enone shown in Figure 15 is rather similar to the problem with the *t*-butyl [3.3.1] enone. The [3.2.1] enone was generated from the ketophosphonate shown. Reaction of the ketophosphonate with sodium hydroxide in water formed the anion that closed to the carbonyl group, and eliminated to form the enone that was trapped by water to give the ketol. Unfortunately, this ketol is sufficiently unstable that we have not been able to trap it. Instead the ketal opens and the only product we isolate at the end of the reaction is the diketone shown. If we trap the [3.2.1] enone with a nucleophile that cannot open up such as methanol, we get the corresponding β-methoxy compound, which is stable.

One might suppose that a better way to do this reaction would be to carry out the Emmons-Horner reaction with an aprotic system such as sodium hydride in DME. Curiously, when we do this we find that the first step of this reaction, the formation of the anion and the aldol closure step, occurs, but the intermediate aldol product does not undergo elimination. This suggests that the alkoxide formed in these reactions probably does not undergo the same type of decomposition that one has come to expect in the conventional Wittig reaction. It appears we have to have a protic solvent. If the intermediate is mixed with water and base, the diketone is formed. If the intermediate is mixed with sodium methoxide and methanol, the methyl ether is formed. Consequently, attempts to effect intramolecular Emmons-Horner reactions in aprotic solvents have not been useful. This same observation also applies to the other [3.2.1] systems we have examined.

Our efforts to obtain suitable precursors have been more successful for the isomeric [3.2.1] enone shown in Figure 16. We can generate the intermediate ketol in one of two ways. A somewhat longer route involves adding the lithio derivative of dimethylsulfone to the lactone shown followed by an aldol closure. We removed the sulfone functionality from the ketol by a standard sodium amalgam reduction. A more efficient way to generate the same compound was from the corresponding ketophosphonate. Generated at -70^{0} from the phosphonate and lactone, shown, this intermediate species was oxidized to the diketone. With water or methanol and base this intermediate again generated the enone. In methanol solution the enone formed the methyl ether, but in water it added water to give the ketol. Again reaction with sodium hydride and DME gave an intermediate aldol that did not eliminate until we added a protic solvent.

The important feature for our discussion is the ability to isolate the ketol and convert it to species with a better leaving group at the β carbon. The groups we have chosen to study are the mesylate and tosylate leaving groups. When these compounds are heated with triethylamine, as shown in Figure 17, there is a slow separation of a precipitate. We have examined that precipitate at various points throughout the reaction. It appears that the initial precipitate is a mixture of both the triethylamine sulfonate salt and some of the corresponding species where enone has added triethylamine and then precipitated as the sulfonate salt. As the reaction proceeds, the adduct evidently decomposes back to enone and other products because at the end of the reaction, none of the adduct remains in the precipitate.

Although either of the two synthetic routes can be used, there is a complication with use of the mesylate that we believe arises from elimination of a sulfene from the methanesulfonate. So the cleaner reaction uses the tosylate to generate the enone. The [3.2.1] enone can be trapped with metha-

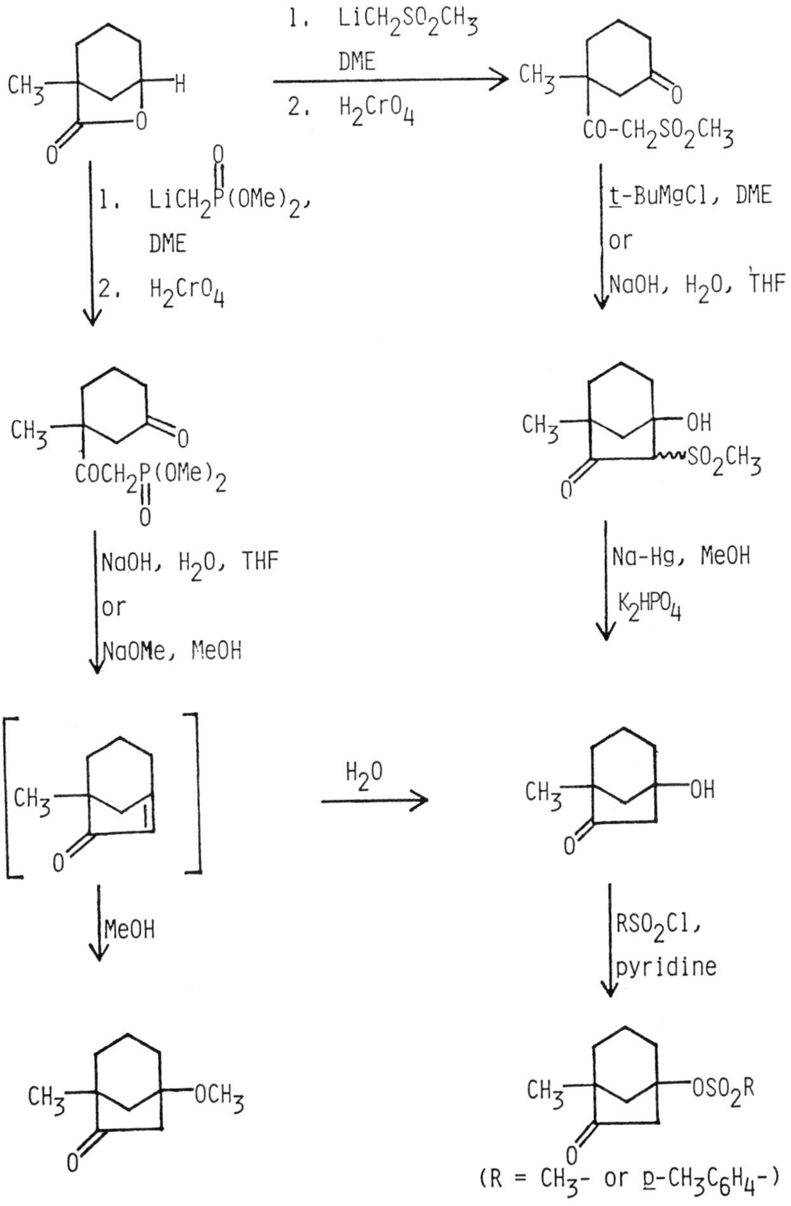

CH_3 —H O O

1. $LiCH_2SO_2CH_3$, DME
2. H_2CrO_4

CH_3 O $CO\text{-}CH_2SO_2CH_3$

1. $LiCH_2P(O)(OMe)_2$, DME
2. H_2CrO_4

t-BuMgCl, DME
or
NaOH, H_2O, THF

CH_3 O $COCH_2P(O)(OMe)_2$

CH_3 OH SO_2CH_3 O

NaOH, H_2O, THF
or
NaOMe, MeOH

Na-Hg, MeOH
K_2HPO_4

[CH_3 O]

H_2O

CH_3 OH O

MeOH

RSO_2Cl,
pyridine

CH_3 OCH_3 O

CH_3 OSO_2R O

(R = CH_3- or *p*-$CH_3C_6H_4$-)

Figure 16. Precursors for the 6-keto bicyclo-[3.2.1]octane system.

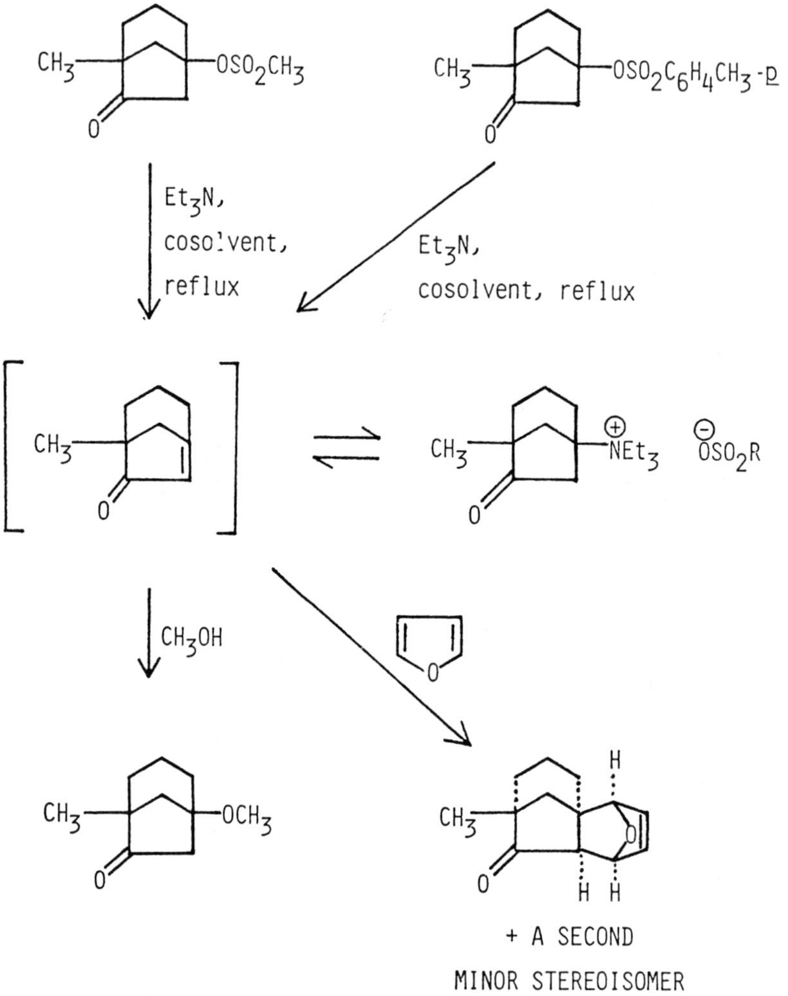

Figure 17. Generation of the 6-keto-bicyclo[3.2.1]enone.

nol to form the methyl ether shown earlier, or it can be trapped with furan to form two stereoisomeric adducts in a mixture of about 80% of the structure shown and about 20% of a second stereoisomer. The stereochemistry of the major adduct was established by crystallography. The important feature in our continuing effort to form the [3.2.1] system is the ability to generate a furan adduct. Obviously, our earlier experience with Walt Trahanovsky's vacuum pyrolysis apparatus suggested that this adduct could be a valuable intermediate for enone generation. So we were curious to see whether this adduct would reverse to form furan and the [3.2.1] enone. Heating the adduct in a sealed tube in an inert solvent, cyclohexane, to 240° formed the product shown in Figure 18.

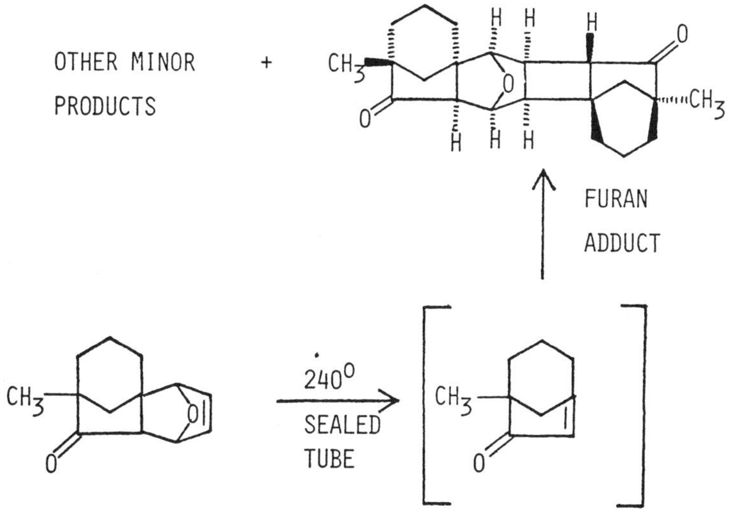

OTHER MINOR PRODUCTS + [structure]

FURAN ADDUCT

240° SEALED TUBE

Figure 18. Alternative generation of the 6-keto-bicyclo[3.2.1]enone.

This is obviously not a short contact time experiment. The mixture was heated in a sealed tube for 6-8 h to get the complete decomposition of the furan adduct. Under those circumstances what we actually isolated was a mixture in which the major product had the structure and stereochemistry shown in Figure 18. Upon examination, you will realize that this structure is the [2+2] cycloadduct from the [3.2.1] enone and the furan adduct. Because the enone is generated in the presence of an excess of the furan adduct, the formation of the observed product is not unexpected.

We consider this observation as reasonable evidence that the [3.2.1] enone can be generated by pyrolysis of the furan adduct. The next step obviously was to send this material to Walt Trahanovsky and have him pyrolyze the material under reduced pressure. They pyrolized the [3.2.1] system and observed only dimer at -50^{0}.

So evidently -50^{0} is not sufficiently cold to keep the [3.2.1] enone. However, it does appear that the vacuum pyrolysis will be a useful procedure.

Discussion

Traylor: It seems very strange that the dioxetane would hold together in the open form. Is it possible that the compound could be a dimer, an eight-membered ring?

House: There is no way I can exclude that. The material is of such a nature that we have not yet been able to successfully recrystallize it or purify it. You can look at things like the mass spectrum, but that does not prove anything. It has a molecular ion of the monomer, but so does the triketone. I really cannot exclude the possibility that it is a dimer or a polymer. One thing that led us initially to think it might be dimeric or polymeric is the fact that it is insoluble in acetonitrile. The thing that might argue against a mixture of different structures is the C-13 NMR spectrum. However, since the C-13 lines are broad, this observation could be consistent with mixtures of stereoisomeric dienes or polymers. We have not been able to recrystallize the material because it decomposes at room temperature in a period of about 2-3 hours. Probably if we worked in a cold room and were very patient we could get some pure material from the crude product. We would never get a crystal structure of it and just what would be the best technique to establish its molecular weight I do not know. Do you have any suggestions for proceding?

Adam: Low temperature osmometry.

House: If we had it pure.

Foote: We made our first dioxetanes from enamines. We observed a very similar phenomenon to what you report, broadened lines. We believed at the time the compounds were dimeric. We spent a lot of work on this with no further clarification. The enamine dioxetanes are of course much less stable. They break down at -30°C, so we really could not isolate them.

House: Well, one of the things we had thought about has been the suggestion that the mode by which the dioxetane decomposses is to proceed by reversible opening to singlet and then slow conversion to triplet which goes on to product. Now I do not know whether the current lore will change that notion or not. But all we would suggest is that what we have seen is not inconsistent with seeing this reversible opening to singlet.

Foote: The energy associated with C-C bond breaking in the diradical is extremely low.

Adam: Once you open up a 1,2-dioxetane, it is very hard to expect that the resulting diradical will close back up again to regenerate the dioxetane, especially in your case in which an additional 10 kcal/mol of strain is present in such a bridgehead structure. Instead of homolytic fission of the peroxide bond, how about heterolytic cleavage of one of the C-O bonds to generate a dipolar intermediate, which is then trapped?

House: You are quite correct that if we open the other bond, there would be more reason for it to reclose. It would be nicer of course in terms of getting hyperfine splitting back in other areas of the molecule. We would have to have a compound pure and know it was pure before we can answer that question.

Adam: I guess you looked for chemiluminescence?

House: Yes, with negative results. We have examined it both in solution and in a melting point tube as it melted. However, an experiment which

Paul Bartlett suggested earlier is one of which I was sort of ignorant; namely, that we should have put some dibromoanthracene in the solution and then looked. That experiment we have not done.

Shea: You were successful in the photochemical [2+2] cycloaddition of the [4.3.0] system. Is there any evidence that photochemically you can induce cis-trans isomerization?

House: Nothing that we have seen. That is a rather messy reaction as regards the number of products and what the genesis of all of them might be we are not certain. Also one of the things that one begins to see is isomerization of the double bond to the back ring of the molecule. Probably there are also some reactions that involve hydrogen atom abstraction from that back ring. There are a lot of products and I certainly cannot exclude it. The products we isolated did not require such isomerization.

Caldwell: If you were suggesting that it is a localized singlet biradical which I thought you said, I do not see how that is going to give you line broadening because you are dealing with diamagnetic species throughout. It is nothing other than site rearrangement.

House: I am told by my nmr colleagues that if one in fact opens the O-O bond there would be hyperfine splitting. I cannot comment whether this is right or wrong.

Herndon: There seem to be nice rationalizations of all the data except for the downfield shift of the β carbon in the parent compound and a lack of shift in the 2-phenyl compound.

House: There does not appear to be a shift in the 2-phenyl compound. We have not made the right model yet to be sure about that.

Herndon: But the model is about as good a model as you can get.

House: We have the phenyl group bound to the carbon-carbon double bond.

Herndon: Yes, but it is two carbon atoms away.

House: I suspect one thing that may be different about the parent system and the phenyl system that would not show up in the force field calculations is that one could get more distortion of the C-C double bond that could be alleviated by a delocalization into the phenyl ring. If the phenyl ring is not there you may get a great deal more distortion at the other carbon atom.

Herndon: Have you tried to make any cycloadducts with electron poor olefins? For example, [2+2] adducts?

House: The olefins we have tried have not added any [2+2] cycloadditions. The ones we have tried include cyclohexene, vinyl acetate, vinyl ethers and enamines.

Herndon: Did you ever try dicyanoethylene?

House: We have not tried dicyanoethylene. The olefin we really wanted to try would be difluorodichloroethylene or dichloroethylene. Very recently this exact experiment has been done with the olefinic bridgehead compounds by Becker.

Paquette: Has there been any work done in an attempt to acquire a series of 3 or 4 of these enones in optically active form to see if there exists a probe one could use such as circular dichroism that would be a more sensitive gauge to the degree of twisting?

House: No. I know of people who discussed the possibility of resolving these enones in an effort to get some idea of what the shape would be. The trouble is that the interesting ones will be tough to resolve. You have to work with them under difficult conditions. You are quite correct in that it would be a way to get information about the shape of the chromophore. We have obviously tried like the devil to get crystalline derivatives of some of these enones. We have a crystalline derivative of the [4.3.1] compound, but we simply have not been able to get adequate crystals of the others. In the [3.3.31] series we do not have any derivative crystals, but we may be able to make some.

Greene: You had several experiments pertaining to triethylamine interactions both with the [3.3.1] phenyl substituted ketone as well as perhaps the dioxetane. You did not formulate them in terms of electron transfer, but triethylamine is fairly good at donating electrons.

House: I guess I have gotten a little gun shy about seeing an electron transfer reaction under every bush. What I thought about it is this. If we get the reduction potential of some enones low enough, it could well be that the reduction potentials of some enones in the range of minus 1.1 to 1.2 volts could be possible. If that is the case then many anions could be quite acceptable reagents to transfer electrons. There is absolutely nothing about the products we isolate that tells us whether that has been a direct nucleophilic addition or whether it has in fact been an electron transfer to form the ion radical, followed by coupling. It is a possibility we have to recognize.

Traylor: One suggestion about the dimer versus monomer. If it were a dimer you might have head-to-tail, head-to-head mixtures in which case you might explain the line broadening in the nmr.

House: It appears there will need to be more than two species. There are no obvious differences in the line widths as you go from line to line.

Traylor: I am talking about the peroxide.

House: We are talking about the same thing. I don't think the material is just a mixture of two compounds. But of course there are a number of stereoisomers possible when you form dimers or trimers. All of these could be present.

Traylor: If you use doubly labeled oxygen, ^{18}O, ^{16}O, you get different distributions. In the dioxetane you get 18-18 only. If it is an 8-membered ring, it could be different.

References

[1] J.R. Wiseman, J. Am. Chem. Soc. 89, 5196 (1967).
[2] J.A. Marshall and H. Faubl, J. Am. Chem. Soc. 89, 5966 (1967).
[3] W. Carruthers and M.I. Qureshi, J. Chem. Soc (D), 832 (1969).
[4] B.G. Cordiner, M.R. Vegar, and R.J. Wells, Tetrahedron Lett., 2285 (1970).
[5] G.L. Buchanan and G. Jamieson, Tetrahedron 28, 1129 (1972).
[6] G.L. Buchanan and G. Jamieson, Tetrahedron 28, 1123 (1972).
[7] W. Carruthers and A. Orridge, J. Chem. Soc. (Perkin I), 2411 (1977).
[8] O. Ermer, Zeitschrift Naturforschung 32b, 837 (1977).

9. SYNTHESIS AND CHEMISTRY OF BRIDGEHEAD ALKENES

Kenneth J. Shea

Department of Chemistry, University of California, Irvine, California 92717

Twisting the two ends of a carbon-carbon double bond in opposite directions results in a torsionally distorted alkene. This distortion will, to a first approximation, result in diminished overlap of the two p orbitals. The torsional distortion is expected to result in an increase in energy of the system and enhanced chemical reactivity. The degree of these effects will be related to the magnitude of the torsional distortion. The molecular species that embody this type of torsional distortion are the trans-cycloalkenes. For purposes of calibration, the smallest isolatable trans-cycloalkene is trans-cyclooctene.

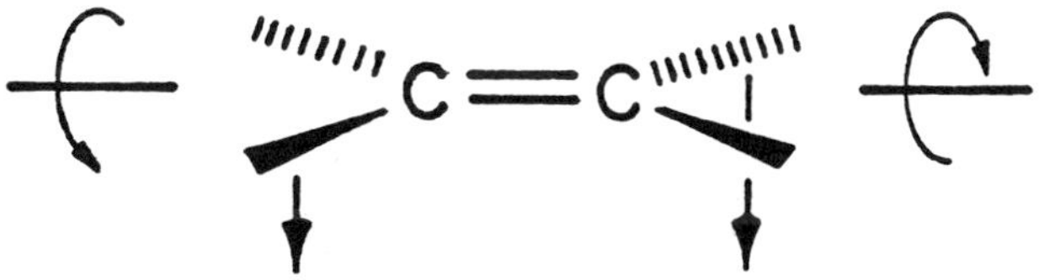

C=C

A subgroup of trans-cycloalkenes has an additional carbocyclic bridge connecting the double bond to the ring. The result of this connection is a bridged bicyclic molecule with the double bond occupying a bridgehead position. These molecules are frequently referred to as anti-Bredt olefins. As was suggested some years ago [1], there is a correspondence between the stability of bridgehead alkenes and the corresponding trans-cyclooctene. The most highly strained isolatable bridgehead alkene contains a trans-cyclooctene ring, eg, bicyclo[3.3.1]non-1-ene.

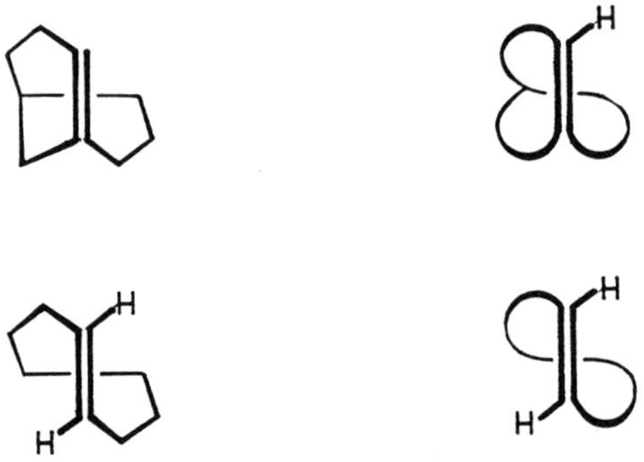

If attention is focused on the structural consequences of the torsional distortion, several very interesting features emerge. A rotation of the ends of the double bond in opposite directions results in a deviation from coplanarity of the two p orbitals. Furthermore, it was suggested by Mock [2], Pople [3], and Allinger [4] that double bonds, when subjected to this type of distortion, will undergo a pyramidilization as indicated in the structure below.

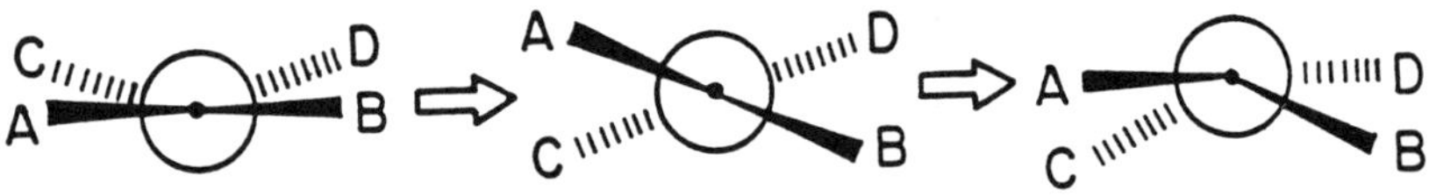

Although the number of structural studies of compounds that contain a torsionally distorted double bond is limited, the suggestion of pyramidalization appears to be substantiated. The magnitude of these distortions can be appreciated by consideration of the X-ray crystal structure of a derivative of trans-cyclooctene-3-ol [5]. Sighting down the carbon-carbon double bond reveals four independent geometrical parameters that are necessary to define the geometry of the double bond.

These entail the angles designated Φ (deviation from coplanarity of the p orbitals) and χ, out of plane bending (pyramidalization). In trans-cyclooctene, these distortions are impressive: Φ ranges from 15 to 22° and χ ranges from 20 to 28°.

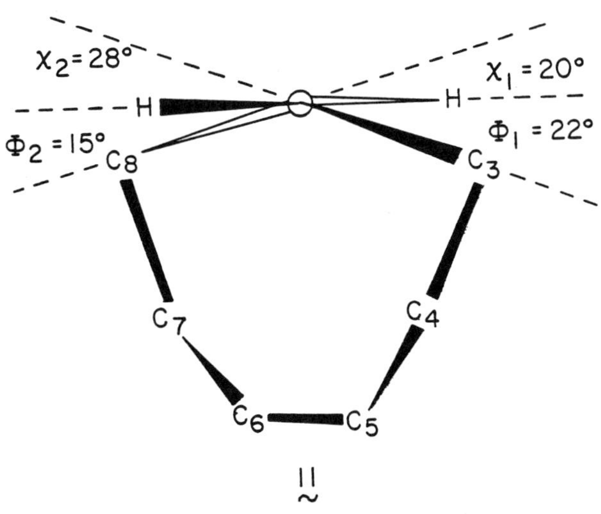

trans-Cyclooctene-3-ol

Although the geometrical distortions in trans-cyclooctene are significant, the energy associated with them, is relatively small. Consider, for example, the total strain energies of trans-cyclooctene and bicyclo[3.3.1]non-1-ene. These values, 22 [6] and 24 [7] kcal/mole, respectively, are relatively modest in the strained molecule arena [8]. If the strain energy associated with the torsionally distorted alkene is artificially partitioned from the total strain energy [9], the torsional strain is indeed very small (7-15 kcal/mole). Thus, one arrives at the very limits of isolatable molecules by investment of only very modest amounts of energy. This fact is of considerable importance in the design of new synthetic entries to molecules that contain these structural features.

To illustrate, consider the enthalpy change of several very common pericyclic reactions, the

STRAIN ENERGY

C=C

	SE (kcal)	OS (kcal)
	22	7.8
	24	12-15

Diels-Alder cycloaddition (ΔH_{rexn}-40 kcal/mol), the Claisen rearrangement (ΔH_{rexn}-16 kcal/mole) and the ene reaction (ΔH_{rexn}-22 kcal/mol). All of these

Reaction Enthalpy

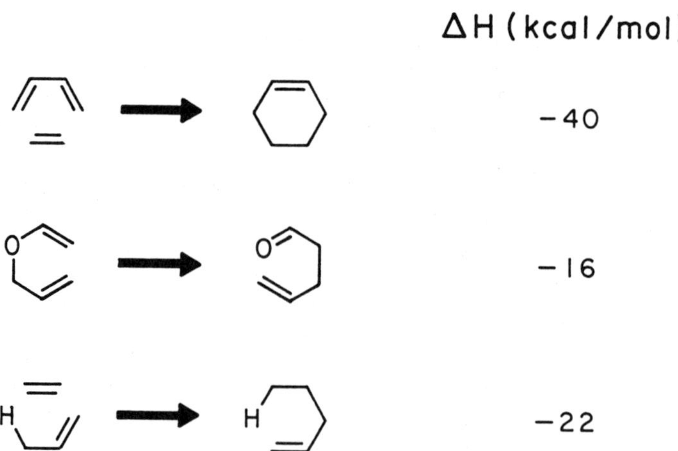

reactions, in principle at least, possess the necessary thermodynamic driving force for the synthesis of molecules that reside at the limits of isolatable torsionally distorted alkenes. Although one does not normally consider pericyclic reactions as a means of synthesizing strained bridgehead alkenes, an awareness of the energetics of these processes renders this approach very plausable. In the sections that follow, we will discuss several recent applications of pericyclic reactions to the synthesis of bridgehead alkenes.

Bridgehead Dienes

Two systems that we have targeted for study are the bridgehead dienes, molecules that contain two bridgehead double bonds "locked" in either a parallel or perpendicular geometry. These molecules can be viewed as derivatives of the two conformational isomers of _trans,trans_-cycloalkadienes

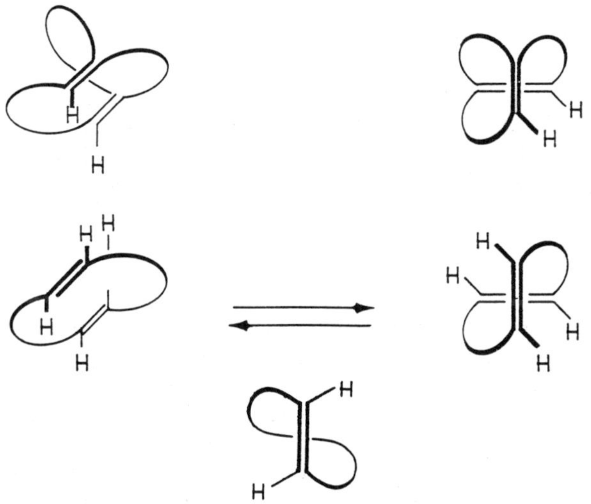

Meso bridgehead dienes. Our strategy for the synthesis of representatives of this class is shown in eq. (1). There are two stereoisomeric boat-like transition states available to _cis_-1,2-divinylcycloalkenes. A (3.3) sigmatropic rearrangement

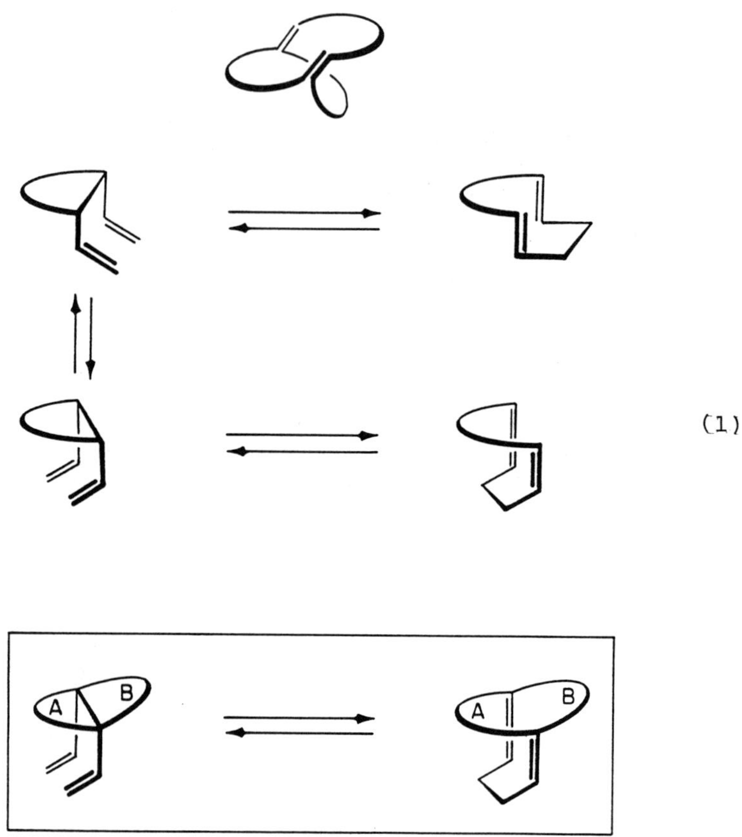

via the endo conformation gives rise to a *cis,cis*-1,5-cycloalkadiene while the *exo* conformation produces a *trans,trans*-1,5-cycloalkadiene. In small and medium ring cycloalkadienes the *cis,cis*-isomer is considerably more stable than the *trans,trans*. The observed sigmatropic rearrangement in the divinyl compound therefore proceeds via endo conformation resulting in formation of the *cis,cis* product.

When *cis*-divinyl groups occupy the bridgehead position of a [n.m.o.]bicycloalkane, rearrangement *must* result in formation of a *trans trans*-1,5-cycloalkadiene. This is illustrated in the figure below where rearrangement is shown to result in a trans,trans-1,5-cycloalkadiene linkage in the B branch (cis,cis- in the A branch). The key question regarding these reactions concerns the position of equilibrium.

Our initial effort involved the generation of 1,4-divinylbicyclo[2.2.0]hexane. At temperatures up to 0°C neither the Cope rearrangement product, nor its subsequent (3.3) sigmatropic rearrangement product was detected [10]. The only organic product isolated is a tetraene (24%) formed by retro [2+2] reaction. We believe this reaction takes precedence over (3.3) sigmatropic rearrangement because of the extremely low bond dissociation energy of the zero bond in the bicyclic divinyl precursor.

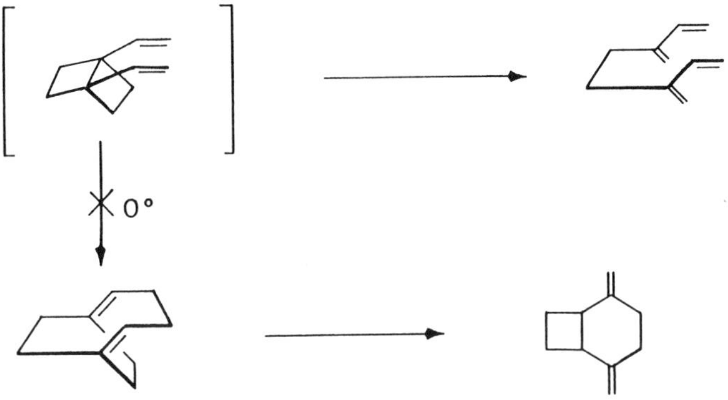

0°

Subsequent investigations revealed the (3.3) sigmatropic reaction pathway provideds a general synthetic entry into meso bridgehead dienes. For example, at room temperature 1,5-divinylbicyclo-[3.2.1]heptane undergoes spontaneous rearrangement to E,E-bicyclo[4.3.2]undeca-1(2),5(6)-diene. Although the gross spectral properties do not distinguish between EE and ZZ isomers, chemical and thermal stability as well as thermodynamic considerations strongly support the assigned structure.

A summary of our results to date are presented below [11]. Several general observations

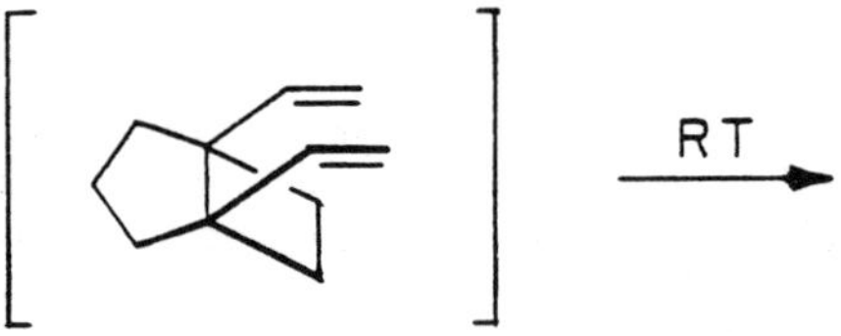

RT

regarding these rearrangements are in order. In all cases examined, the equilibrium for these thermal rearrangements lies exclusively on the side of bridgehead dienes. Furthermore, the rate of the rearrangements seems to parallel the strain energy in the bicyclic alkane starting material.

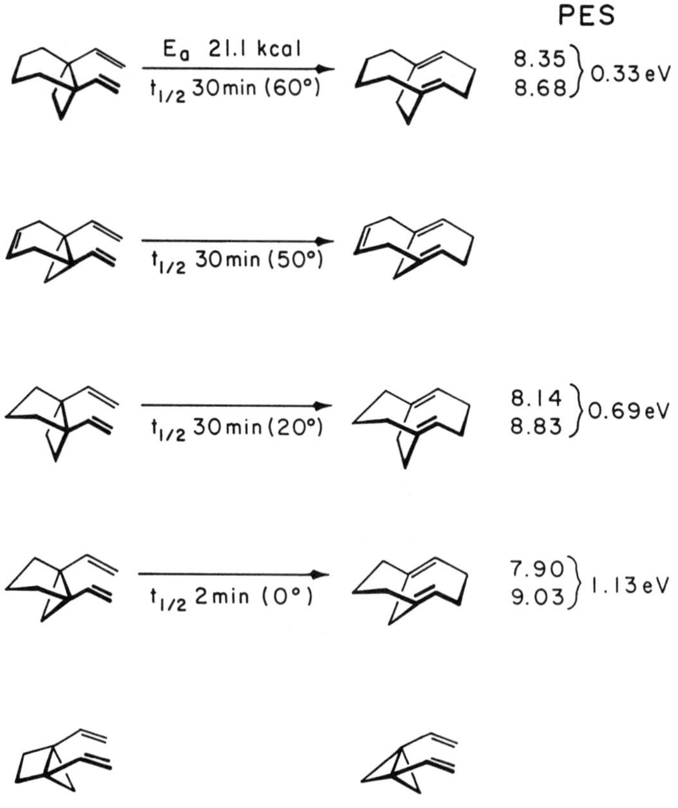

PES

E_a 21.1 kcal, $t_{1/2}$ 30min (60°) — 8.35, 8.68 } 0.33 eV

$t_{1/2}$ 30min (50°)

$t_{1/2}$ 30min (20°) — 8.14, 8.83 } 0.69 eV (2)

$t_{1/2}$ 2min (0°) — 7.90, 9.03 } 1.13 eV

Each of the bridgehead dienes contains two double bonds stacked in a parallel array. In order to obtain some measure of the combined through space-through bond interaction between these two groups we have examined the photoelectron spectra of several of these molecules (eq. 2). All systems exhibit a splitting between the π energy levels. Interestingly, the midpoint of the π levels remains relatively constant (8.5 eV) while the magnitude of energy level splitting increases as the double bonds are squeezed closer together. The 1.1 eV split in the 1,5-nonadiene derivative represents a substantial perturbation of the π orbitals (comare, for example, the splitting of 0.8 eV in norbornadiene) [12].

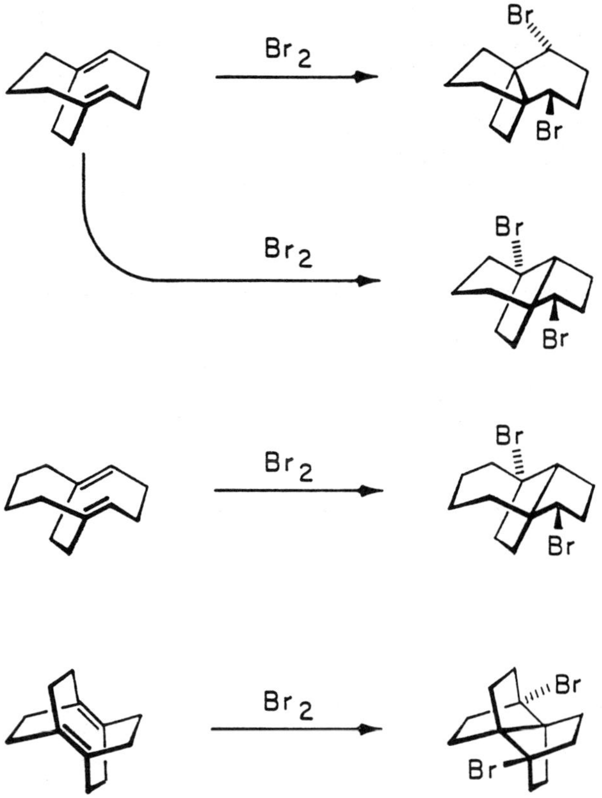

(3)

Further evidence for the interaction of the two π systems comes from a study of their reaction with electrophiles. The bridgehead dienes (eq. 3) take up an equivalent of Br_2 to yield a single dibromide adduct. The structure of the adduct, as revealed by C-13 NMR, is quite clearly the unsymmetrical dibromide. These observations are very similar to that of Wiberg and co-workers [13] in the bromination of the symmetrical bridgehead diene shown below. The chemistry of these three bridgehead dienes parallel that of *cis,cis*-1,5-cyclooctadiene, a molecule known to undergo transannular ring closure to produce bicyclo[3.3.0]octane derivatives [14].

d,l-Bridgehead Dienes. We rely once again on a pericyclic reaction for a synthetic entry into a new class of d,l bridgehead dienes. Our interest in this area was piqued by an unexpected observation made during the course of a mechanistic investigation. We noted that at 190°, 1-vinyl-4-methylenebicyclo[3.2.0]heptane undergoes isomerization to 2,6-bismethylene bicyclo[3.3.0]octane. Our proposed (formal) mechanism for the isomeri-

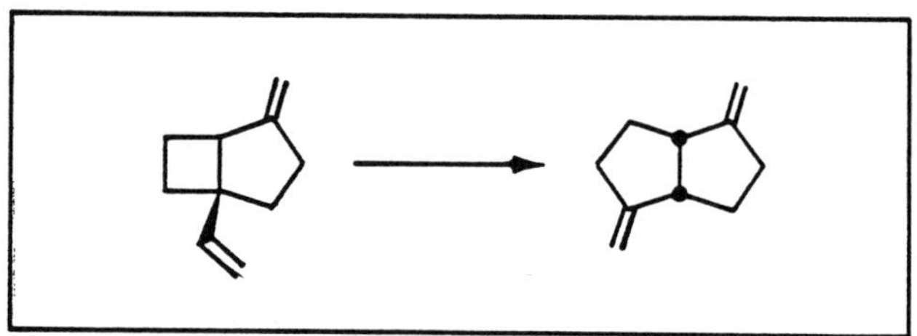

(4)

zation involves a two-step sequence of (3.3) sigmatropic rearrangements (eq. 4).

The proposed intermediate in this transformation is a d,l-bridgehead diene, a derivative of trans, trans-1,5-cyclooctadiene "locked" in the criss-cross conformation by the ethano bridge. The bridgehead diene does not accumulate during the reaction. but undergoes a second (3.3) sigmatropic rearrangement resulting in formation of the observed reaction product. After careful examination of this reaction. we posed the following question: Could pericyclic reactions, such as the one illustrated above, serve as a general synthetic entry into d,l-bridgehead dienes? In the figure below, the key substructural features are given; these encompass an exocyclic double bond joined to a vinyl group through a fused ring junction. A formal (3.3) sigmatropic shift results in bridgehead diene formation. The position of equilibrium, of course, will dictate the utility of this strategy.

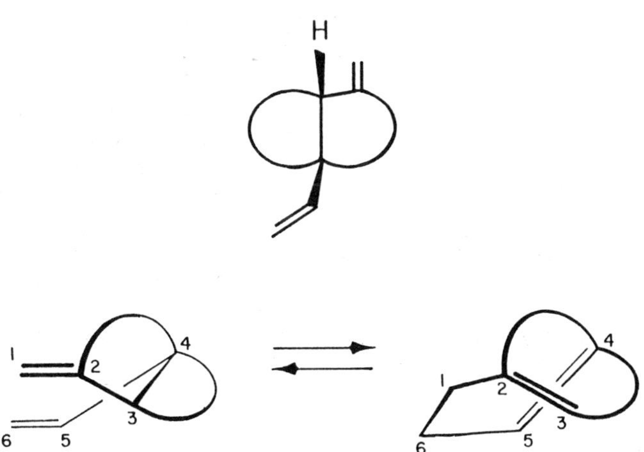

Selected bicyclic dienes were synthesized and subjected to thermolysis [15]. For example, when the bicyclo[3.3.0] derivative was heated (22h, 300°C) a single product was recovered (>77% yield). As indicated in eq. (5), bridgehead diene was not observed, rather a product that contained the hydrindane skeleton was found.

300°C
22 hrs
77%

(5)

Despite the fact that bridgehead diene was not observed, the rearrangement was viewed as an encouraging sign since it adds further support to our proposed mechanism for thermal rearrangement. The hydrindane product is viewed as arising from a sequence of formal (3.3) sigmatropic rearrangements shown in the equation above. The first sigmatropic shift produces a bridgehead diene. The molecule contains a trans,trans-1,5-cyclononadiene "locked" in the criss-cross conformation. Apparently, this too does not accumulate during the reaction, but undergoes the second rearrangement to yield the thermodynamically more stable hydrindane in a sequence of reactions very similar to the rearrangement of the bicyclo[3.2.0]heptane derivative.

This same pattern of reactivity is exhibited by the hydrindane derivatives shown in eq. (6). After 24 hr at 300 °C, an equilibrium was established between starting material and the bismethyleneperhydroazulene, the equilibrium ratio was 93:7. Here, too, a similar sequence of (3.3) sigmatropic rearrangements accounts for the equilibrium, and once again, the intermediate bridgehead diene does not accumulate during the reaction. It apparently lies at higher energy than either product or starting material.

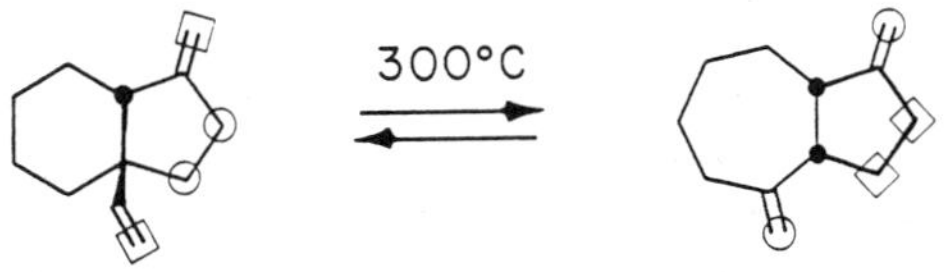

300°C

93 7

(6)

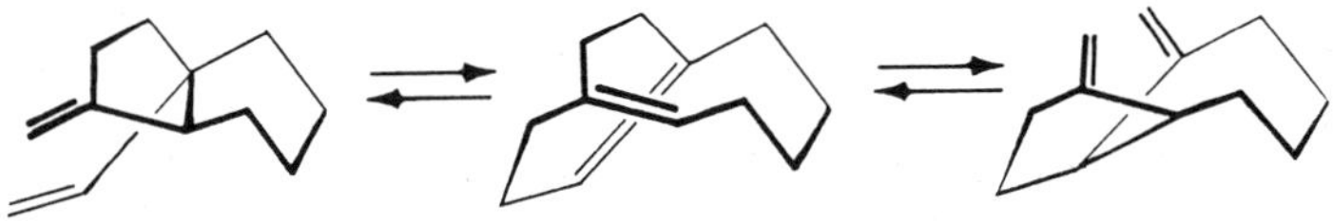

In the three preceding reactions d,l-bridgehead dienes are reaction intermediates. In each case, their isolation was thwarted by a subsequent sigmatropic rearrangement. In order to prevent this subsequent rearrangement, modification of the ring size is necessary. The system selected to remedy

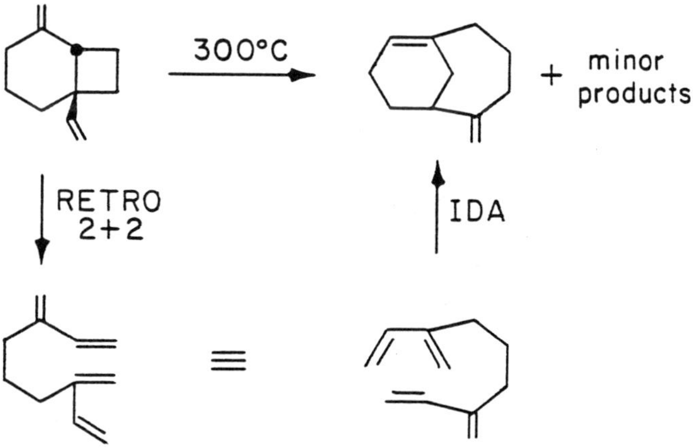

300°C

+ minor products

RETRO 2+2

IDA

(7)

this problem, a bicyclo[4.2.0]octane derivative, is shown in eq. (7). When neat samples of the bicyclic diene were heated (300°C, 3hr), the major product isolated from the reaction mixture (55%) contained a bridgehead double bond, but only one! Proton and C-13 NMR also revealed an exocyclic methylene group. Furthermore, the reaction mixture contained higher molecular weight polymeric material (33%) as well as starting material and several minor products. The structure of the bridgehead olefin was established by independent synthesis. We can account for the formation of this product by the sequence of reactions shown above. Cyclobutane fragmentation (retro 2+2) results in formation of a tetraene. This species then undergoes intramolecular Diels-Alder cycloadditions to produce the observed bridgehead alkene product.

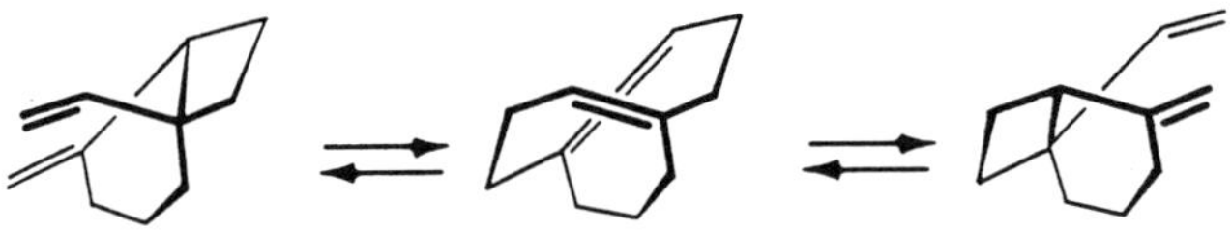

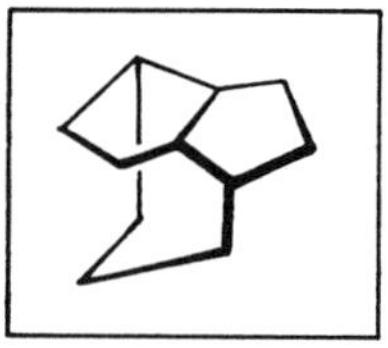

Although once again the dominant reaction pathway did not result in bridgehead diene formation, we have isolated a minor reaction product (5%) that we believe is the hoped-for bridgehead diene. Both ^{1}H and C-13 NMR spectra are consistent with the assigned structure. Interestingly, both catalytic and homogeneous reductions do <u>not</u> result in formation of bicyclo[3.3.3]undecane, rather only one equivalent is taken up to give an unsymmetrical tricyclic alkene, the independent synthesis of which is in progress. An important conclusion to be drawn from these results is that (3.3) sigmatropic rearrangements offer a viable synthetic entry into novel classes of bridgehead dienes.

Bridgehead Alkenes

The past ten years has witnessed an enormous resurgence of interest in the Diels-Alder reaction, both from a synthetic as well as mechanistic standpoint. One of the more exciting developments in this area has been the successful extension of this reaction to the intramolecular domain. Prior to our entry into this area, virtually all intramolecular Diels-Alder reactions involved union of the diene and dienophile at the 1-position of the diene. This substitution pattern gives rise to a fused bicyclic ring system.

TYPE I

TYPE II

We recognized that there was a variation on this theme, ie, union of diene and dienophile at the two position. Cycloaddition in this case gives a bridged bicyclic molecule with a double bond at the bridgehead position. This represents yet another example of a pericyclic reaction serving as a new synthetic entry into molecules that contain a torsionally distorted double bond. At the start of this program, the practical aspects of this problem were unknown. For example, where did the equilibrium lie and what sort of activation energy barrier could we expect?

We were very pleasantly surprised rather early on in this area to learn that some of the most highly strained isolatable bridgehead alkenes were available by the Diels-Alder route [16]. For example, 3-methylene-1,7-octadiene, is converted into bicyclo[3.3.1]nonene in the gas phase at atmospheric pressure at 400°. This is a flow pyrolysis experiment, and the contact or residence time is adjusted by the rate of N_2 flow. Typical residence times range from 5-20 sec.

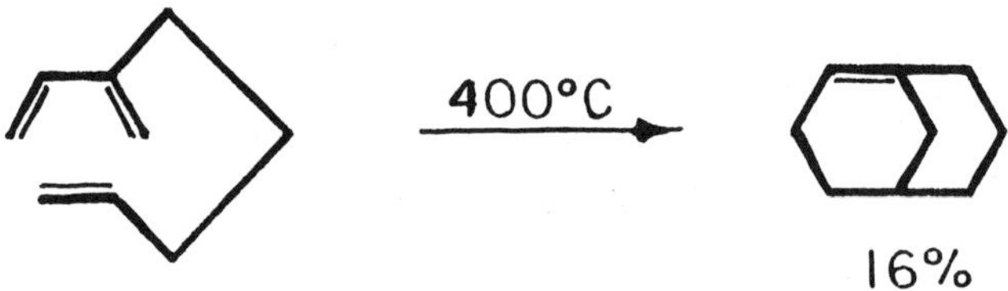

Figure 1 is a plot of the pyrolysate composition as a function of temperature (16 sec. contact time). These reactions are very clean, and the pyrolysate consists of only starting material and product. The cycloaddition becomes kinetically important only at temperatures in excess of 360°C.

We have surveyed a variety of hydrocarbon trienes to establish if the cycloadditions are general. They indeed are, and the reactions reported represent optimized temperatures and contact times [17]. From both kinetic and thermodynamic investigations of the cycloaddition reactions, we have been able to evaluate the energetics of the cycloaddition step. This information is summarized in eqs. (8) and (9). Interestingly, even for formation of the highly strained bicyclo[3.3.1]non-1-ene, the Diels-Alder reaction is exothermic by 18 kcal/mol.

We have examined a variety of triene substitution patterns, and have found that the intra-

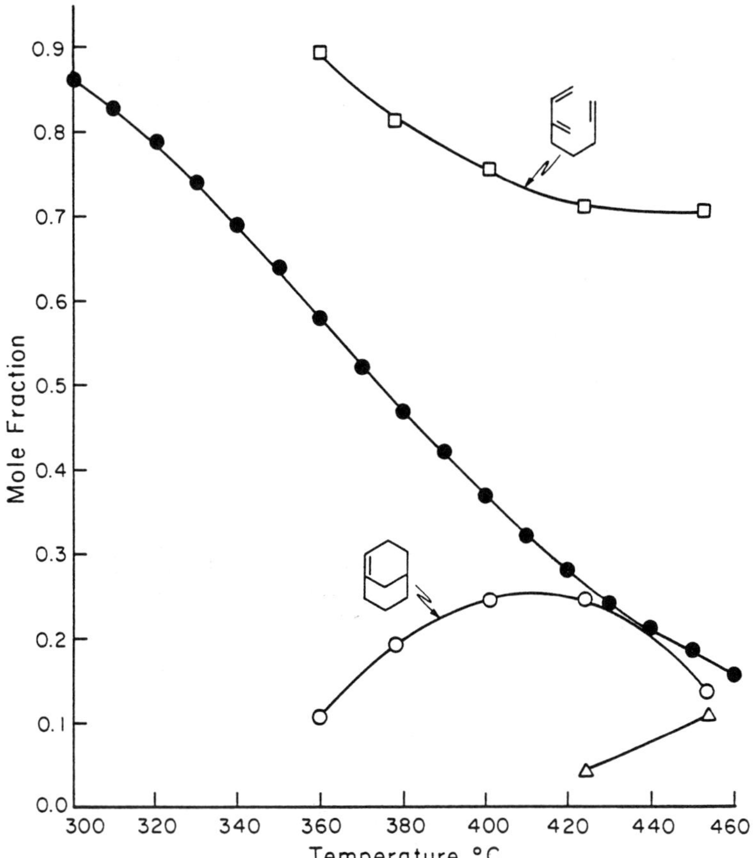

Figure 1. Pyrolysate composition as a function of temperature.

molecular Diels-Alder reaction provides a very high yield regio- and stereospecific synthesis of bridgehead alkenes. A significant portion of our effort is now directed toward exploring their synthetic potential [18,19,20].

Several areas that are currently under investigation are discussed in the following sections.

Bridgehead Enol Lactones

A potentially valuable variation of the intramolecular Diels-Alder reaction involves incorpo-

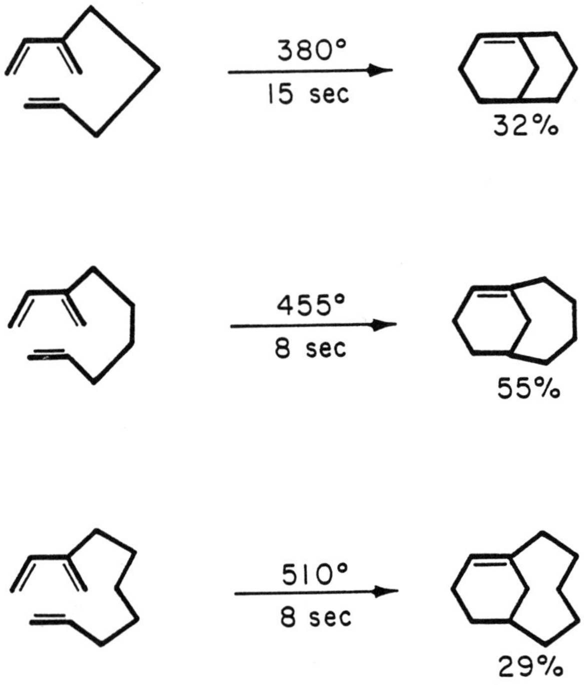

380°
15 sec
32%

455°
8 sec
55% (8)

510°
8 sec
29%

Energetics of Diels–Alder Cycloadditions

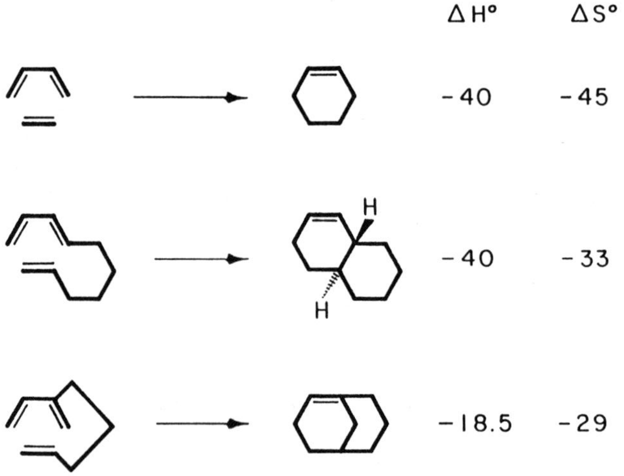

		$\Delta H°$	$\Delta S°$
		-40	-45
	H H	-40	-33
		-18.5	-29

(9)

ration of the diene component as part of an enol ester. Should this compound undergo cycloaddition, the product would contain a bridgehead double bond that is part of an enol ester, in this case, a bridgehead enol lactone.

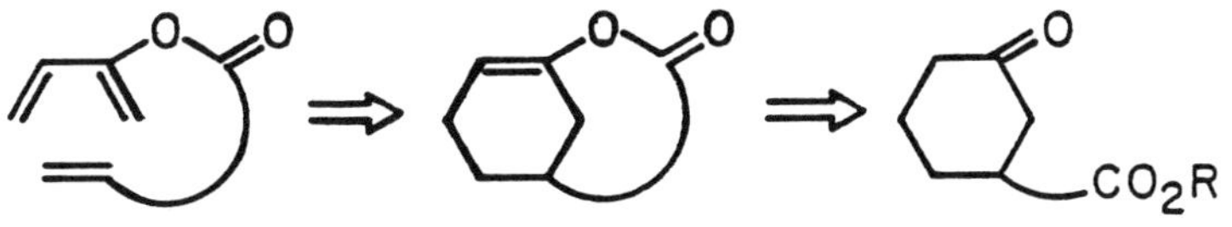

1 2 3

From a mechanistic standpoint, these experiments are interesting since they will provide an opportunity to assess the influence of a torsionally distorted double bond on the chemical properties of enol esters.

A second reason for our interest in these systems has a more synthetic-oriented goal. The hydrolysis of the bridgehead enol lactone unmasks a β-substituted cyclohexanone. This is a rather unorthodox entry into cyclohexanones, and, since the cycloaddition step is concerted, it affords the opportunity to control the relative stereochemistry at positions remote from the carbonyl.

The synthesis of a representative of this class was accomplished, and the results of the thermolysis of this dienol ester are shown in eq. (10). When 0.1 M solutions of the triene are heated in benzene containing a trace of inhibitor at -185°C for 18 h, the bridgehead enol lactone is isolated in over 90% yield. It is a crystalline, hydroscopic

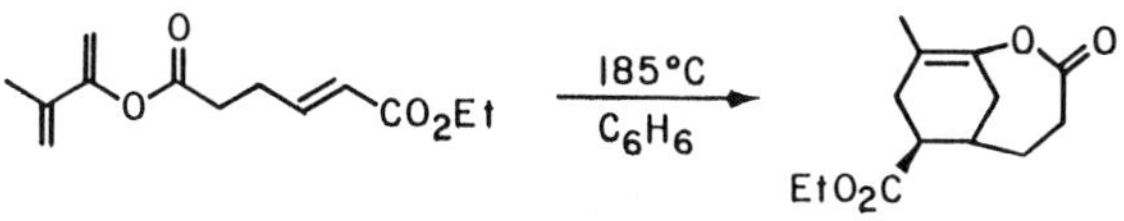

CO_2Et → 185°C, C_6H_6 → EtO_2C (10)

solid that has been completely characterized spectroscopically. Not suprisingly, under both mild acid or base conditions (aq EtOH) the bridgehead enol lactone is converted in quantitative yield to an epimeric mixture of cyclohexanone derivatives.

The kinetic product distribution for both acid and base enthanolysis reactions is remarkably similar. These results are best accommodated by a common reaction pathway involving O-acyl cleavage to produce a cyclohexane enolate or enol in a high energy boat conformation. There is a subsequent competition between proton transfer and conformational relaxation to a more stable conformer of the enolate.

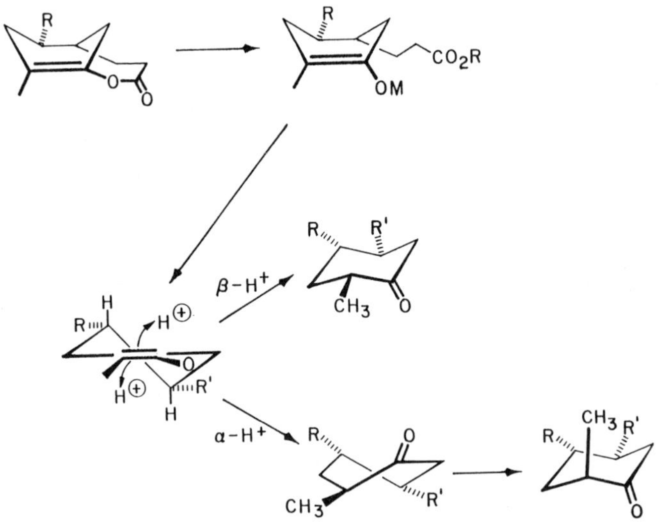

In order to understand the high stereospecificity in the proton transfer step, we propose the following scenario: After O-acyl cleavage the boat conformation relaxes to one of the two quasi-chair conformations of the cyclohexane enolate anion. Proton transfer can take place from either the β-face to give the observed major product or from the α-face to give the minor product.

Under conditions where the two products are not equilibrated, the reaction proceeds by 90% β attack, 10% α attack. This protonation reaction is highly stereospecific: higher, for example, than alkylation reactions of cyclohexane enolates. The stereochemical outcome can be explained in terms of axial proton delivery to yield the most stable chair-conformation of the cyclohexanone derivative. Axial delivery from the α-face results in a high energy twist boat conformaton.

An interesting difference emerges between simple enol esters and bridgehead enol esters. Under mildly acidic conditions, simple enol esters hydrolyze by rate-determining carbon protonation. The mechanistic cross-over is presumably a result, in part at least, of the instability of the resulting bridgehead carbonium ion.

Scheme 4

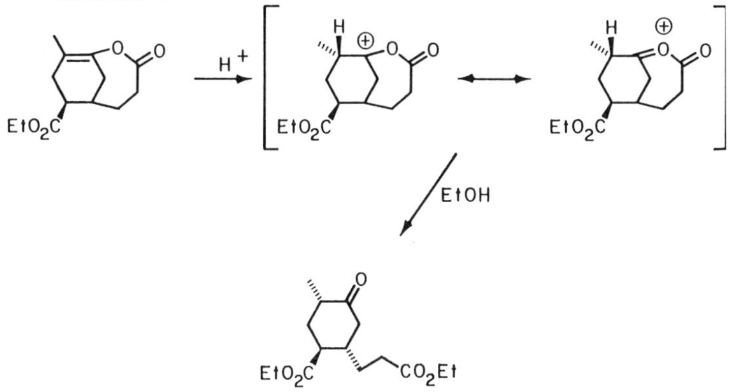

H^+ EtOH

EtO_2C CO_2Et

Diels-Alder Route to Bridgehead Dienes

Intramolecular Diels-Alder cycloaddition of dienes substituted at the 2-position with acetylenic groups offers a potential entry into bridgehead dienes, more specifically 1,5-bridged-1,4-cyclohexadienes. Consideration of the energetics of the parent reaction, butadiene and acetylene, reveals that the process is considerably more exothermic (14 kcal/mol) than is the ethylene plus butadiene reaction.

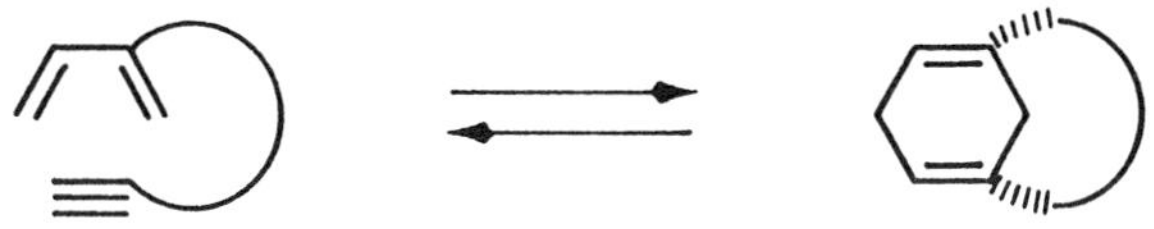

One of the first systems investigated proved to be less straightforward than we had hoped. For synthetic reasons, we selected the oxadienyne shown in eq. (11). The pyrolysis (gas phase, 425° C, 5 s) proceeds smoothly to give a single major product, a diene aldehyde.

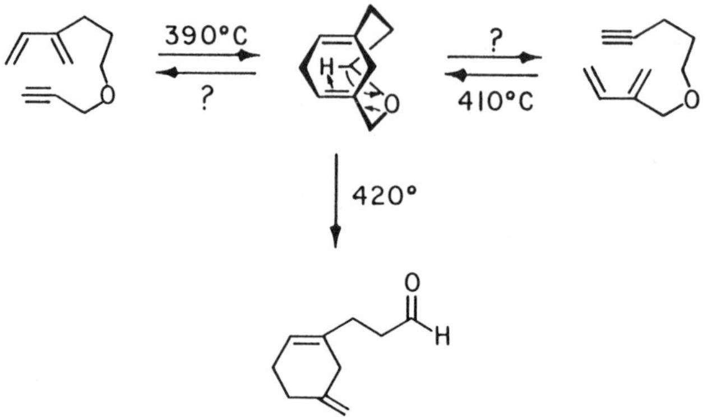

(11)

Our explanation for the observation is shown in eq. (11). The first step in this sequence involves intramolecular Diels-Alder cycloaddition to give the bridgehead diene. Apparently, under the reaction conditions, the adduct is not stable and undergoes a retro heteroene reaction to yield the observed diene aldehyde. Inspection of the molecular model of this compound shows that the CH_2 group is tucked under the ring, and the hydrogen that must be transferred is very accessible to the carbon-carbon double bond.

Support for this proposal comes from a study of the pyrolysis of the isomeric dienyne. When it is subjected to the pyrolysis condition, we observe formation of the same diene aldehyde. This suggests a common intermediate for these reactions, the bridgehead diene. We have not, however, observed the oxygen scrambling reaction. This implies that the retro Diels-Alder reaction is not competitive under these conditions. Careful examination of the pyrolysis mixture reveals that the bridgehead diene does accumulate under the reaction conditons, and it has, in fact, been isolated and completely characterized.

We have also established that deuterium substitution has an influence on the retro heteroene step. The influence of deuterium on product ratio is shown below.

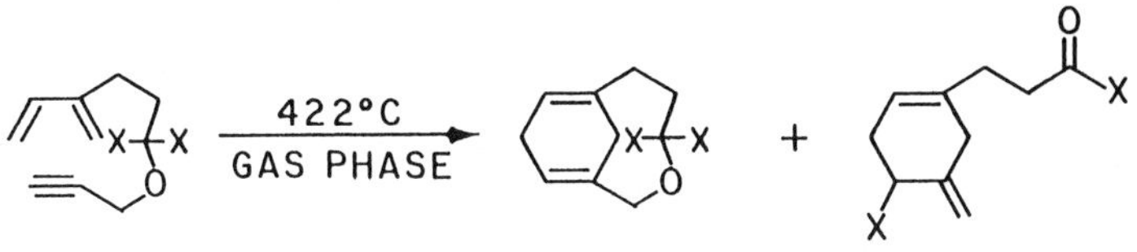

422°C
GAS PHASE

X		
H	1	8.9
D	1	2.1

A primary deuterium isotope effect is observed in the retro heteroene step. Thus, substitution of deuterium for hydrogen significantly alters the final product distribution by diminishing the kinetic lability of the bridgehead diene intermediate.

Subsequent investigations of a variety of dienynes reveal that the intramolcular Diels-Alder cycloaddition is a very general synthetic entry into bridgehead dienes, and with appropriately activated substituents it can result in near quantitative synthesis of bridgehead dienes.

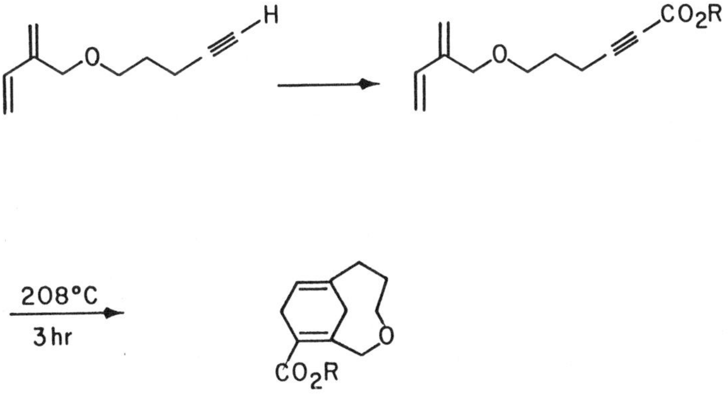

H

CO_2R

208°C
3hr

CO_2R

98%

Conclusion

We have demonstrated that pericyclic reactions offer a very general and valuable synthetic entry into a wide variety of molecules containing a torsionally distorted carbon-carbon double bond.

Acknowledgements. I am particularly grateful to my co-workers in this research whose names are included in the references. I also wish to thank Professor Don Aue from the University of California, Santa Barbara for the photoelectron spectra, and I acknowledge the financial support of the National Science Foundation and the Petroleum Research Foundation.

Discussion

Lin: In the condensation pyrolysis reaction, the temperature was something like 300-400° C. What about lower temperatures of maybe 160° with metals or metal surfaces? You might expect many different types of products.

Shea: I am not aware of any successful efforts involving catalyzed intramolecular Diels-Alder reactions on pure metal surfaces.

Diels-Alder reactions are, however, responsive to several types of catalysis; perhaps the most successful known have been Lewis acids. We were reluctant to pursue this line of research because of the complications that we perceived that could arise between the Lewis acids and the bridgehead double bonds. These concerns were not well founded. As it turns out, many of the high temperature Diels-Alder cycloaddition reactions discussed on the preceding slides can be run at room temperature in the presence of Lewis acid catalysts.

For example, cycloaddition of the triene ester shown below requires 12 hr at 180° C. This same reaction is complete in 1 hr at 22° C when run in the presence of 1 equivalent of diethylaluminum chloride.

Herndon: You talked about type I and type II reactions. Of course there is another type, type III which would lead to a paracyclophane derivative. Do you have any examples of these?

Shea: Two regioisomeric transition states are available to type II intramolecular Diels-Alder cycloadditions. The factors that are usually deemed important in establishing regiochemical preference in intramolecular Diels-Alder cycloadditions are secondary orbital effects and the cumulative effects of torsion, strain and non-bonded interactions associated with the bridge joining diene and dienophile. As is usually the case, these latter factors dominate. As in the example cited above, the meta regioisomer is the only reaction product. It is only when the chain joining the diene and the dienophile spans six atoms or more do we observe formation of the para regioisomer. For example

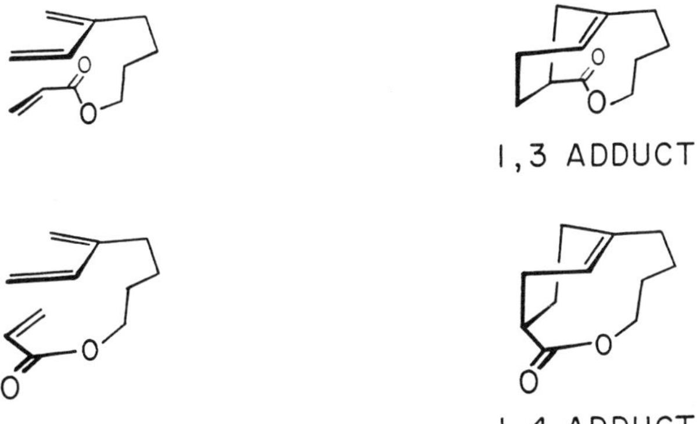

Herndon: The Lewis acid would of course even favor the para orientation to a larger extent.

Shea: The above reaction was examined in the presence of Lewis acids. The isomer distribution changes from 90:10 (180 °C) to 81:19 (22 °C, Et_2AlCl). Thus, in the presence of Lewis acids, there is a shift in the regioselectivity towards the para regioisomer. This shift is consistent with that predicted by FMO analysis. The magnitude of the shift, however, is not large and may also be accounted for by a slight temperature dependence of the regiochemistry.

Traylor: Does the meta-connected cyclohexadiene undergo addition with closure or photo-closure in any way?

Shea: The meta-bridged cyclohexadienes have been designed as potential photochemical precursors to the as yet unknown *syn*-tricyclohexane ring system. These experiments are presently underway, but we have no results to report.

Traylor: Have you added bromine or other electrophiles?

Shea: No. We have not yet explored this reaction.

References

[1] J.R. Wiseman and W.A. Pletcher, *J. Am. Chem. Soc.* *92*, 956 (1970).

[2] W.L. Mock, *Tetrahedron Lett.*, 475 (1972).

[3] L. Radom, J.A. Pople and W.L. Mock, *Tetrahedron* ᴵ*ett.*, 479 (1972).

[4] N.L. Allinger and J.T. Sprague, *J. Am. Chem. Soc.* *94*, 5734 (1972).

[5] O. Ermer, *Angew. Chem. Int. Ed.* *17*, 604 (1974).

[6] D.W. Rogers, H. Voitkengerg and N.L. Allinger, *J. Org. Chem.* *43*, 360 (1978).

[7] P.M. Lesko and R.B. Turner, *J. Am. Chem. Soc.* *90*, 6888 (1968).

[8] A. Greenberg and J.F. Liebman, "Strained Organic Molecules", Academic Press, New York, 1978.
[9] W.F. Maier and P.v.R. Schleyer, J. Am. Chem. Soc. 103, 1891 (1981).
[10] J.R. Wiseman and J.J. Vanderbilt, J. Am. Chem. Soc. 100, 7730 (1978).
[11] K.J. Shea, S. Wise, P.D. Beauchamp, S. Nguyen, and A. Greeley, manuscript in preparation.
[12] K.J. Shea and D. Aue, unpublished results; E. Heilbronner and J.P. Maier, "Electron Spectroscopy, Theory, Techniques and Applications," C.R. Brundle and A.S. Baker (Ed.), Academic Press, N.Y., 1973.
[13] M.C. Matturro, R.D. Adams, and K.B. Wiberg, J.C.S. Chem. Comm., 878 (1981).
[14] G. Haufe and M. Muhlstadt, Z. Chem. 19, 170 (1979).
[15] K.J. Shea, P. Beauchamp, P.D. Davis, and A. Greeley, manuscript in preparation.
[16] K.J. Shea and S. Wise, J. Am. Chem. Soc. 100, 6519 (1978).
[17] K.J. Shea, S. Wise, and L. Burke, manuscript in preparation.
[18] (a) K.J. Shea and S. Wise, Tetrahedron Lett., 1011 (1979). (b) K.J. Shea, Tetrahedron 36, 1683 (1980).
[19] K.J. Shea, P. Beauchamp, and R. Lind, J. Am. Chem. Soc. 102, 4544 (1980).
[20] K.J. Shea, S. Wise, L.D. Burke, P.D. Davis, J.W. Gilman, and A.C. Greeley, J. Am. Chem. Soc. 104, 5708 (1982).
[21] K.J. Shea and E. Wada, J. Am. Chem. Soc. 104, 5715 (1982).
[22] K.J. Shea and L.D. Burke, manuscript in preparation.

10. SINGLET OXYGEN: NEW AND OLD MYSTERIES

Christopher S. Foote, C.-L. Gu, L. Manring, J.-J. Liang, R. Kanner, J. Boyd, M. L. Kacher, M. Kramer, and P. Ogilby

Department of Chemistry and Biochemistry
University of California, Los Angeles, California, 90024

This paper will discuss the photooxidation of three classes of compounds: sulfides, an acyclic diene, and substituted indenes. Although sulfides are not, of course, π systems, there are many instructive analogies between their reactions and those of other systems, and it is useful to consider them together.

The photooxidation of sulfides has been known for some time; it was first described by Schenck's group [1]; the initial mechanistic work was done by Foote and Peters [2,3]. The reaction is smooth under many conditions and proceeds to give the sulfoxide. One mole of oxygen is taken up and 2 moles of sulfoxide are formed; both atoms of oxygen are used.

$$2R_2S \xrightarrow[{}^1O_2]{} 2R_2SO$$

(also phosphorus and other atoms)

There is an intermediate in this reaction; we are not yet sure of its structure. Three possible ways of writing such an intermediate are shown below:

$$R_2\overset{+}{S}\text{-O-}\overset{-}{O} \qquad R_2S\!\!<\!\!\begin{matrix}O\\|\\O\end{matrix} \qquad R_2\dot{S}\text{-O-O}\cdot$$

This intermediate reacts with a second sulfide to give the sulfoxide. One of the ways of detecting the intermediate is to trap it. For example, diphenyl sulfide is essentially unreactive toward singlet oxygen, but if it is added to photo-oxidizing diethyl sulfide, diphenyl sulfoxide is produced along with diethyl sulfoxide.

$$Ph_2S \xrightarrow[{}^1O_2]{} N.R.$$

$$Et_2S + Ph_2S \xrightarrow[{}^1O_2]{} Et_2SO + Ph_2SO$$

Peters and Foote carried out a kinetic study which established that competitive trapping of the intermediate by the two sulfides was occurring in methanol [2,3]. The behavior they observed was consistent with scheme 1.

Scheme 1:

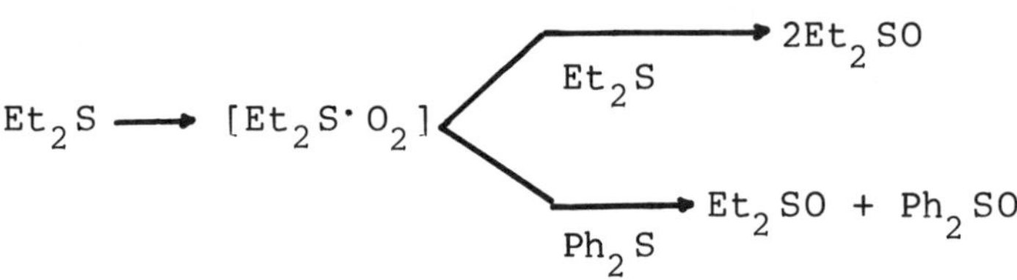

However, there are some strange things about this reaction. Scheme 1 describes the reaction well in methanol; in aprotic solvents, however, the overall rate at which singlet oxygen is removed is about the same as in methanol, but product formation is about 50-fold less. It turns out that there is a very large amount of quenching of singlet oxygen under these conditions; roughly 98% of the interactions with singlet oxygen result in its deactivation and only about 2% lead to product [2,3].

Mysteriously, the reaction in aprotic solvents speeds up dramatically at low temperatures; it turns out that this effect is due to the fact that, although the initial rate of interaction with singlet oxygen is constant, the quenching reaction is suppressed in favor of product formation [4]. These observations are summarized below:

In aprotic solvent:

1. Inefficient at room temperature.
2. Quenching suppressed at lower temperature.
3. Stoichiometry: two sulfides per mole O_2.

There are several other features of the reaction which are not easy to understand. One is associated with the kinetics of the quenching reaction. If there were an intermediate which was either reacting with the sulfide to give products or decaying to starting materials, one would expect that the higher the sulfide concentration, the lower the fraction of quenching would be, since the intermediate would be trapped before it could decay. However, that was not the case; the fraction of product formation was independent of sulfide concentration. The only way to fit these observations into this simple scheme was to add a sulfide-dependent quenching step, for which there was no rationale (Scheme 2).

Scheme 2:

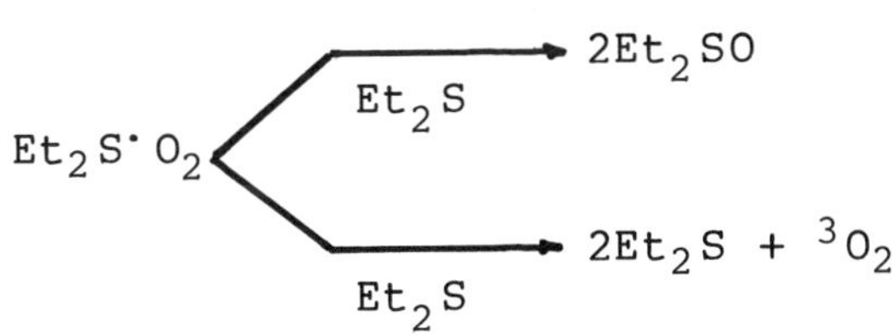

The only virtue of the sulfide-dependent quenching step was that it satisfied the kinetics. A few years ago we began to reexamine the reaction. We noticed that under some conditions, there was a reasonable amount of the sulfone produced in addition to the sulfoxide. We thought one reason might be that some of the sulfoxide product was also competing for the reactive intermediate. We found that diphenyl sulfoxide is also inert to singlet oxygen. However, if it is added to photooxidizing diethyl sulfide, it is oxidized to the sulfone very rapidly and traps the intermediate (much more efficiently, in fact, than diphenyl sulfide). In a given period of time, for example, about five times as much intermediate is trapped by diphenyl sulfoxide in methanol as by diphenyl sulfide. The numbers by each step below represent the relative trapping efficiencies of each of the traps in methanol.

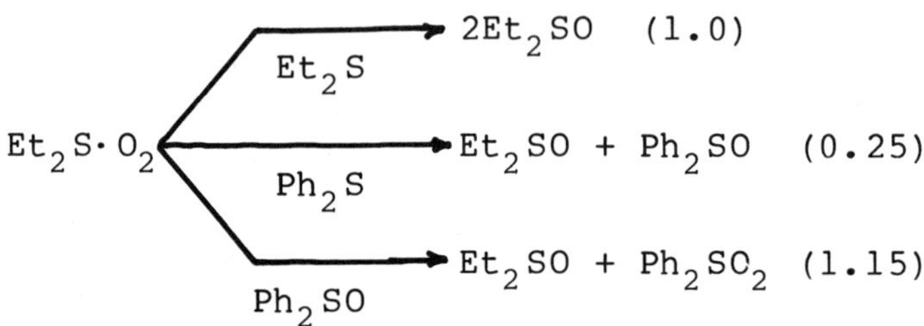

In benzene, the ratio is even larger. In fact, diphenyl sulfoxide (and other sulfoxides) are considerably better traps than diethyl sulfide itself, which accounts for the fact that in aprotic solvents there is so much sulfone formed: the product sulfoxide does indeed trap the intermediate [5].

At about this time, two studies of the efficiency of trapping of the intermediate by various compounds appeared. Ando's group reported that when diaryl sulfides are used as trapping agents, they act as <u>nucleophiles</u> [6]. Sawaki and Ogata independently reported that sulfoxides function as <u>electrophiles</u> toward the intermediate [7].

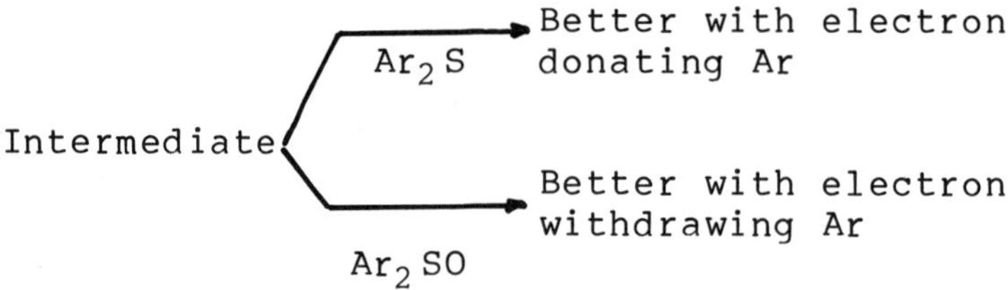

These observations led us to wonder whether there might be more than one intermediate. The kinetics we determined previously for trapping by diphenyl sulfide in methanol were consistent with

Scheme 1, in which there is an intermediate which either reacts with diethyl sulfide to give two Et_2SO or with Ph_2S to give Et_2SO and Ph_2SO. The experimental plot obtained for diphenyl sulfide is Figure 1, which fits Scheme 1 well.

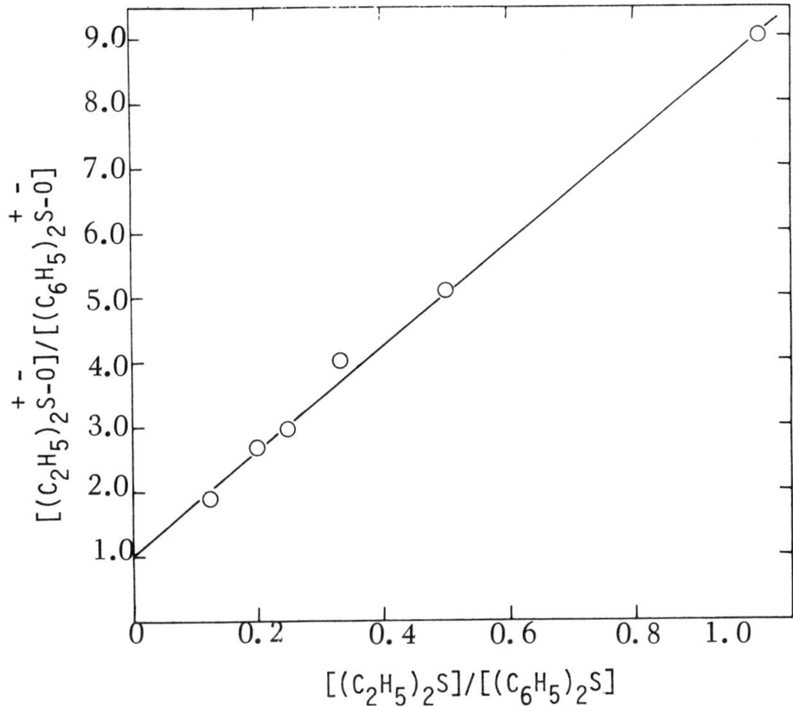

Figure 1. Photooxidation of mixtures of Et_2S and Ph_2S.

We decided to study the kinetics of trapping by diphenyl sulfoxide. To do this, we varied the concentration of both the diethyl sulfide and the diphenyl sulfoxide. The kinetic scheme described above would give the following relationship between the sulfone product, diethyl sulfide, and diphenyl sulfoxide:

$$\frac{Et_2SO}{Ph_2SO} = 1 + \frac{2\underline{k}_1 (Et_2S)}{\underline{k}_2 (Ph_2SO)}$$

The ratio of Et_2SO to Ph_2SO_2 is dependent upon both Et_2S and Ph_2SO. This is a perfectly competitive situation. We carried out this reaction in methanol, and also in benzene and other aprotic solvents. The results were strikingly different. In methanol, exactly the behavior expected was observed. Using three different diethyl sulfide concentrations, the plot of the reciprocal of the product ratio versus the reciprocal of the sulfoxide concentration gave good straight lines (Figure 2).

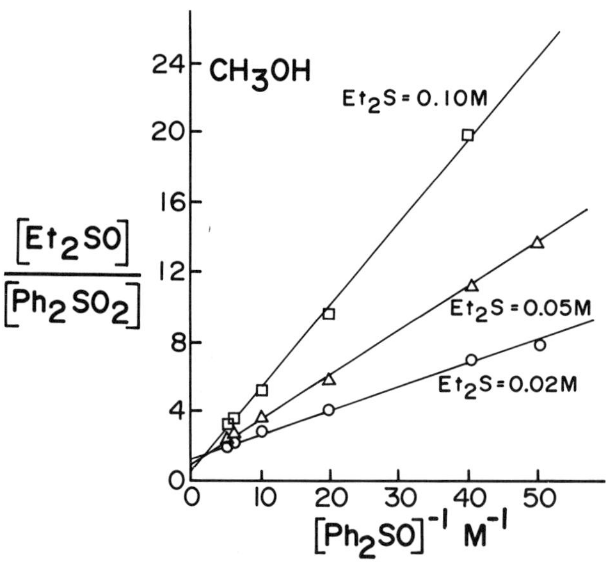

Figure 2. Diphenylsulfoxide trapping, methanol.

The slope was dependent upon diethyl sulfide, as it should be. However, when we did the same experiments in benzene, the slope of the line was absolutely independent of sulfide concentration (Figure 3). This plot is not consistent with the competitive trapping seen in Figures 1 and 2.

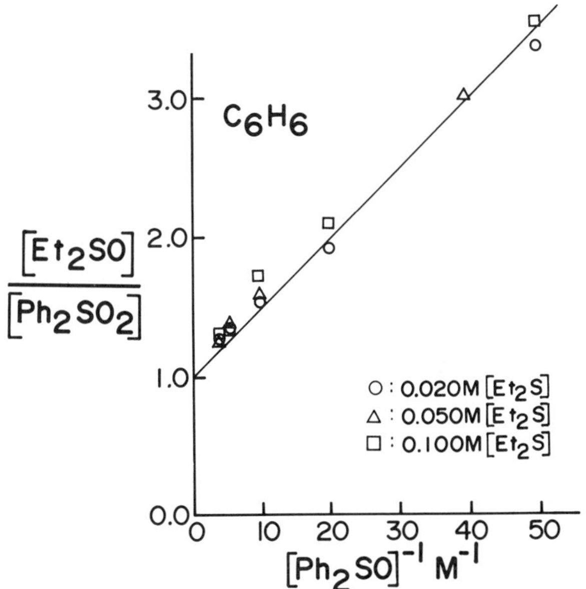

Figure 3. Diphenylsulfoxide trapping in benzene.

Many different explanations for this behavior have been considered [5]; the simplest is that there is not one but two intermediates. For example, if there is one intermediate, which reacts with Ph_2SO to give Et_2SO and Ph_2SO_2 but then decays by some first-order process to an intermediate (Y) which is no longer trapped by Ph_2SO but reacts with Et_2S (scheme 4), the kinetics reduce to the expression shown below:

Scheme 4:

$$Et_2S \xrightarrow{^1O_2} Et_2S\cdot O_2 \xrightarrow[\underline{k}_1]{Ph_2SO} Et_2SO + Ph_2SO_2$$

$$Et_2S\cdot O_2 \xrightarrow{\underline{k}_2} Y \xrightarrow[Et_2S]{} 2Et_2SO$$

$$\frac{Et_2SO}{Ph_2SO_2} = 1 + \frac{2\underline{k}_2}{\underline{k}_1(Ph_2SO)}$$

The slope of the line depends upon Ph_2SO but not on Et_2S. This scheme predicts a kinetic plot exactly like that observed in Figure 3. The amount of intermediate which is removed by diphenyl sulfoxide cannot be restored by increasing the sulfide concentration because all of the inhibition has occurred at an earlier stage.

Clearly, there are two different kinetic schemes. The reaction in methanol involves competitive quenching; in benzene, the inhibition is noncompetitive.

Diphenyl sulfide was also found to <u>suppress</u> the quenching of singlet oxygen in methanol. This fact can easily be accounted for by a slight expansion of the kinetic scheme (Scheme 5). Because of the large amount of quenching in benzene ($\underline{k}_q/\underline{k}_2 = 50$), adding diphenyl sulfoxide actually <u>increases</u> the total amount of Et_2SO formed, since 1 mole is formed by the $\underline{k}_1$ pathway. Scheme 5 shows other interesting features. The amount of quenching which occurs in aprotic solvent depends on the amount of added sulfoxide trap, but not on sulfide. The sulfoxide diverts the intermediate which can quench to product. Diphenyl sulfide does not. Clearly, the sulfoxide is trapping at an earlier stage than diphenyl sulfide. The nucleophilic intermediate is trapped by the sulfoxide, the electrophilic intermediate is trapped by the sulfide; this is consistent with the results both of Ando [6] and of Sawaki and Ogata [7].

Scheme 5:

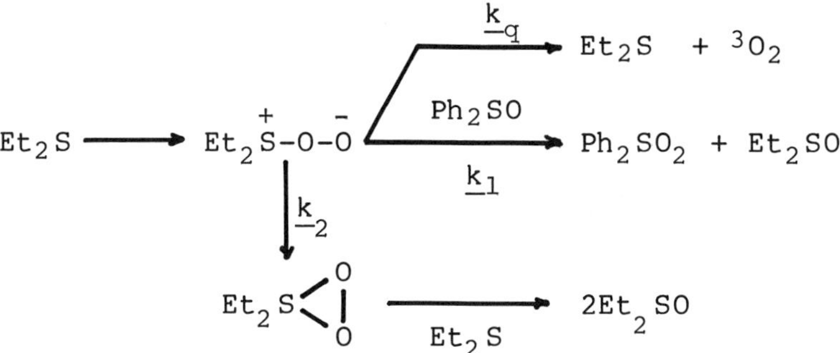

Although it is clear that there are two intermediates, it is not so clear what their structures are. The structures written in Scheme 5 are at least understandable in terms of nucleophilicity and electrophilicity in aprotic solvent, since the persulfoxide structure might be expected to react as a nucleophile with the sulfoxide. It is clear that the sulfoxide trapping competes with the quenching step. In aprotic solvents, where nothing acts to stabilize the intermediate, the first intermediate collapses to the second intermediate. A reasonable structure for the second intermediate would be the cyclic sulfuran (thiadioxirane) which might be expected to act as an electrophile. The structures for the intermediates in Scheme 5 are reasonable ones and have precedent in corresponding structures suggested for the carbonyl oxide and dioxirane isomers [8].

Scheme 5 immediately suggested a further test: we had never used diphenyl sulfide in aprotic solvents. If the scheme is correct, diphenyl sulfide should trap the second intermediate and should be competitive with diethyl sulfide. This behavior is precisely what is observed (Figure 4), confirming the suggestion that there are two intermediates. One acts as a nucleophile, reacts with sulfoxides, and quenches, while the other reacts as an electrophile and reacts with sulfides.

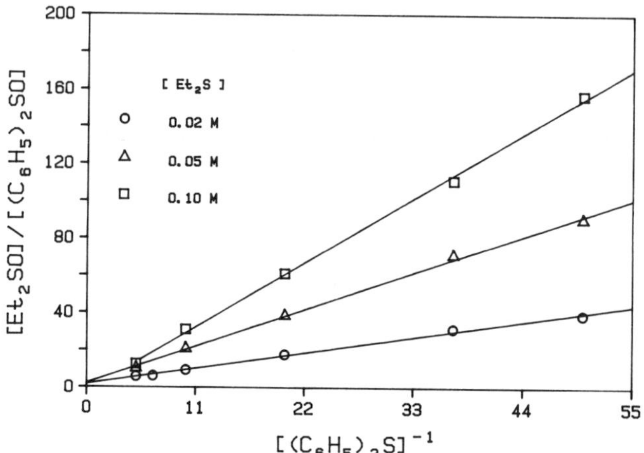

Figure 4. Diphenylsulfide trapping in benzene.

The situation in methanol is quite different. There is no quenching and both sulfoxide and sulfides are competitive. There are at least two possibilities to be considered for this situation. Methanol may stabilize the first intermediate by hydrogen bonding to keep it from closing. The hydrogen bonding would remove some negative charge and make the intermediate more electrophilic so that it can react with either the sulfides or diphenyl sulfoxide. The second possibility, perhaps the more reasonable, is that methanol actually adds to the intermediate to give a peroxysulfurane. J.C. Martin's group has isolated some similar compounds. This intermediate should act as a peracid and would be expected to react with both the sulfoxide and the sulfide.

Kinetic proofs are never very satisfactory; however, they are all that is available in this case. To summarize: this system has a very unusual quenching in aprotic solvents and none in protic solvents; at low temperature, the quenching disappears. There are two distinct intermediates in aprotic solvents, but only one in protic.

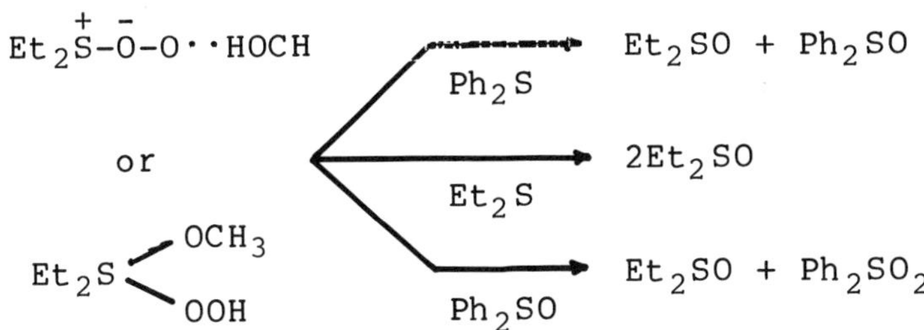

This situation is reminiscent of some earlier results in the indene series. There is a similarly complicated situation with indene, which reacts quite slowly in aprotic solvent at room temperature, but more efficiently in protic solvents or at low temperature [9]. We began to wonder whether the same type of situation as in the case of the sulfides caused the behavior in this system.

If indene (or substituted indenes) are photooxidized at -80 C in aprotic solvent, there are nearly quantitative yields of a very strange product (I) with 2 moles of oxygen added [9,10]. However, when indene is photooxidized in methanol, instead of the endoperoxide, a dioxetane (II) is formed [11].

Subsequently, we worked out the chemistry of these products; it is fairly clear that the initial product in aprotic solvents is an endoperoxide (III). This endoperoxide rearranges to diepoxide IV in what is almost certainly an electron transfer from excited dye, because it is very sensitizer dependent [12]. This rearrangement can also be carried out with other electron donors in model endoperoxide systems.

The less electron donating sensitizers allow the initial product to be trapped as a dienoperoxide (V) [13]. With other sensitizers, rearrangement occurs to give the diepoxide (IV), which is then trapped with a second mole of singlet oxygen to give I. We have been able to detect the initial endoperoxide III by carrying out the reaction to fairly low conversions at low temperature. Since this product is more reactive than the starting material, it never builds up to high concentration. By using Fourier transform NMR (both proton and

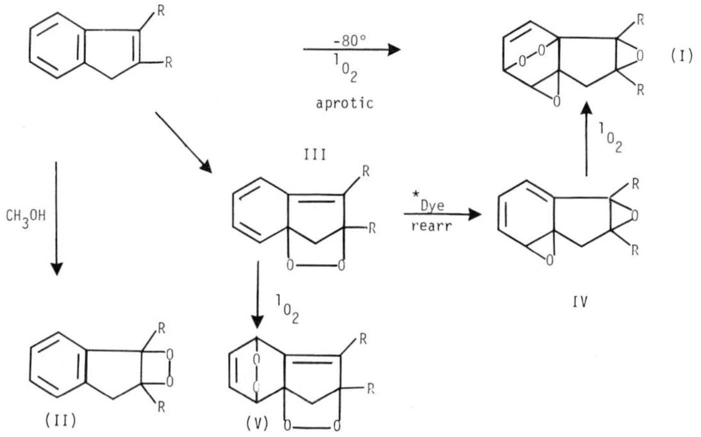

C-13 at low temperatures, we could see the formation of a small amount of product which has vinyl hydrogen resonances in the region expected for the monoendoperoxide. However, we can't be absolutely sure of the structure since we have no way of isolating this compound.

To summarize: Indene reacts slowly and inefficiently in aprotic solvents and it reacts faster at lower temperatures. The reaction is much faster in protic solvents but gives a different product.

Subsequently, we made some quantitative studies on the temperature effect with indenes; we determined the total interaction rate (the sum of the reaction, k_r, and quenching, k_q processes for 1-methyl-2-phenylindene (MPI) and 1,2-diphenylindene (DPI), and the total reaction rate with MPI. As in the case of the sulfides, the temperature dependence is due to quenching. In fact, the rate of removal of singlet oxygen is almost independent of temperature for both substrates, being very slightly slower at low temperature than at 25°C. However, the fraction of the reaction of MPI which leads to product, $\underline{k}_r/(\underline{k}_r + \underline{k}_q)$, goes up by a factor of 10 when the temperature is lowered to -50°C (Table 1).

Table 1. Effect of Temperature on Reaction of 1-Methyl-2-phenylindene (MPI) and 1,2-Diphenylindene (DPI)

T,°C	Substrate	$\underline{k}_r+\underline{k}_q$,$M^{-1}s^{-1}(x10^{-6})$	$\underline{k}_r/(\underline{k}_r+\underline{k}_q)$
25	MPI	1.5	0.11
-49	MPI	1.0	1.0*
25	DPI	8.4	
-25	DPI	8.2	

*Determined at -78°C

This situation is reminiscent of the sulfides; we suspect that the intermediate is simply thermolyzed at room temperature, regenerating starting material, since the bands characteristic of the endoperoxide disappear on warming. We have so little intermediate that we cannot be certain whether it regenerates starting material or produces some ill-defined product. It is likely that the initial Diels-Alder reaction from this compound is quite efficient, but simply undergoes retro Diels-Alder reaction at higher temperatures. If this is true, however, the yield of singlet oxygen cannot be quantitative in the reverse reaction, or it would not result in quenching.

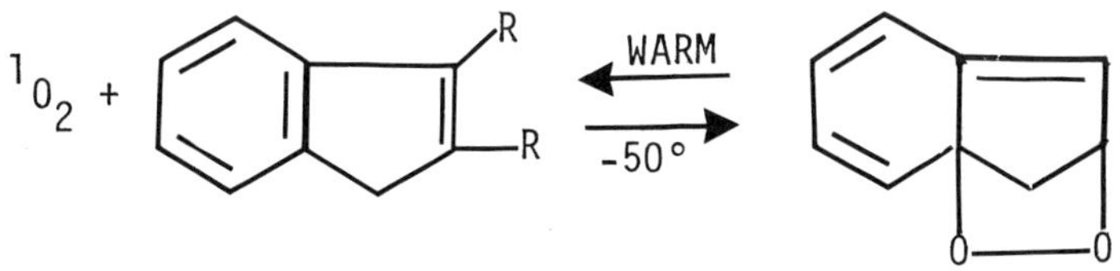

We have no evidence whatsoever for what is creating the solvent effect in the indene reaction. One rationale is that there is a charge-transfer complex, a zwitterion, or a perepoxide intermediate formed. Hydrogen bonding solvents divert the intermediate in one direction and aprotic solvents divert it in another; however, we have no evidence for this suggestion, nor is a good way of testing it apparent.

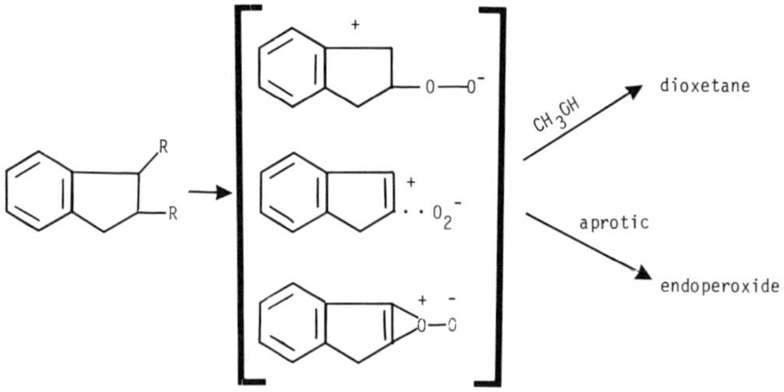

Dimethylhexadiene (DMHD) is another compound that has been of interest to us for some time. We studied this compound many years ago, and Hasty and Kearns published a communication on its reaction [14], which was reinterpreted by Gollnick and Kuhn [15]. This compound is a diene which cannot adopt a _s_-cis conformation. Upon photolysis, some endoperoxide is formed, but only in trace amounts. The major product is a dioxetane; the cleavage product may be isolated depending upon the conditions. There is also a large amount of the ene product, and in methanol there are products of methanol addition to an initial intermediate. We were of the opinion originally that the methoxyhydroperoxides had not been demonstrated to be singlet oxygen products. We thought that they might be formed from an electron transfer reaction via a radical cation intermediate. However, this turns out not to be the case: we carried out quenching studies which showed that all the products are quenched in parallel by singlet oxygen quenchers. It seems likely that all are actually derived from singlet oxygen.

OOH (ene) OOH OCH_3

O—O endoperoxide (traces) OOH OCH_3

O—O dioxetane methanol adducts

In this system, there is also a strong effect of solvent. In methanol, there is a large amount of dioxetane and a relatively small amount of ene product compared to the reaction in aprotic solvents. Recently, we developed a method of measuring rates of singlet oxygen reactions conveniently using direct detection of the luminescence of singlet oxygen [16]. The singlet oxygen luminesces very weakly, because it is almost all undergoing radiationless deactivation. Nevertheless it is not hard to detect the luminescence in a time-resolved mode using a laser to excite the dye and a germanium photodiode [16,17].

Figure 5 shows the intensity as a function of monochromator setting, demonstrating that the luminescence is the 1.27μ emission from singlet oxygen. Figure 6 shows the luminescence as a function of time, and Figure 7 shows the logarithm of the intensity as a function of time, demonstrating excellent first-order decay kinetics. This technique provides a very direct and useful method of measuring the rate constants of reactions involving singlet oxygen and gives the rate constant directly without interpretation.

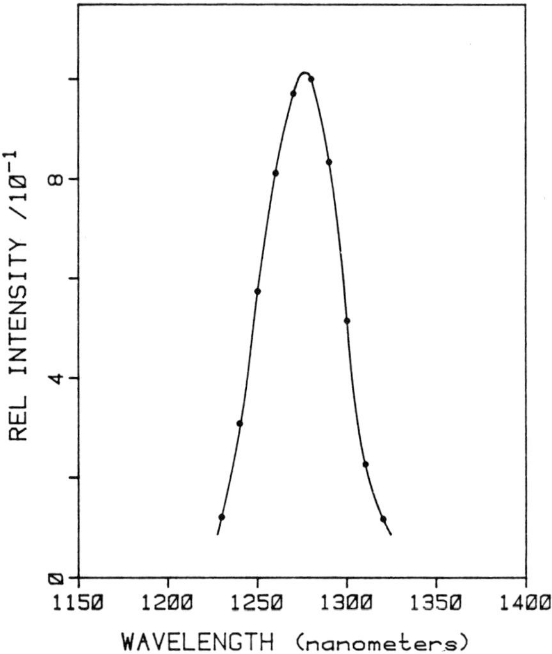

Figure 5. Intensity of $^1O_2$ luminescence as a function of wavelength.

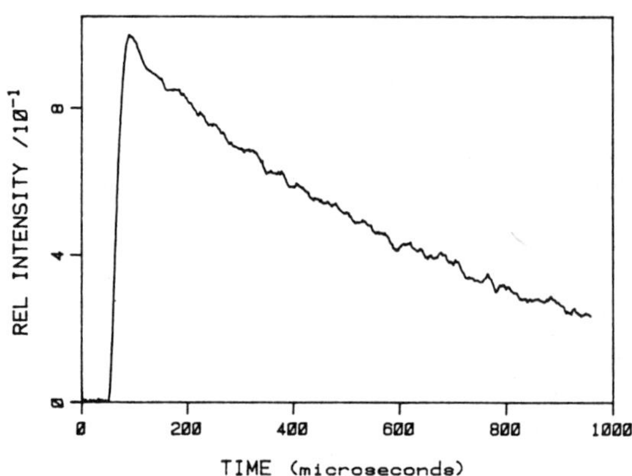

Figure 6. $^1O_2$ luminescence as a function of time.

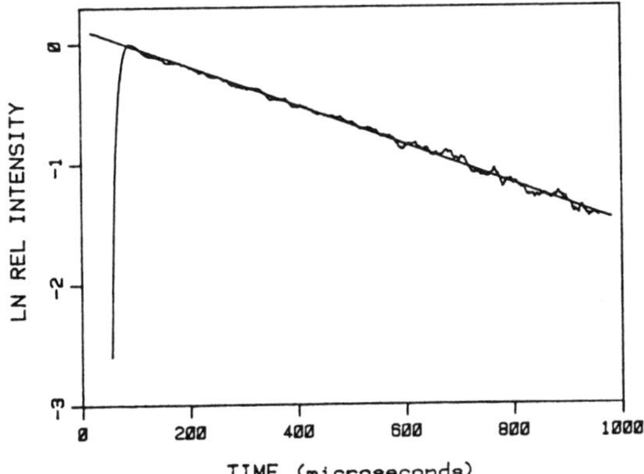

Figure 7. Ln(luminescence) as a function of time.

We measured reaction rates in the dimethylhexadiene system using this technique, hoping to be able to answer the question of whether dioxetane and ene product are formed by totally separate mechanisms with separate transition states or whether there is a common intermediate which solvent diverts to one course or the other.

Table 2. Rates of Interaction and Reaction of DMHD with $^1O_2$. All rates in $M^{-1}s^{-1}$ and x 10^{-6}.

Solvent	$k_r + K_q$	k_r^c	k_r^d
CH_3OH	2.8[a]	3.4	2.5
CH_3COCH_3	3.9[b]	0.38	0.69
CH_3CN	5.2[a], 6.3[b]	0.76	1.6
C_6H_6	3.6[b]	---	<0.25
CH_2Cl_2	5.2	0.60	1.4

a. Calculated from $k_d/(k_r + k_q)$
b. Determined directly, see Ref. [16].
c. Ref. [14]. Determined by competition with methylcyclohexene.
d. This work. Determined by competition with 2M2P, adjusted for solvent effects on 2M2P rates.

Table 2 shows the rates of interaction of DMHD in different solvents with singlet oxygen; the total rate of removal is $k_r + k_q$. The total reaction rate (k_r) was also measured by competition with methylcyclohexene [14] or 2-methyl-2-pentene.

In solvents as diverse as acetonitrile, methylene chloride, and acetone, there is no substantial change in overall rate of deactivation of singlet oxygen (reaction + quenching); however, in methanol, there is a large amount of dioxetane formed, although the rate of $^1O_2$ deactivation remains nearly constant (Table 3.). This is similar to the results in the indene case, where protic solvents seem to favor dioxetane.

Table 3. Products from DMHD in Different Solvents and k_r, Relative to Methylcyclohexene (MCH).

Solvent	Dioxetane/ene[14]	k_{rel}, MCH
CH_3CN	0.01	6.3
CH_2Cl_2	0.1	5.0
Acetone	0.2	3.2
CH_3OH	2.6	28.0

The ene process is comparatively independent of conditions, but the dioxetane formation and the quenching processes seem to be competing in some way; as one process increases, the other decreases. It is very hard to do good product analyses in this system, which makes quantitative kinetics difficult.

The analogy between a perepoxide and a persulfoxide occurred to us, and we wondered whether it would be possible to trap the intermediate in this reaction. However, using diphenyl sulfide and diphenyl sulfoxide at concentrations as high as 0.50 M, we were never able to intercept an intermediate. Although the analogy is appealing, the reactive intermediate must be too short-lived to trap.

The effects of temperature on the reactions of DMHD are shown in Table 4. The rate of the product formation ($\underline{k}_r$) and overall reaction ($\underline{k}_r + \underline{k}_q$) are both nearly independent of temperature in acetone. The dimethylhexadiene case thus differs from the sulfide and indene cases in that the fraction of quenching is nearly independent of temperature; the appealing analogy between the different systems begins to break down.

Table 4. Effect of Temperature on the Reaction of DMHD in Acetone. (All rate constants $\underline{M}^{-1}s^{-1} \times 10^{-6}$).

T, °C	$\underline{k}_r + \underline{k}_q$	$\underline{k}_r$
25	3.9	0.69
-78	3.1*	0.63

*Determined at -68°C.

We wondered whether the reaction with this s-trans diene was a special case because of the restricted conformation. However, the ratio of the total rates of reaction of dimethylhexadiene and 2-methyl-2-pentene is independent of solvent and temperature, which suggests that the initial interaction with singlet oxygen is very similar.

Another interesting observation is that the activation enthalpies of dimethylhexadiene and the two indenes are zero within experimental error; they may actually be slightly negative (Table 5).

Table 5. Activation Parameters

Olefin	ΔH, kcal	ΔS, e.u.
DMHD	-0.2	-29
MPI	-1	-34
DPI	-0.5	-34

Almost the only way to have a negative enthalpy of activation is to have a complex preceding the transition state for the reaction; such negative activation enthalpies have been observed for certain Diels-Alder [18] and carbene-olefin [19] reactions, for example, and have been interpreted as implying that a charge transfer-complex may be involved. In the case of the singlet oxygen reaction, it is quite likely that a charge-transfer complex is formed (technically, an exciplex). In all cases, the entropies of activation are large and negative, typical for singlet oxygen reactions.

To summarize the facts we have learned about dimethylhexadiene:

1. No activation enthalpy
2. No effect of temperature on quenching
3. No effect of temperature on product distribution in acetone
4. Ene reaction rate unaffected by solvent
5. Dioxetane increases, quenching decreases in methanol

This story is not yet complete; a tentative rationalization for these results is that there are two different processes: one leading to the ene product, and one which gives some intermediate "X"; these two paths are unaffected by temperature and solvent. However, X can partition to give either quenching or dioxetane depending on solvent. However, without detailed kinetics, this hypothesis cannot be tested thoroughly.

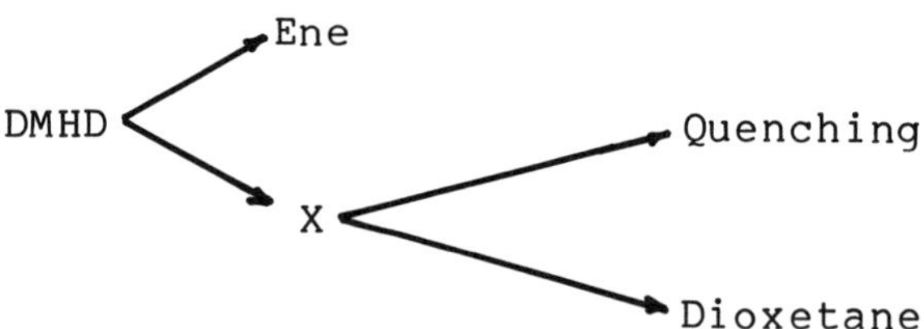

As for the structure of X, many possibilities come to mind; three are shown below: a charge-transfer complex, a perepoxide, and a zwitterion.

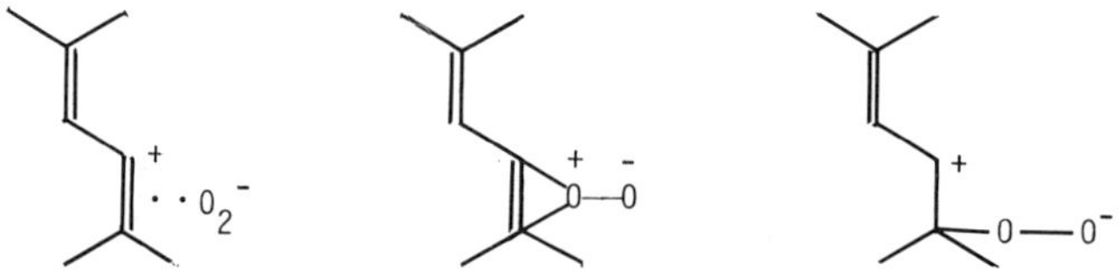

Of these, the first two may not really differ; indeed, they all probably represent a continuum, depending on solvent and structure. The zwitterion would be especially favored by resonance stabilization in both the indene and dimethylhexadiene cases and should be strongly influenced by hydrogen bonding solvents. The zwitterion formulation is fully compatible with the formation of methanol adducts in both the indene and dimethylhexadiene systems; in fact, the corrected structures of the methanol adducts from indenes strongly suggest a zwitterion intermediate [20]. Unfortunately, our inability to trap intermediates in these systems limits our present ability to characterize them.

Acknowledgement. This work has been supported for many years by NSF and NIH grants.

Discussion

Perrin: Is there any direct conversion to the sulfone at very low concentrations of sulfide? Why doesn't the three-membered ring open up directly?

Foote: There is a small pathway that goes directly to sulfone. This reaction is about as comnlex as it can be. It is very difficult to study that because it is inefficient and you get only a small amount. It seems to come from the initial intermediate, but it is hard to distinguish it from the much larger amount of sulfone that you get by trapping the intermediate. We have studied it at low conversions at the limit of GC detection, and almost certainly it is a direct process for the conversion of that intermediate to the sulfone.

Traylor: The three-membered ring, could it be dimeric?

Foote: It could be dimeric, but there is no concentration effect.

Traylor: It would have to be a photolysis rate effect.

Foote: We have done it under a wide variety of conditions, and we get a consistent set of kinetics. We have no special insight as to the nature of these intermediates. We know there are two and the structures we have written are the most logical.

Adam: Can you do other oxidations with those intermediates? For example, pyridine?

Foote: No. We probably haven't looked at this as hard as we should. We looked initially at electron rich and electron poor olefins. We are not able to detect any oxygen transfer. Of course the electron rich olefins are being oxidized by singlet oxygen directly. The electron poor ones provide only negative evidence.

Adam: I sense quite a bit of similarity between your intermediate and the dioxirane versus carbonyl oxide problem.

Foote: Yes. They should be quite similar. So far we have not detected any adducts, but we probably have not looked as hard as we should.

Adam: Can you oxidize pyridine to an N-oxide?

Foote: I would strongly doubt it, but we haven't tried.

Adam: Pyridine does distinguish between the carbonyl oxide and the dioxetane. Dioxetane oxidizes pyridine and the carbonyl oxide does not.

Foote: I wasn't aware of that. Who did it?

Adam: Griesbaum [JCS Chem. Comm., 920 (1966)] showed that carbonyl oxides, generated in ozonizations, do not oxidize pyridine into its N-oxide, while Edwards showed [J. Org. Chem. 46, 1681 (1981)] that dioxiranes, generated from acetone and caroate, afford pyridine N-oxide. We have repeated these studies and confirmed them, but it is not published yet.

Greene: In the reactions of the indenes, in some cases you had the phenyl-substituted indenes. In others did you have both a methyl in one position and a phenyl in the other?

Foote: We did a large number of different groups. Most of the kinetics have been with methylphenylindene which is more soluble at low temperatures.

Greene: What is the product of that? Are you not getting attack at the methyl?

Foote: No, essentially none. This kind of chemistry mentioned in the talk is almost all which is going on.

Herndon: Do you ever see a Diels-Alder reaction, a 1,4-addition, with any open chain diene?

Foote: Yes. Under the proper conditions; if the dienes can get *s-cis* they react well.

Houk: An interesting point about the diene scheme you wrote near the end is, of course, that in all of the deuterium and isotope effects your intermediate "X" comes before the ene reaction.

Foote: Yes, I know. Perhaps there is a "Y" back here and maybe "X" is something else. Possibly a zwitterion. This is a reasonable case to have a zwitterion.

Houk: I was thinking of placing a charge transfer complex where you have the zwitterion.

Foote: Yes that would be consistent with our other systems. There seems to be a constant proportion of "X" formed and something grabs it in methanol. It might be a zwitterion, or it might be a perepoxide which decides whether it goes back to oxygen or gives dioxetane. The "X" also probably is the intermediate which adds methanol.

Bartlett: Does singlet oxygen do anything interesting with disulfides?

Foote: Yes. Murray showed they are converted to the thiosulfinates. One of the sulfur atoms is oxidized to an S-O and the other is inert.

Bartlett: The second oxygen disappears again?

Foote: Yes, exactly the same as in the sulfide cases. The intermediate oxidizes another mole of sulfide. The kinetics look very similar to the sulfides. Oae has shown that these intermediates are useful in distinguishing between electrophilic oxidants and nucleophilic oxidants. We are going back to try one of these thiolsulfinates to see which one of the sulfurs gets oxidized. Nucleophiles attack the sulfur that already has an oxygen. Electrophilic oxidants attack the one that doesn't have an oxygen.

Perrin: Can I get something straight with respect to the sulfoxide versus the sulfide? Are they completely exclusive? One attacks only after the first intermediate and the other attacks only after the second?

Foote: Unfortunately, we only have the steady-state kinetics. You can write other schemes that will satisfy them, but they all have to have two different intermediates. Secondly, since the quenching is inhibited by sulfoxide only and not by sulfides, we think the intermediate that reacts with sulfoxide comes first.

Perrin: But does the second intermediate react only with sulfide? Does it exclude the possibility of reacting with sulfoxide?

Foote: It probably would react with sulfoxide under higher concentrations. The sulfide will react with the first species, but it has to be at much higher concentrations to do it because the intermediate is so much more reactive with sulfoxide.

Greene: What do you see of a dioxetane when you use a monoene? When you look at the solvent dependence of singlet oxygen, are you seeing anything?

Foote: We thought we would try some simple enes and we went to perfluoroisopropyl alcohol to see if we could make them give dioxetanes, but we could never get a trace, even at the 1% level. In fact, the simple olefins which give dioxetane are limited to unusual ones like adamantylideneadamantene, bisnorbornylidene and bisfluorenylidene and one or two others. Otherwise, you need electron rich olefins, which suggested to us that we had a charge transfer complex. I think a zwitterion is equally compatible, although we were never able to trap one. We looked at the enamine case and used many zwitterion traps and we never trapped them.

Traylor: Did you say that the oxygen from the sulfur complex goes back into solution?

Foote: Yes, the quenching clearly regenerates oxygen. We have not tried to detect that by labeling experiments.

References

[1] G.O. Schenck and C.H. Krauch, Chem. Ber. 96, 517 (1963); K. Gollnick, Advan. Photochem. 6, 1 (1968).
[2] C.S. Foote and J.W. Peters, IUPAC Special Lectures 4, 129 (1971).
[3] C.S. Foote and J.W. Peters, J. Am. Chem. Soc. 93, 3795 (1971).
[4] C.-L. Gu and C.S. Foote, J. Am. Chem. Soc. (1982), in press.
[5] C.-L. Gu, C.S. Foote, and M.L. Kacher, J. Am. Chem. Soc. 103, 5949 (1981).
[6] W. Ando, Y. Kabe, and H. Miyazaki, Photochem. Photobiol. 31, 191 (1980).
[7] Y. Sawaki and Y. Ogata, J. Am. Chem. Soc. 103, 5947 (1981).
[8] J.O. Edwards, R.H. Pater, R. Curci, and F. Difuria, Photochem. Photbiol. 30, 63 (1979).
[9] C.S. Foote, S. Mazur, P.A. Burns, and D. Lerdal, J. Am. Chem. Soc. 95, 586 (1973).
[10] P.A. Burns, C.S. Foote, and S. Mazur, J. Org. Chem. 41 899 (1976).
[11] P.A. Burns and C.S. Foote, J. Am. Chem. Soc. 95. 4339 (1974).
[12] J.D. Boyd, C.S. Foote, and D.K. Imagawa, J. Am. Chem. Soc. 102, 3641 (1980).
[13] J.D. Boyd and C.S. Foote, J. Am. Chem. Soc. 101, 6758 (1979).

[14] N.M. Hasty and D.R. Kearns, J. Am. Chem. Soc. 95, 3380 (1973).
[15] K. Gollnick and H.J. Kuhn in "Singlet Oyxgen", H.H. Wasserman, R.W. Murray, eds. Academic Press, Inc., New York (1979) Chapter 3.
[16] P.R. Ogilby and C.S. Foote, J. Am. Chem. Soc. 104, 2069 (1982).
[17] J.R. Hurst, J.D. Macdonald, and G.B. Schuster, J. Am. Chem. Soc. 104, 2065 (1982).
[18] V.D. Kiselev and J.G. Miller, J. Am. Chem. Soc. 97, 4036 (1975).
[19] N.J. Turro, G.F. Lehr, J.A. Butcher, R.A. Moss, and W. Guo, J. Am. Chem. Soc. 104, 1754 (1982).
[20] T. Hatsui and H. Takeshita, Bull. Chem. Soc. (Japan) 53, 2655 (1980).

11. CONFORMATIONAL MOBILITY AND PHOTOCHEMISTRY

Peter J. Wagner

Department of Chemistry, Michigan State University
East Lansing, Michigan 48824

Many excited-state reactions occur at least as rapidly as processes whereby molecules change their shapes. This coincidence of rates for quite different competing processes produces effects which can be considered from several viewpoints. One can consider how lack of conformational mobility might limit the photoreactions of certain classes of compounds. Alternatively, one can consider how photoreactions might be studied to monitor rates and efficiencies of conformational change. Finally, one can consider how knowledge of conformational restrictions might provide information about orientational requirements for interaction of two groups. This talk will explore this approach for simple electron transfer reactions.

Most intermolecular reactions, which necessarily involve two distinct functional groups, have intramolecular counterparts. Therefore there should be analogies between the kinetics of bimolecular and unimolecular bifunctional reactions. Conformational effects are of primary importance in intramolecular reactions, wherein one part of a molecule might be close enough to another part for the two to interact.

Scheme 1 shows the parallel between diffusion together of two molecules and rotation together (by whatever combination of bond rotations is required) of two functional groups in the same molecule. Two boundary conditions develop in each case, corresponding to reaction of the two moieties A and B being either much slower or much faster than their movement apart.

The necessity for diffusion together of two

Scheme 1. Bifunctional reactions

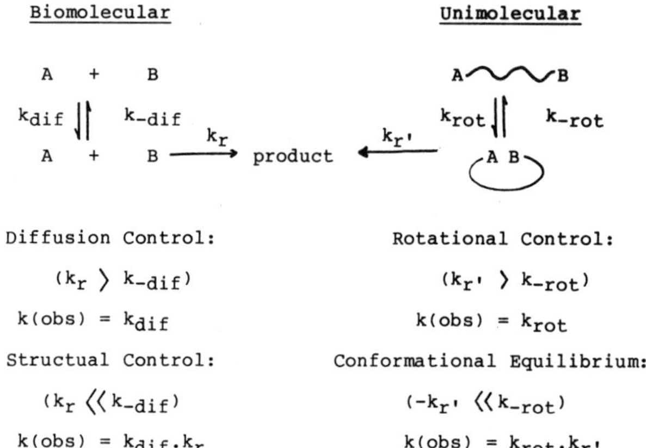

slowly reacting molecules usually can be tacitly ignored. The observed rate constant can be correlated directly with the structures of the reactants. However, the observed rate constant for a "slow" intramolecular reaction also includes a conformational equilibrium constant (k_{rot}/k_{-rot}) which may vary substantially with structure [1].

Diffusion-controlled bimolecular reactions, especially of reactive intermediates, represent a well-known phenomenon. Rotationally controlled reactions represent the intramolecular equivalent and should be expected whenever the bimolecular reaction between A and B is known to be diffusion-controlled. The phenomenon is restricted to highly energetic molecules, such as those formed by rapid absorption of light.

Scheme 2 presents the situation for excited-state reactions. F and U represent conformations which are favorable and unfavorable, respectively, for a given intramolecular interaction. Three separate boundary conditions arise, depending on how rates of excited-state conformational change compete with chemical and physical decay rates of the excited state.

Scheme 2. Competition between triplet reaction and bond rotation

$$\begin{array}{ccccc} F & \underset{k_d}{\overset{h\nu}{\rightleftharpoons}} & F^* & \xrightarrow{k_r} & \text{Biradical} \\ k_{UF}\uparrow\downarrow k_{FU} & & k_{UF^*}\uparrow\downarrow k_{FU^*} & & \downarrow P \\ U & \underset{k_{d'}}{\overset{h\nu}{\rightleftharpoons}} & U^* & & \text{Product} \end{array}$$

Conformational Equilibrium: $(k_{FU^*}, k_{UF^*} \gg k_r, k_d)$

$$k(obs) = k_r k_{UF^*}/(k_{FU^*} + k_{UF^*}) = X_{F^*} k_r$$

$$\Phi = \frac{k_r X_{F^*} P}{k_r X_{F^*} + k_d X_{F^*} + k_{d'} X_{U^*}}$$

Ground State Control: $(k_{FU^*}, k_{UF^*} \ll k_r, k_d)$

$$k(obs) = k_r$$

$$\Phi = k_R X_F P/(k_R + k_d)$$

Rotational Control: $(k_{UF^*}\ k_d : k_{FU^*} \ll k_r)$

$$k_1(obs) = k_r \qquad k_2(obs) = k_{UF^*}$$

$$\Phi = \frac{k_R X_F P}{k_R + k_d} + \frac{k_{UF} X_{U^*} P}{k_{UF^*} + k_{d'}}$$

With relatively slow excited-state reactions, conformational equilibrium is established before reaction. Observed rate constants include excited-state conformational equilibrium constants. There is only one kinetically distinct excited state even if two different conformations undergo two different reactions.

With fast excited-state reactions and/or slow conformational change, the quantum efficiency for the intramolecular reaction of interest is limited by the ground-state population of favorable conformations. There are two kinetically distinct excited states, only one of which (F*) leads to the reaction of interest. This situation is called "ground state control" [2] and applies only to reactions induced by instantaneous insertion of energy.

Finally, if conformational change and decay are competitive, rotation-controlled reactions can occur. Note that again there are two kinetically distinct excited states and that the reaction of interest arises from both but determines the lifetime of only one.

It is worthwhile to consider explicitly how observed rate constants and quantum yields depend on actual reaction rate constants k_r under the different boundary conditions. Comparing first the two extreme situations, we see that k(obs) equals k_r under conditions of ground-state control, while the maximum quantum yield (ie, when $k_r >> k_d$) equals X(gs). Under equilibrium conditions k(obs) is only a fraction of k_r, while the maximum quantum yield can be unity. The two situations can be distinguished by measuring both Φ and k(obs) while varying k_r in a known fashion.

Rotational control manifests itself dramatically, at least when ($k_r + k_d$) is measurably different from ($k_{UF}* + k_{d'}$), as two kinetically distinct decay lifetimes and thus two different k(obs) values, one the ground-state control component, the other the rate of conformational change. The rotationally controlled k(obs) is totally independent of k_r as long as k_r is sufficiently fast that $k_{FU}*$ does not become competitive.

In this presentation I will discuss conmationally sensitive photoreactions of ketones. The choice is not arbitrary, since ketones are the only compounds whose photochemistry has been so fully studied and which occur in such structural variety that examples of all three kinetics boundary conditions have been firmly established.

Triplet excited ketones undergo only three common bifunctional reactions: hydrogen atom abstraction, charge transfer (leading to exciplexes or radical-ion pairs), and electronic energy transfer [3]. Each of these processes can occur internally in suitable bifunctional molecules.

Of these three reactions, hydrogen abstraction is the best understood geometrically: the carbonyl oxygen must approach within O-H bonding distance of the C-bonded H atom. Since the other two processes do not involve any bonding-rebonding, they can occur over variable distances. Therefore I will first discuss some hydrogen abstraction reactions which exemplify the various kinetic boundary conditions for intramolecular reactions and then will return to the question of orientations in n-π interactions.

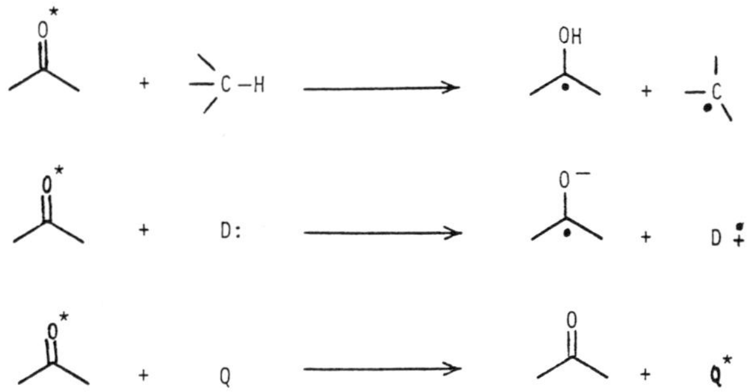

Conformational Equilibrium

Rate constants for bimolecular hydrogen atom abstraction by triplet ketones are much smaller than diffusion-controlled ones [4]. Consequently ketones with any conformational mobility are expected to attain conformational equilibrium before their excited states undergo intramolecular reaction.

Triplet ketones obey the general rule that, in flexible acyclic systems, 1,5-hydrogen atom transfers are the most rapid [5]. For example, γ-hydrogen abstraction (which leads to Norrish type II reactions [6]) is some 20 times faster than the equivalent δ-hydrogen abstraction in acyclic ketones [7]. The transition states for the two processes are cyclic, containing six or seven atoms. The latter geometry is disfavored both statistically and energetically. The preference for γ-hydrogen abstraction reflects the unique ability of a six-membered cyclic transition state to achieve a strain-free chair-like geometry [7].

$$k_\gamma(\text{obs}) = X_F \cdot k_\gamma^0 \qquad (1)$$

The observed rate constant for a given γ-hydrogen abstraction contains the factor X_F which describes the equilibrium fractional population of favorable conformations, ones from which γ-hydrogen abstraction can take place with the intrinsic rate constant k^o.

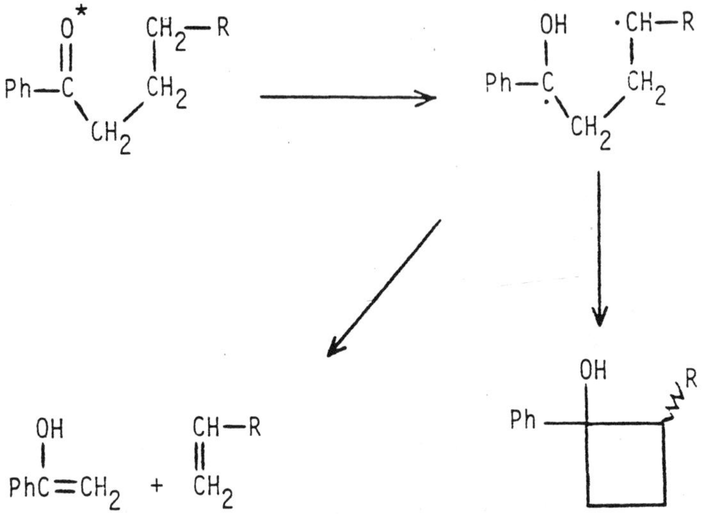

Benzoylcyclobutanes undergo type II reactions in low quantum efficiency and display very long triplet lifetimes [8]. This situation was originally ascribed to a very low equilibrium population of the pseudoaxial conformer, with the pseudoequatorial conformer having no accessible γ-hydrogens. Triplet exo-5-benzoylbicyclo[2.1.1]hexane, an appropriate model for the reactive conformer of benzocyclobutane, reacts with a rate constant of 5 x 10^9, sec^{-1} [9], as expected for a molecule fixed into a favorable conformation [10].

Alexander and Uliana clarified the situation by studying several benzoylcyclobutanes with ring substituents known to change triplet reactivity [11]. Table 1 lists their results which show a linear relationship between quantum yield and intrinsic triplet reactivity. This finding eliminates the possibility of ground-state control, in which case quantum yields would have been independent of k_H (Scheme 2). It is unlikely that n,π* excitation of the pendant benzoyl group affects the ring puckering significantly, so that the derived value also applies to the ground state. Thus study of a photo-

U $\underset{k_{FU}}{\overset{k_{UF}}{\rightleftharpoons}}$ F

reaction has provided a measure of a conformational equilibrium constant too small to determine by more classic techniques. Other examples will be provided later.

Table 1. Comparison of Quantum Yield with Excited-State Reactivity

X	k_r ($R=C_4H_9$)	Φ (R-$\underline{C}$-C_4H_7)
CF_3	28×10^7	0.09
F	15	0.05
H	12	0.04
CH_3	1	0.001
OCH_3	0.05	0.0001

Ground-State Control

Several benzoylcyclohexane derivatives have provided the most clear-cut examples of ground-state control for any photoreaction. Lewis provided the first example in the behavior of 1-benzoyl-1-methyl cyclohexane [12]. Alpha cleavage and γ-hydrogen abstraction are competing triplet reactions, but they occur from kinetically distinct triplets. The classic steady state experimental evidence for such a situation is that the two different products are quenched with distinctly different efficiencies. The measured rate constant for α cleavage is 2 x 10^9 sec^{-1} and that for γ-hydrogen abstraction is 6 x 10^8 sec^{-1}.

O* Ph X 10^5 X O* Ph $k=2x10^7 s^{-1}$ $6x10^8$ (CH_2) $2x10^{10}$ (N-Me) O PhC· + X X OH Ph

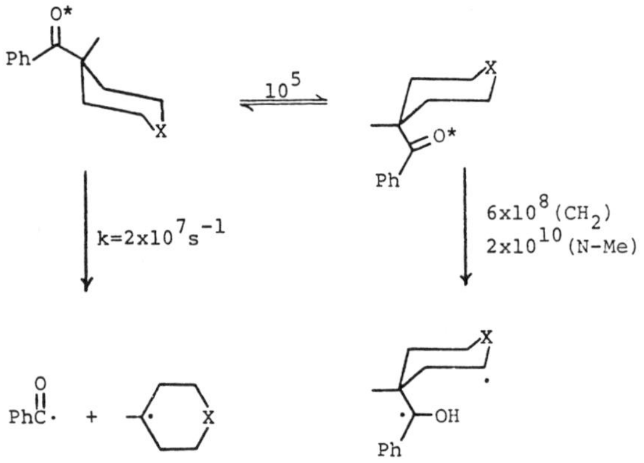

We later observed the same general phenomenon in 4-benzoyl-4-methylpiperidines [13]. The presence of the amine function complicates the photochemistry because the radicals produced by α cleavage attack both reactant and cyclic product. The chemical yields observed under various conditions are given below. The presence of sufficient thiol to trap all benzoyl and alkyl radicals allows for a material balance near 100%. Benzaldehyde, the α-cleavage product, represents some 75% of the total products, the same fraction observed for the benzoylcyclohexane. In the presence of sufficient triplet quencher, the only product is that corresponding to type II cyclization. Stern-Volmer plots are similar to those for the cyclohexane, except that cyclic alcohol formation is barely quenchable.

hν	PhCHO	+
benzene	12%	12%
" + 0.4M RSH	72%	27%
" + 1M $C_{10}H_8$	0	100%

As Lewis has noted for benzoylcyclohexane, the two triplet reactions both proceed much faster than the $\sim 10^5$ sec^{-1} ring inversion of cyclohexane. This inequality must also hold for the piperidine. The conformer with benzoyl axial can undergo internal hydrogen abstraction rapidly; its epimer cannot and, therefore, undergoes the slower α cleavage. The two ground-state conformations are in rapid equilibrium, with some 75% of the molecules having an equatorial benzoyl group. Upon excitation, that 75% cleave into radicals before any ring inversion can occur. This readily quenchable cleavage reaction can be totally suppressed by modest concentrations of quencher, so that only the conformer with axial benzoyl reacts. Hydrogen abstraction by this conformer is much faster in piperidine than in the corresponding cyclohexane. The rate enhancement reflects a striking stereoelectronic effect, the lone pair on the nitrogen being held perfectly trans-periplanar to the axial γ C-H bonds.

Our conclusions provided a nice further example of Lewis' discovery. However, our motive in studying the piperidines was to determine the effect of the nitrogen on the triplet lifetimes. The longer lived conformation has the same 50-nsec triplet lifetime whether or not the nitrogen is present. This observation indicates that any intramolecular quenching of the triplet carbonyl by the remote amine function occurs at a rate significantly slower than 10^7 sec^{-1}. The importance of this conclusion will be discussed below.

Rotationally Controlled Hydrogen Abstraction

The most clear-cut example of rotation control in a photochemical reaction is provided by photoenolization of ortho-alkylphenyl ketones [14,15]. This basic photochromic system remained mechanistically confusing until it was realized that two kinetically and conformationally distinct triplets were involved [14]. Sensitization studies revealed that ketones such as ortho-methylacetophenone and ortho-methylbenzophenone produce two distinct triplets, one with a subnanosecond lifetime and another with a 30-nsec lifetime (in benzene). Recent flash kinetic work has verified these conclusions [16]. That these different triplets correspond to syn and anti conformers was indicated by the behavior of 8-methyl-1-tetralone, which is locked into a syn conformation and which displays only a nanosecond triplet [14].

Enolization of the syn triplet corresponds to γ-hydrogen abstraction from a geometrically perfect conformation, with $X_F \sim 1$. That the decay of the longer lived triplet corresponds to anti → syn rotation was deduced from the structure independence of the rate constant. Thus valerophenones with ortho-CH_3, CD_3, and CH_2CH_3 groups all undergo competitive type II elimination and enolization, the latter occurring at the same rate in all three, as indicated in Table 2. The rate-determining step obviously does not involve H-atom abstraction, in contrast to the competing γ-hydrogen abstraction where rate constants vary with bond energy. Likewise, additional ring substituents, which slow down hydrogen abstraction by stabilizing the ketone π,π^* triplet, do not affect the rate of enolization of the longer lived triplet. Stern-Volmer quenching studies indicate that enolization of ortho-methylacetophenone and ortho-methylbenzophenone occurs from both long-lived and short-lived triplets [15]. Therefore, since the longer lived triplets do undergo hydrogen transfer, but not in the rate-determining step, the structure independent rate-determining step for enolization is deduced to be bond rotation.

It is worth emphasizing that the bond rotation

Table 2. Photoreactivity of X-$PhCOCH_2CH_2R$ in Benzene

X	R	Φ_{II}^{max}	$k_{\gamma-H}$-$10^7 s^{-1}$	-k_{enol}
o-CH_3	CH_3	0.026	0.4	3.0
o-CH_3	CH_2CH_3	0.10	2.7	2.9
o-CH_3	$CH(CH_3)_2$	0.16	14	4.0
o-$CH_2\dot{C}H_3$	CH_2CH_3	0.07	3.0	2.3
2,4-$(CH_3)_2$	$CH(CH_3)_2$	0.08	1.0	2.6
2,4-$(CD_3)_2$	$CH(CH_3)_2$	0.12	2.0	2.7

rate constant of 3 x 10^9 sec^{-1} deduced from these studies is slower than in the ground state because of the additional conjugation between phenyl ring and carbonyl group found in the excited state.

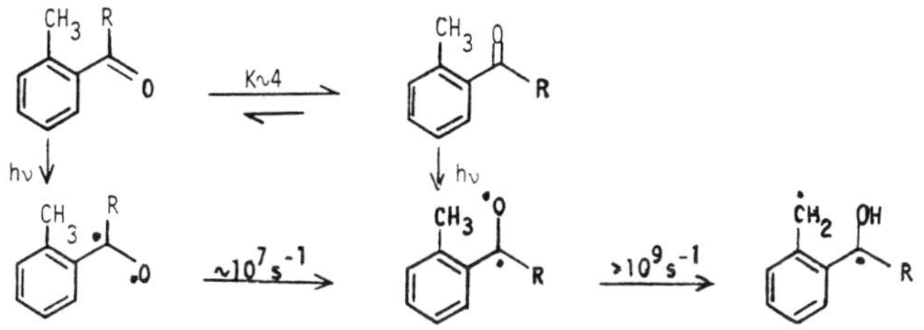

Since these ortho-alkyl ketones are subject to ground-state control, the fraction of the absorbed light which produces the longer lived triplet represents an otherwise inaccessible measure of the ground-state equilibration between syn and anti conformers. Sensitization and flash studies both indicate that 10-35% of the absorbed light produces the longer lived triplet, the exact fraction depending on structure.

Remote Hydrogen Abstraction

Efficient intramolecular hydrogen abstraction can occur from positions other than γ when there are no γ hydrogens or when conformational rigidity prevents access to γ hydrogens. The latter situation has not received much study but has been exploited spectacularly by Paquette in his synthesis of dodecahedrane [17].

Table 3. δ-H Abstraction from ortho-Alkoxy Ketones.

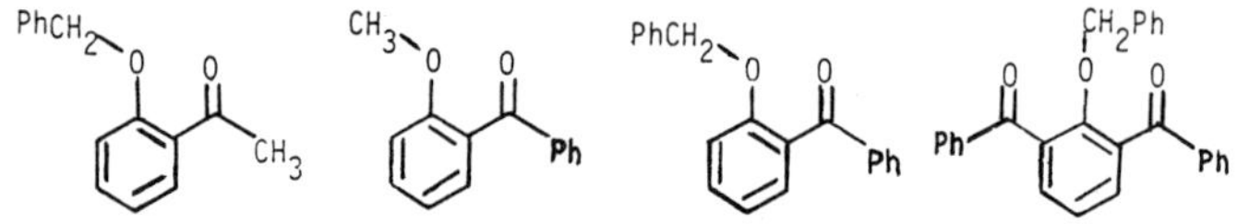

Φ	0.10	0.40	0.95	0.78
k_H	0.02	0.13	1.8	20 x 10^7

We have recently measured conformational equilibrium constants for a case of δ hydrogen abstraction [18]. Irradiation of several ortho-alkoxyphenyl ketones is known to produce benzohydrofuranols, presumably by cyclization of 1,5-diradicals [19]. Table 3 compares the behavior of several such ketones. The measured triplet lifetimes and high quantum yields indicate that k(obs) for δ-hydrogen abstraction is proportional to the intrinsic k_H values, the methoxybenzophenone being less reactive than the benzyloxybenzophenone, and the acetophenone (with a π,π* lowest triplet) being even less reactive. Therefore the triplet states achieve dynamic conformational equilibrium before decaying. In the dibenzoyl ether, there is a carbonyl group near the benzyl C-H bonds no matter which way the alkoxy group is twisted. Its enhanced triplet reactivity confirms that the unreactive anti rotamer is normally favored. Since we have shown independently that meta-acyl substitution barely affects triplet ketone reactivity [20], the 10-fold rate enhancement afforded by the extra benzoyl group suggests an equilibrium constant of 0.10 for rotation of the alkoxy group about the phenyl-O bond.

It is noteworthy that syn-anti isomerism about the phenyl-carbonyl bond is not important here as it is for the _ortho_-alkylketones. Anti → syn rotation still presumably occurs with a rate on the order of $3 \times 10^7 sec^{-1}$ in the triplet, but the effective rate of hydrogen abstraction is barely competitive with rotation back to anti. Combined rotational preferences provide too small a fraction of discrete syn-syn or anti-anti conformers for observation.

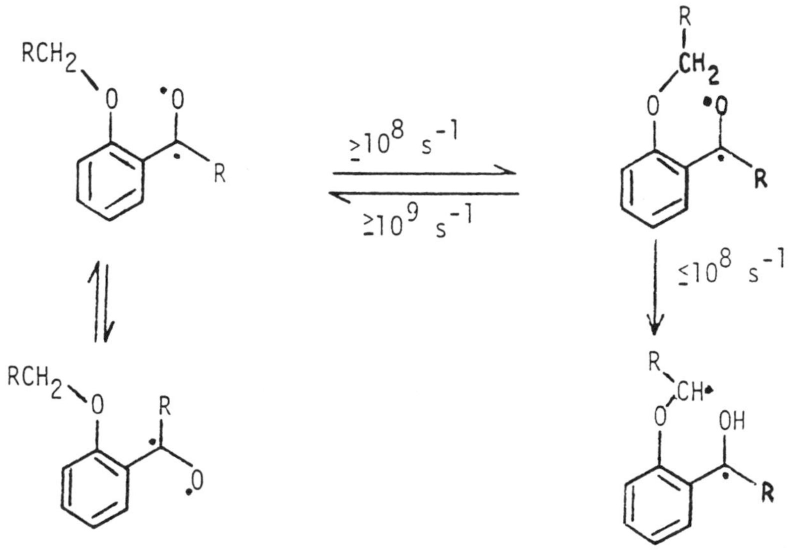

RCH2 O •O R ≥10^8 s^-1 ≥10^9 s^-1 R CH2 O •O R ≤10^8 s^-1 RCH2 O R •O R CH• O OH R

Intramolecular Charge-Transfer Interactions

A wide variety of intramolecular charge- or electron-transfer processes has been studied, some for the express purpose of deducing orientational requirements for such processes. No clear picture has emerged of the relative orientational requirements of donor and acceptor during electron- or charge-transfer interactions. Many of the systems studied have involved donors and acceptors with extended π-systems; in such cases sufficient orbital overlap for rapid reaction apparently is possible over a large range of orientations and distances. Therefore we have purposely studied some n-orbital systems where overlap is more limited.

Photochemistry of PHCO(CH2)$_n$X

Several laboratories have looked at intramolecular interactions in α-substituted ω-benzoylalkanes. Table 4 summarizes our results for X = thioalkoxy [21], dimethylamino [22] and vinyl [23]. The primary interest in such studies is observing how the rate constant for internal quenching of triplet ketone varies with n and with the nature of X.

Table 4. Rate Constants (10^7 sec^{-1}) and Effective Molarities for Internal Quenching in Triplet $PhCO(CH_2)_nX$ in benzene at 25°

n/X	SBu	NMe_2	$CH{=}CH_2$
1	130(3)	<10(-)	-
2	460(9)	480(1.6)	20(100)
3	240(5)	740(2.8)	<10(<10)
4	14(0.5)	50(0.2)	-
5	<2(-)	20(0.1)	-
k_2 [a]	30	300	0.8

Numbers in parentheses are effective molarities.
[a] corresponding bimolecular rate constant, $10^7 M^{-1}s^{-1}$.

Both sulfides and tertiary amines quench triplet ketones by charge-transfer (CT) interactions [24, 25]. The amines display 10-fold larger bimolecular quenching rate constants because of their lower oxidation potentials [26,27]. The simplest picture of such CT quenching involves the formation of an exciplex with significant overlap of the donor HOMO with the acceptor LUMO [26-29]. In these particular cases, both orbitals are highly directed and highly localized n orbitals: the lone pair on S or N acts as donor, the half-vacant carbonyl n orbital in the n,π* benzoyl triplet as acceptor. Since such overlap resembles σ-bond formation, the molecule must attain a conformation something like that of a cycloalkane for CT quenching to occur. How exact the resemblance is will depend on how closely the donor and acceptor orbitals must approach each other.

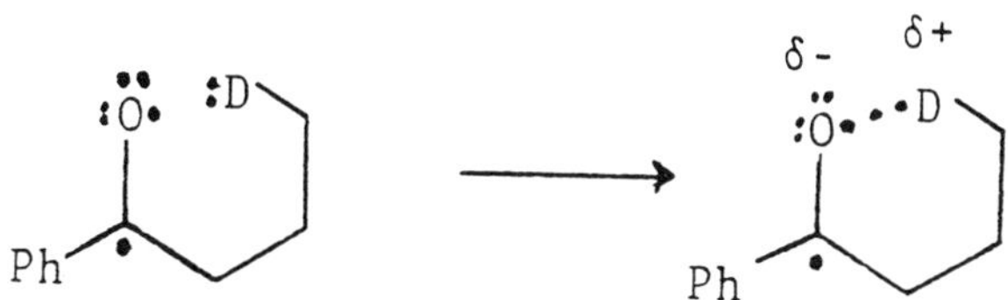

The number of atoms in these cyclic interactions equals (n+3). Therefore it is significant that both donors display the fastest quenching when n = 2 or 3, reflecting the expected ease of formation of 5- and 6-membered rings. As n gets larger, k decreases dramatically, reflecting the expected decreased probability of forming medium-sized rings. In both cases, k is also lower when n = 2, as expected for 4-membered ring formation, but there is no disappearance of reactivity as noted for β-hydrogen abstraction. It is interesting that the n=3/n=4 rate ratios are the same as that for γ/δ-hydrogen abstraction. The similarity suggests that entropic effects dominate for both processes when n > 3.

The fact that the kinetics so accurately parallel known ground-state conformational preferences says two important things about these CT interactions. The simple model about HOMO and LUMO overlap must be substantially correct; and the two n orbitals must overlap enough that the system actually feels some of the strain of the corresponding cycloalkane. The latter conclusion can be deduced independently from the steric effects observed in CT quenching [13,27,29]. Of course, orbital size dictates the distance required for effective overlap. In this regard, the differences between RS and R_3N are quite significant.

The thiyl group is actually a more effective quencher than the amino group, especially when attached α to the carbonyl, where RS is a rapid quencher despite the four-atom ring involved. This difference is probably due to the longer "reach" of the sulfur's 3p lone pair compared to the nitrogen's sp^3 lone-pair orbital. In fact, k(intra)/-k(inter) for SR remains larger than for NMe_2 for n = 2-4. This ratio represents the "effective molarity" [30] of a quencher and is known to vary with the geometry of the interaction [31], in this case increasing with the spatial reach of the donor orbital. The enhanced internal reactivity of RS seems to fall with increasing n, suggesting ever increasing importance of entropic effects. Unfortunately, we could not measure k values for SR when n > 5 because the competing Norrish type II reaction used to monitor triplet decay is too efficient.

Ph (N) vs. Ph (S)

$2p\text{-}sp^3$
poor overlap

2p-3p
good overlap

When X is vinyl, overlap with π orbitals becomes important in quenching. The interaction again is CT in nature, alkenes lying on the same linear free energy plot which correlates quenching rate constants of amines and sulfides with their oxidation potentials [37]. Bimolecular quenching is not intrinsically rapid for 1-alkenes [27]. The only case where intramolecular quenching is rapid is for β-vinylketones. There is a sharp contrast between the behavior of the n donors and of vinyl when n changes from 2 to 3.

The unique internal quenching ability of vinyl groups when attached β is strikingly similar to the relative efficiencies of remote double-bond participation in ester solvolyses, where the 5-hexenyl system shows much greater participation than either the 4-pentenyl [33] or the 6-heptenyl [34]. Just as for double-bond participation in solvolysis [35], double-bond quenching of triplet ketones is accelerated multiplicatively by each additional alkyl group on the double bond [23,27,36]. Both

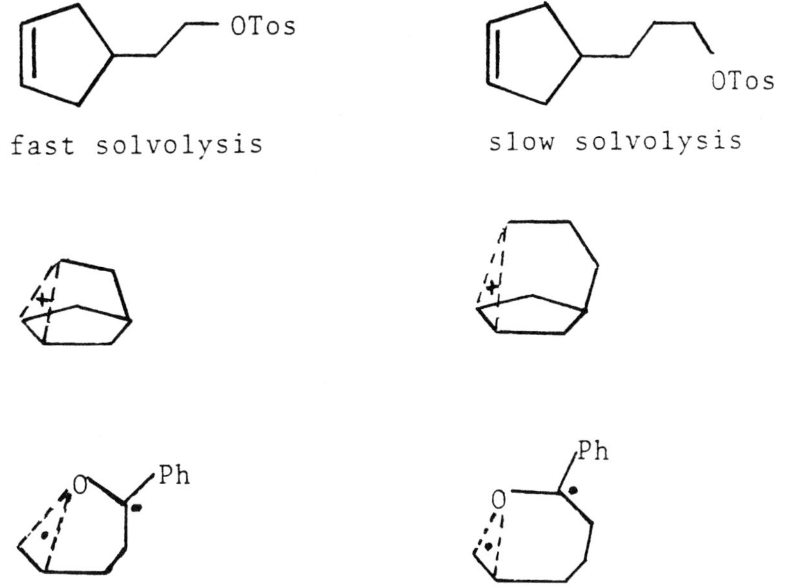

processes involve fairly symmetric overlap of the electron-deficient reaction center with the π orbital of the double bond rather than bonding to one end. Therefore the CT complex responsible for intramolecular triplet quenching must assume a sort of bicyclic geometry, the [3.1.0] structure being easier to form than the [4.1.0], just as for bridged carbocations. The strong parallel with known ground-state behavior again indicates the closeness required between donor and acceptor centers for effective quenching.

Regioelectronic Control of Charge-Transfer Quenching

If the half-empty n orbitals in n,π* triplets are indeed as localized as the above considerations suggest, and if substantial HOMO-LUMO overlap is required for charge-transfer quenching, then internal CT quenching may not occur in molecules which cannot assume any conformation suitable for charge transfer interaction (ie, X_F = 0). The 4-benzoyl-piperidine discussed above [13] is one of only a few such cases. Such selectivity may be called "regioelectronic" in analogy to the accepted usage of the terms "regioselective," "stereoselective," and "stereoelectronic".

We recently reported a striking contrast in the regioselectivity of internal charge-transfer quenching in some aminoketones depending on whether the ketone possesses an n,π* or a π,π* lowest triplet [37]. When R = phenyl, the aminoketone has an n,π* triplet and shows long-lived phosphorescence comparable to that of the model methyl ester. However, when R = alkyl, the lowest triplet is π,π* and internal quenching is very rapid, $\underline{k} > 5 \times 10^8$ sec^{-1}. Since the half-empty HOMO of the excited acceptor is a π orbital heavily localized on the benzene ring, overlap is possible with the donor lone-pair orbital on nitrogen. In the benzophenone derivative, $\underline{k} \leq 10^5$ sec^{-1}; the nitrogen lone pair cannot reach the carbonyl for any kind of effective overlap. The situation here is comparable to that found by Winnik for ω-unsaturated esters of benzoylbenzoic acid [38].

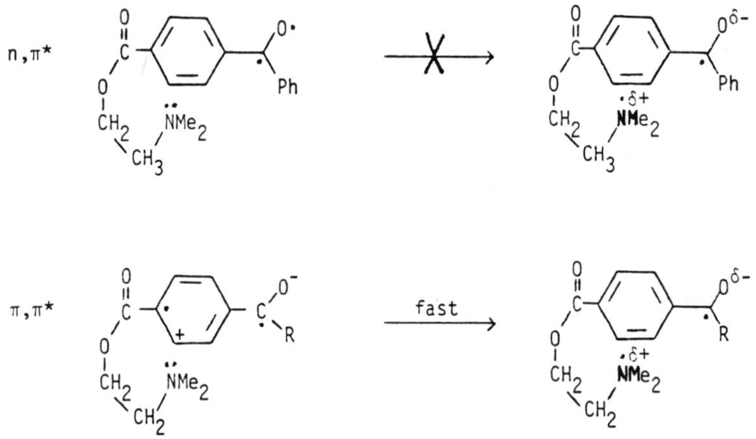

Concluding Remarks

Intramolecular hydrogen abstraction by triplet excited ketones has provided examples of all the limiting conditions involving competitive excited-state reaction and bond rotation. Application of the conformational principles so learned to intramolecular charge-transfer processes has provided a clearer picture of the extent of orbital overlap required for efficient donor-acceptor interaction.

Discussion

House: In the charge transfer reactions with the amino ketone, do the materials simply relax back to starting materials or do you get some chemistry and some products?

Wagner: In the bimolecular case, one gets fairly efficient photoreduction. There is enough freedom of movement that a hydrogen atom on the α carbon can be plucked off by the negatively charged oxygen. You get radicals. In these internal reactions we have never seen products through the charge transfer route. Sometimes we get competing direct hydrogen abstraction which you can factor out. We published, that apparently the geometry of what you get (called an exoplex) has the n-orbital of the nitrogen overlapping with the carbonyl n orbital. It is tightly held and stays that way. The hydrogen atoms on the carbon α to the nitrogen are held away from the oxygen and there is no way to get a proton transfer over to the oxygen. The radiationless decay from these exoplexes is always fast and seems to dominate. From the point of view of kinetics it is fortunate there is not a lot of this other stuff going on, but from a synthetic viewpoint nothing much useful happens.

Caldwell: There is an experiment crying out to be done. I presume you are at it. If you take the dialkylamino group in the phosphorescence experiment, and replace it with an electron acceptor the quenching group should now be a good one for going after the electron-rich triplet benzophenone π system. In this case you should see much-enhanced quenching. It would be a nice demonstration of the electron-richness of the benzophenone triplet LUMO. Are you doing anything along those lines?

Wagner: Not precisely. One could have a cyano-olefin out there. We are looking at having a stilbene out there where one now has an energy transfer phenomenon, but those molecules have proved to be difficult to isolate. It is a good suggestion.

Plummer: When you had the orthodibenzoyl substituent, was that rate constant corrected for the statistical factor of 2?

Wagner: No.

Houk: What about the ionization potential of sulfur versus nitrogen?

Wagner: Dialkyl sulfides have higher ionization potentials than trialkyl amines and in a bimolecular sense the rate constant for quenching is always an order of magnitude slower than by the amine. The same is true of oxidation potentials in as much as they can be measured. They are never reversible so one is guessing.

References

[1] S. Winstein and N.J. Holness, J. Am. Chem. Soc. 77, 5562 (1955).

[2] (a) W.G. Dauben and M.S. Kellogg, J. Am. Chem. Soc. 102, 4456 (1980). (b) J.E. Baldwin and S. M. Krueger, J. Am. Chem. Soc. 91, 6414 (1969); and references therein.

[3] P.J. Wagner, Topics Curr. Chem. 66, 1 (1976).

[4] (a) J.C. Scaiano, J. Photochem. 2, 168 (1973/1974). (b) L. Giering, M. Berger, and C. Steel, J. Am. Chem. Soc. 96, 953 (1974).

[5] R.H. Hesse, Advan. Free Rad. Chem. 1, 83 (1969).

[6] P.J. Wagner, Acc. Chem. Res. 4, 168 (1971).

[7] P.J. Wagner, P.A. Kelso, A.E. Kemppainen, and R.G. Zepp, J. Am. Chem. Soc. 94, 7500 (1972).

[8] A. Padwa, E. Alexander, and M. Niemcyzk, J. Am. Chem. Soc. 91, 465 (1969).

[9] A. Padwa and W. Eisenberg, J. Am. Chem. Soc. 94, 5859 (1972).

[10] F.D. Lewis, R.W. Johnson, and D.R. Kory, J. Am. Chem. Soc. 96, 6100 (1974).

[11] E.C. Alexander and J.A. Uliana, J. Am. Chem. Soc. 96, 5644 (1974).

[12] F.D. Lewis, R.W. Johnson, and D.E. Johnson, J. Am. Chem. Soc. 96, 6090 (1974).

[13] P.J. Wagner and B.J. Scheve, J. Am. Chem. Soc. 99, 1858 (1977).

[14] P.J. Wagner and C.-P. Chen, J. Am. Chem. Soc. 98, 239 (1976).

[15] P.J. Wagner, Pure Appl. Chem. 49, 259 (1977).

[16] (a) R. Haag, J. Wirz, and P.J. Wagner, Helv. Chim. Acta 60, 2595 (1977). (b) P.K. Das, M.V. Encinas, R.D. Small, and J.C. Scaiano, J. Am. Chem. Soc. 101, 6965 (1979).

[17] L.A. Paquette and D.W. Balogh, J. Am. Chem. Soc. 104, 774 (1982); R.J. Ternansky, D.W. Balogh and L.A. Paquette, J. Am. Chem. Soc. 104, 4503 (1982).

[18] M.A. Meador and P.J. Wagner, manuscript in preparation.

[19] (a) G.R. Leppin and S. Zannucci, J. Org. Chem. 36, 1808 (1971). (b) S.P. Pappas and R.D. Zehr, J. Am. Chem. Soc. 93, 7112 (1971).

[20] P.J. Wagner and E.J. Siebert, J. Am. Chem. Soc. 103, 7329 (1981).

[21] M.J. Lindstrom, Ph.D. Thesis, Michigan State University, 1978.

[22] (a) P.J. Wagner, T. Jellinek, and A.E. Kemppainen, J. Am. Chem. Soc. 94, 7512 (1972). (b) W. Mueller, unpublished results.

[23] H. Morrison, V. Tisdale, P.J. Wagner, and K.C. Liu, J. Am. Chem. Soc. 97, 7189 (1975).

[24] S.G. Cohen and J.B. Guttenplan, Chem. Comm., 247 (1969).

[25] S.G. Cohen, A. Parola, and G.H. Parsons, Chem. Rev. 73, 141 (1973).

[26] J.B. Guttenplan and S.G. Cohen, J. Am. Chem. Soc. 94, 4040 (1972).

[27] I.E. Kochevcar and P.J. Wagner, J. Am. Chem. Soc. 94, 3859 (1972).

[28] N.J. Turro, J.C. Dalton, G. Farrington, M. Niemczyk, and D.M. Pond, J. Am. Chem. Soc. 92, 6978 (1970).

[29] R.A. Caldwell and R.L. Creed, Acc. Chem. Res. 13, 45 (1980).

[30] H. Morawetz, Pure Appl. Chem. 38, 267 (1964).

[31] G. Illuminati, L. Mandolini, and B. Masci, J. Am. Chem. Soc. 99, 6308 (1977).
[32] P.J. Wagner and A.E. Kemppainen, J. Am. Chem. Soc. 94, 7495 (1972).
[33] P.D. Bartlett, S. Bank, R.J. Crawford, and G. H. Schmid, J. Am. Chem. Soc. 87, 1288 (1965).
[34] P.D. Bartlett and G.D. Sargent, J. Am. Chem. Soc. 97, 1297 (1965).
[35] P.D. Bartlett, W.D. Closson, and T.J. Cogdell, J. Am. Chem. Soc. 97, 1308 (1965); P.D. Bartlett, W.S. Trahonovsky, D.A. Bolon and G.H. Schmid, J. Am. Chem. Soc. 97, 1314 (1965).
[36] R.A. Caldwell, G.W. Sovoccol, and R.P. Gajewski J. Am. Chem. Soc. 95, 2549 (1973).
[37] P.J. Wagner and E.J. Siebert, J. Am. Chem. Soc. 103, 7335 (1981).
[38] A. Mar and M.A. Winnik, Chem. Phys. Lett. 77, 73 (1981).

12. CYCLIZATIVE CONDENSATION AND ADDITION REACTIONS OF INDOLES AND INDENES

Wayland E. Noland [1]

School of Chemistry, University of Minnesota,
207 Pleasant St. SE, Minneapolis MN 55455

I would like to divide the presentation into two parts:

1. Cycloaddition reactions of indoles with ketones under electrophilic conditions utilizing protic solvents in acid
2. A discussion of cycloadditions of indenes

Shown in eq. (1) is the condensation that was first reported by Scholtz in 1915 [2]. It involves the

2-methylindole + $CH_3\overset{O}{\overset{\|}{C}}CH_2R$ → (1. HCl, EtOH, 25°; 2. NaOH) product (H_3C, CH_2R, R, H_3C)

R	mp °C	% Yield	from
H	182-183	82	acetone
		67	$(CH_3)_2C{=}CH\overset{O}{\overset{\|}{C}}CH_3$
		88	$(CH_3)_2C{=}CH\overset{O}{\overset{\|}{C}}CH{=}C(CH_3)_2$
		91	[structure]
H	229-231 (monoacetyl derivative)	50	[structure]
CH_3	197-198	15	$CH_3\overset{O}{\overset{\|}{C}}CH_2CH_3$
	197-199	20	

(1)

reaction of 2-methylindole with acetone or methyl ethyl ketone in the presence of HCl in ethanol. This led to the 2:2 condensation products with a loss of two molecules of water. Scholtz did not formulate it in this manner; the structure is that proposed by my group in 1961 [3]. The location of the olefinic double bond was based on intuition. A subsequent NMR spectrum confirmed the structure of the product from the indole-acetone reaction.

I want to emphasize the number of sources of acetone equivalents which lead to the same type of product: acetone itself, mesityl oxide, and phorone. With phorone reversal of the aldol condensation generates the acetone for the reaction. It is probable that mesityl oxide also may undergo reversal under these conditions. The product from the addition of 2-methylindole to mesityl oxide may also be used as a precursor; however, we rather doubt that it is a true intermediate in the acetone reaction. Bisindole which also contains an acetone equivalent breaks down under these acidic conditions and probably reverses to a 1:1 product. We believe this is the primary precursor which through dimerization forms the cyclic product.

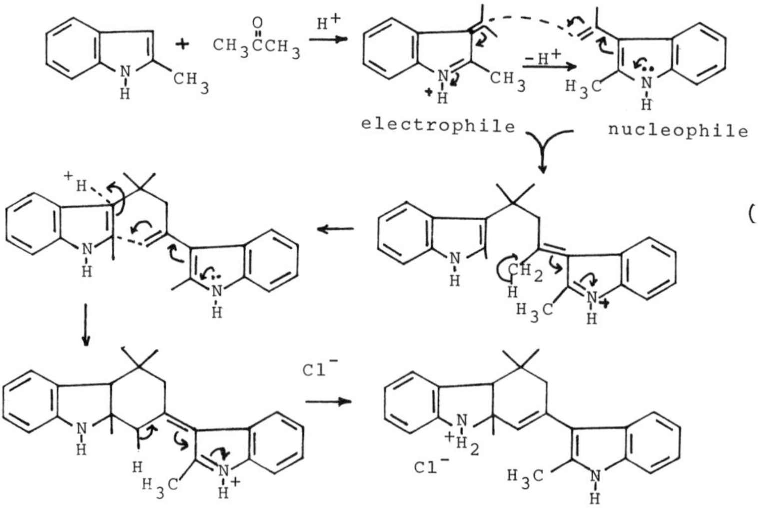

electrophile nucleophile

(2)

The suggested mechanism (eq. 2) involves condensation of the indole with acetone to give the 1:1 dehydration product. The compound is electrophilic, and a proton on the nitrogen enhances this property. Proton loss is suggested to occur from the equivalent of the position of the vinylogous imonium ion. This is really a vinylogous enamine-imonium ion equilibrium situation or something similar to it. This generates a nucleophile and an electrophile in the same medium that can condense to form the intermediate which is on the way to cyclization. A nucleophilic center is generated via the vinylindole configuration. The 2-position of the indole is made electrophilic by protonation of the 3-position which causes the final ring closure to take place. The loss of the proton conceivably could have occurred from the carbazole 1-position or the 3-position; however, it appears to be the 1-isomer. This leads to the monobasic condensation product which then is tied up in the form of the hydrochloride. The precipitation of these products as hydrochlorides is a primary method of isolation. It may also affect the product distribution by removal of this material from an (at least partially) equilibrated system. When the free base is obtained by neutralization of the hydrochloride, these compounds show a great deal of stability as long as there is an alkyl group at the 2-position of the original left-hand indole. In the absence of an alkyl group and with hydrogen at the 2-position the compounds are easily autoxidized.

Figure 1 shows the corresponding compound derived from indole which has a vulnerable hydrogen at the 2-position. Scholtz originally proposed the symmetrical structure at the top [2]; however, this structure does not account for the monobasic character of the cyclization product. In 1961 we proposed placement of the double bond as shown in the middle structure. Without NMR facilities we could not distinguish between the two possible structures [3]. The correct structure is that shown in the lower part of the diagram [1a].

The mechanism scheme (eq. 3) is exactly the same for the formation of the product from indole as from 2-alkylindoles. The branch point for the formation of the double bond in the two possible positions would be at the point in the lower left-center of the scheme (eq. 3). The structure at the lower left is the correct one; the one on the lower right is not isolated.

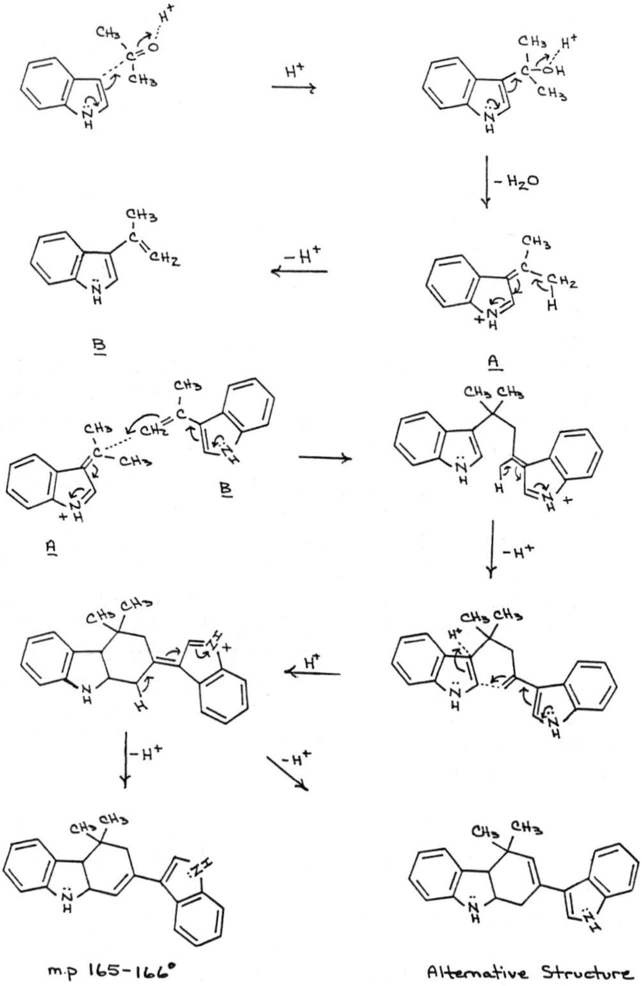

H^+

$-H_2O$

$-H^+$

B

A

(3)

A

B

$-H^+$

H^+

$-H^+$

$-H^+$

m.p 165–166°

Alternative Structure

2 [indole] + 2 $CH_3\overset{O}{\overset{\|}{C}}CH_3$ excess

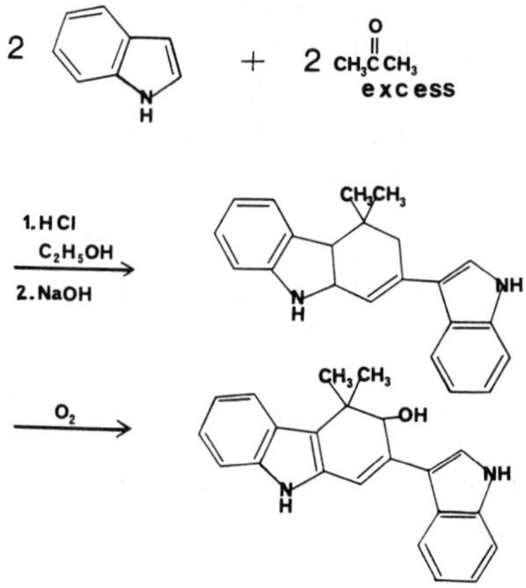

(4)

1. HCl C_2H_5OH

2. NaOH

O_2

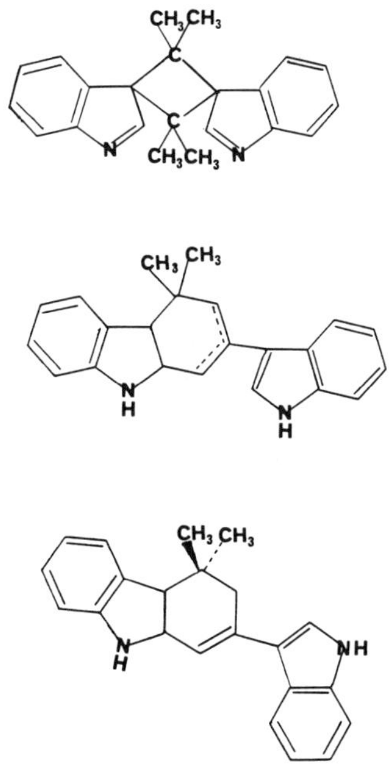

Figure 1. Proposed structures for the monobase.

When the indole does not contain a 2-alkyl group, a number of reactions occur during the workup and isolation (eq. 4) [1b,3c]. Oxidation takes place with the formation first of a double bond at the carbazole 9a,4a-position. In the case of an indole with an N-H substituent further oxidation occurs. In contrast indoles with N-methyl substituents do not undergo as readily the second oxidation step, and we have isolated no such products. In the indole case, further autoxidation takes place in the allylic position, affording the corresponding alcohol. NMR and mass spectral data were used in the structural elucidation. The molecular ion was the base peak in the mass spectrum of the alcohol. Loss of the substituent at the carbazole 4-position is a common feature of the fragmentation patterns of these compounds. Loss of water is another important fragmentation.

Eq. (5) shows a routine autoxidation mechanism by which we can arrive at the hydroperoxide in the carbazole 9a-position. I am not sure if this is

Autoxidation Mechanism

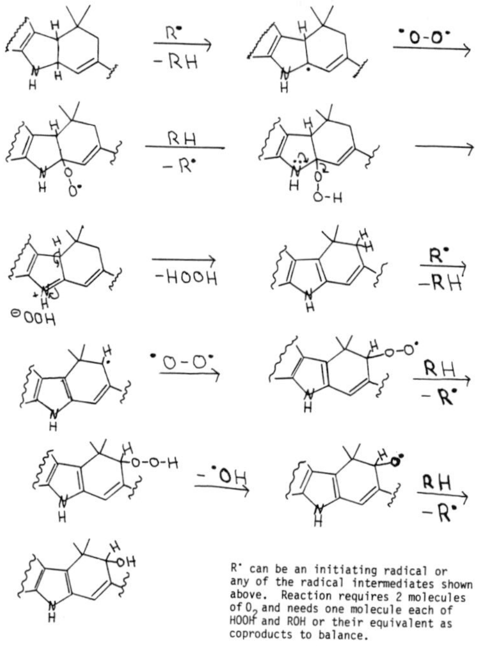

R• can be an initiating radical or any of the radical intermediates shown above. Reaction requires 2 molecules of O_2 and needs one molecule each of HOOH and ROH or their equivalent as coproducts to balance.

(5)

correct, but the mechanism is based upon the classical assumption that the allylic, tertiary, and α hydrogen is extremely vulnerable to autoxidative attack. A free radical chain reaction can convert it to the hydroperoxide which can undergo elimination of the elements of hydrogen peroxide. Loss of the proton from the 4a-position restores the indole character of the nucleus. A system is produced which has conjugation through the left-hand aromatic ring to the hypothetical radical at the 3-position. This radical site can react with oxygen and give a hydroperoxide. Interestingly enough we have not isolated any ketonic product corresponding to this alcohol. This may be related to the fact that indoles are sufficiently good reducing agents that the hydroperoxide is reduced to the alcohol. Acetylation of the alcohol was carried out as part of the structure determination, and a triacetyl derivative was formed ultimately. The UV data given in Table 1 confirm the high degree of conjugation. The long wavelength band at 385 nm is quite intense. Acetylation of nitrogen diminishes the intensity, and the λmax of the long wavelength transition is consistent with the loss of electron availability on the nitrogen.

Table 1. UV Data for Oxidation Product

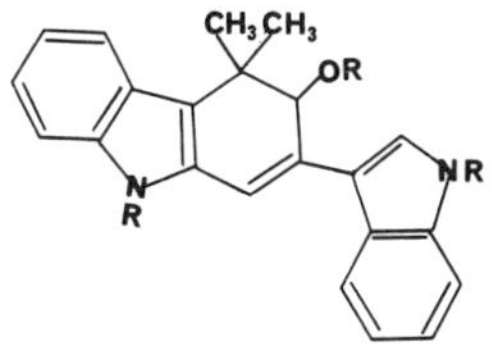

UV in 95% EtOH

R=H		R=$COCH_3$	
λ max(nm)	log E	λ max(nm)	log E
223	4.21	216	3.97
250	3.98	233	3.93
261*	3.86	265*	3.56
278	3.71	293*	3.48
385	4.38	298	3.50
		333*	3.43
		371	3.72

*inflection

The NMR spectrum is consistent with the formulation given for the triacetyl derivative. The most helpful feature was the observation that under too vigorous acetylating conditions a carbazole was formed. We describe this as analogous to a dienone-phenol rearrangement in which the acetate group is lost by ionization. A Wagner-Meerwein shift of a methyl group occurs from the 4- to the 3-position and a corresponding loss of a proton produces aromatization. The corresponding reaction with N-methyl- and N-ethylindole gives rise to monobases, which, if they can be isolated at all, are extremely susceptible to autoxidation. The reaction leads to the formation of the double bond. We have not isolated products involving autoxidation of the remaining allylic position. Even during methylation in liquid ammonia of the monobase derived from indole, the aromatized product was obtained [3c]. The autoxidation occurred during the workup or during the methylation, more likely during the workup.

R^1	R^5	% Yield	mp °C
H	H	46-54	201.5-203
CH_3	H	11-31	181-182
H	CH_3	64	214-216

(6)

Without a 2-alkyl substituent and under conditions of milder acidity the product obtained is also a 2:2 condensation product minus two molecules of water. The compound is not monobasic but is essentially neutral like indole. In 1961 we proposed symmetrical structure 2 (eq. 6) as the product [4]. After NMR became available we became interested in whether the product contained indole nitrogens in a syn or anti relationship. The NMR spectrum showed an AB pattern which indicated that both structures 2 and 3 (eq. 6) were incorrect. We now believe that the products have a five-membered ring structure of the type 1 [1d,e]. The methylene group in the rigid system provides the AB pattern in the NMR. This type of product is obtained not only from acetone but also from butanone (ethyl ethyl ketone). We have additional examples of other ketones in the 1-methylindole series.

There is a question concerning the regiochemistry of substitution on the carbocyclic five-membered ring. On mechanistic grounds we predict the product to be the upper structure in eq. (7), but there is the possibility that a subsequent Plancher rearrangement might occur. For indole compounds in acid the substituents at the 2- and 3-positions exchange *via* a Wagner-Meerwein rearrangement. This leaves open the possibility that the lower structure could be the one actually isolated. The indole product exhibits a wide melting range. A high resolution NMR spectrum indicates two isomers which we believe could only be regioisomers; however, we were not able to separate them. In the

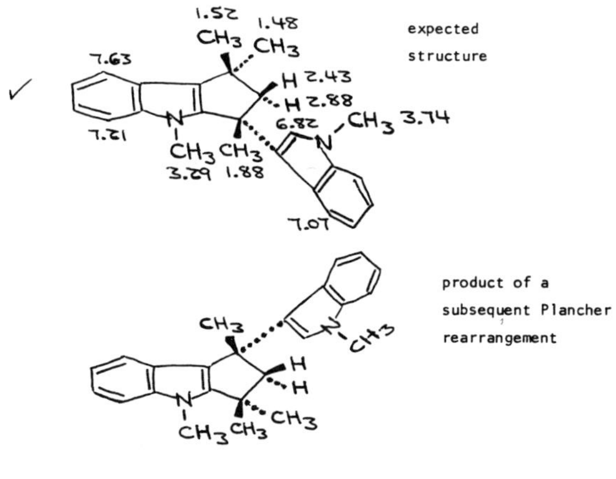

(7)

✓ Confirmed by nuclear Overhauser enhancements (NOE).

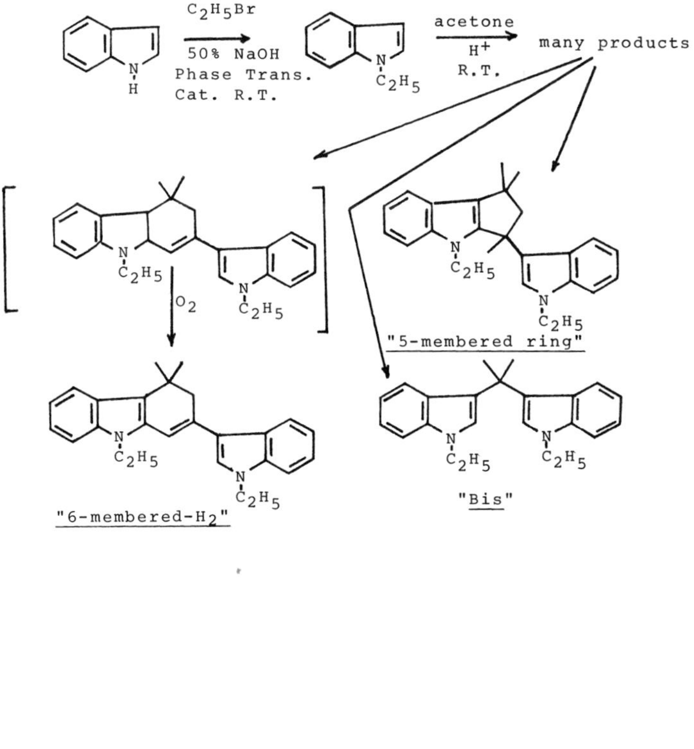

(8)

indole case it is possible the Plancher rearrangement does occur to a sufficient extent to give a mixture which results in two sets of AB patterns in the NMR spectrum. This is not the case for indole compounds with 1- or 5-substituents. A nuclear Overhauser enhancement (NOE) experiment confirmed the structural assignment of the 1-methylindole product which we obtained in a pure form.

In the case of 1-ethylindole and acetone, we have been able to isolate from the same reaction mixture all of the products shown in eq. (8) [1c]. These include the autoxidized derivative of the monobase which is the six-membered ring type of product already described. We obtained the pure product by taking the crude material and bubbling air through it to complete the autoxidation. We also obtained the five-membered ring product under the same conditions; this was the first case where we were able to obtain both products at once. In previous cases we obtained the monobasic products simply by filtering off the precipitated hydrochlorides. In addition to the two cyclized products the ubiquitous bisindole was obtained. This is the most common type of derivative obtained from the reaction of aldehydes and ketones with indoles. It is formed when you don't want it and is an excellent high-yield crystalline derivative of an indole. The problem is usually one of obtaining something else.

The mechanism scheme for formation of bisindole (eq. 9) is similar to that for formation of the six-membered ring product described earlier. The mechanistic departure comes at the point where the attacking nucleophile, instead of being a 1:1 product in the vinylindole form from a previous condensation of an indole with a ketone, is the indole itself. The indole has not had the opportunity to condense with a ketone and to be tied up as the 1:1 product. In general, conditions of lower acidity favor the formation of a bisindole. Conditions of high acidity increase the rate of initial condensation with the ketone so that all of the indole is tied up as the 1:1 product. This leads to further products derived from the 1:1 condensation intermediates. After condensation has occurred, deprotonation at the 3-position gives the final bisindole product.

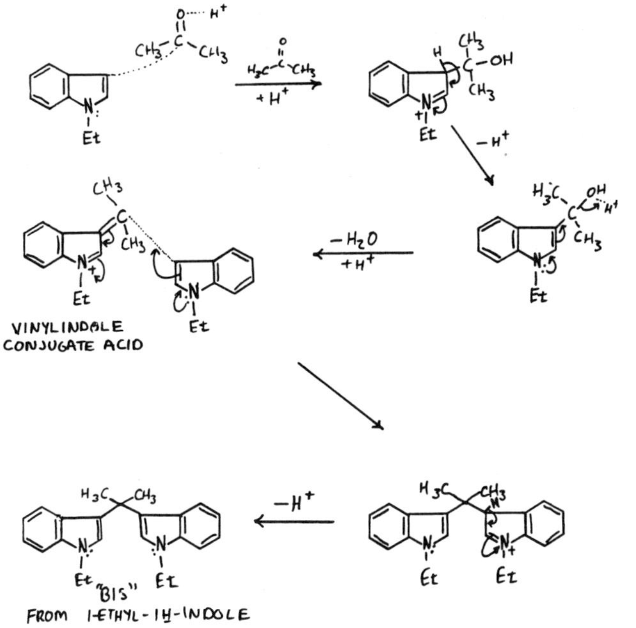

(9)

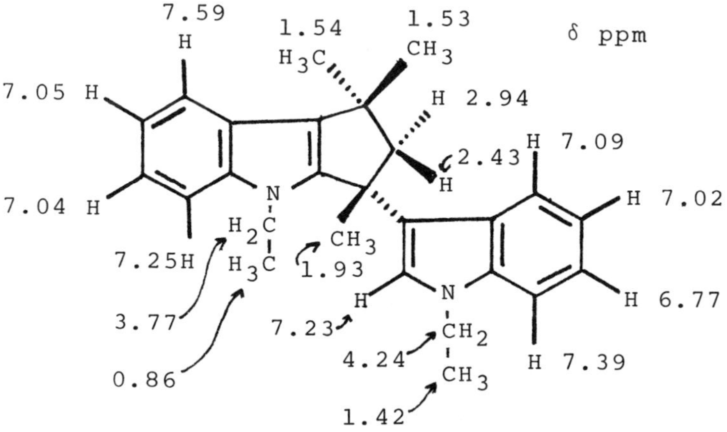

Figure 2. NMR assignment for five-membered ring product.

The NMR analysis of the five-membered ring product from 1-ethylindole is shown in Figure 2. The hydrogen of the ring methylene group cis to the pendent indole is downfield. We think this is due to the hydrogen lying in the deshielding plane of the pendent indole nucleus. The *N*-alkyl substituent in the left-hand indole ring is shielded relative to the *N*-alkyl substituent on the pendent indole nucleus. The trend seems to follow through to some extent for the protons of the aromatic rings of the indole nuclei.

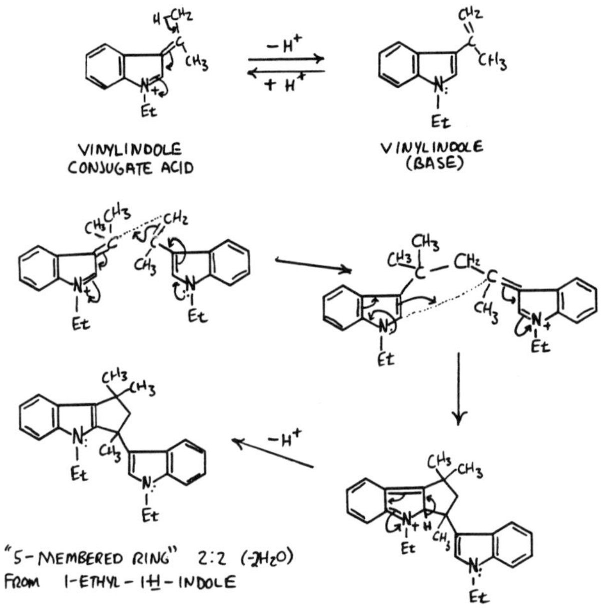

VINYLINDOLE CONJUGATE ACID

$-H^+$ / $+H^+$

VINYLINDOLE (BASE)

$-H^+$

"5-MEMBERED RING" 2:2 ($-2H_2O$) FROM 1-ETHYL-1H-INDOLE

(10)

The suggested mechanism for the formation of the five-membered ring condensation product (eq. 10) is exactly the same as previously discussed for the six-membered ring product. That is, we form the 1:1 product, and it tautomerizes to the vinylindole, which is the nucleophilic component. This material dimerizes through the vinylindole, affording an intermediate which is exactly the same up to this point as was obtained for formation of the six-membered ring product. However, instead of tautomerizing by loss of the proton to create a nucleophilic vinylindole, the same center remains electrophilic, and the nucleophile is provided by the 2-position of the left-hand indole. We have switched the polarities in the cyclization process. The result of this is the formation, after loss of the proton from the 2-position, of the five-membered ring product with a pendent indole and a pendent alkyl substituent remaining at the site of ring closure. A ring methylene group remains which produces the AB pattern in the NMR spectrum.

CH_3; 2.1-1.2 N HCl; H_2O, EtOH, 50°C, 4 h, then 25°C

Ar = RC_6H_4

added dropwise in EtOH solution

1: 2:1 ($-H_2O$) bisindole

2: 2:2 ($-2H_2O$) 5-ring

Product precipitated in 1-3 days except for R=p-NH_2, which required neutralization with $NaHCO_3$

R	type	yield, %	mp °C
p-NH_2	**1**	61	203 -204
p-OCH_3	**2**	20	273 -274
p-CH_3	**2**	10	265.5-266
m-OCH_3	**2**	26	215 -216
p-Cl	**2**	10	284.5-285
p-NO_2	**1**	16	266 -267

(11)

In the case of acetophenone (eq. 11) the options available to the system are limited to the formation of the bisindole and the five-membered ring product. We wanted to look at the general behavior of the NMR AB pattern derived from the ring methylene group and to see how substituents on the acetophenone aromatic ring influenced the relative formation of bisindole and cyclization products. At both ends of the substitution spectrum, ie, p-amino and p-nitro, bisindole was the only crystalline product isolated [1g]. At the intermediate stages of substitution the five-membered ring product is obtained primarily. In the strongly acidic media it is likely that the p-amino substituent is protonated. Accordingly, it is not surprising that its behavior is more like that of the nitro substituent.

The AB pattern and the NMR coupling constants (Table 2) which are usually in the neighborhood of 12-13 Hz are characteristic of the ring methylene group. The NOE evidence indicates that the two aromatic rings derived from the acetophenone are in a cis relationship. The smaller alkyl substituent and the larger indole substituent are on the opposite side of the ring which is what one might expect.

Table 2. NMR AB Patterns and Coupling Constants

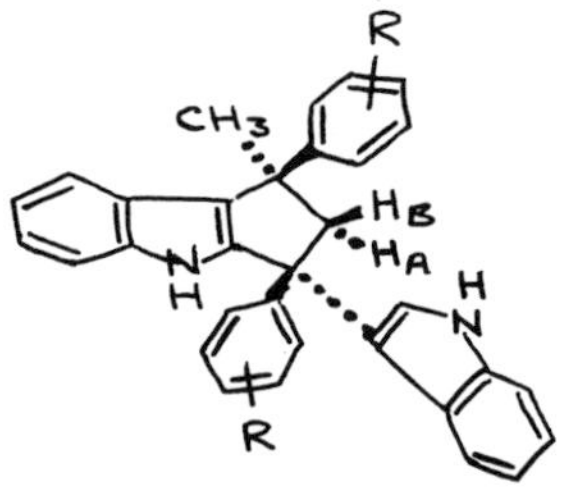

R	δ_A (ppm)	δ_B (ppm)	J_{AB} (Hz)
p-OCH_3	3.55	3.51	13.2
p-CH_3	3.55	3.39	13.5
m-OCH_3	3.65	3.57	13.3
p-Cl	3.73	3.52	11.5

These assignments were confirmed by NOE. To overcome solubility limitations, it was necessary to alkylate the NH groups. The stereochemistry discussed previously was derived via examination of the NMR spectrum of the fully methylated derivative in chloroform-d_6 solution.

One can use more highly substituted ketones such as methyl isobutyl ketone and obtain the five-membered ring condensation product (eq. 12). Thanks to NMR we can now reformulate it [1f]. Methyl benzyl ketone (phenylacetone) gave the corresponding product in low yields [1e]. The melting point was surprisingly low.

Cyclic ketones such as cyclopentanone and cyclohexanone are incorporated into a ring system which

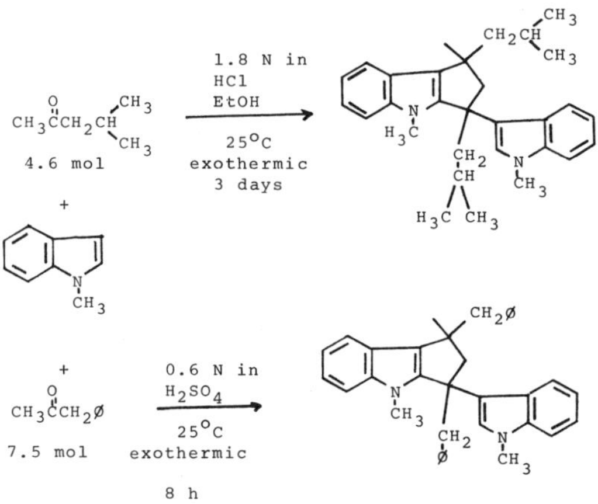

(12)

includes the five-membered ring (eq. 13). Support for this structural assignment lies partly in the location of the unique methine proton which can be found in the NMR [1e].

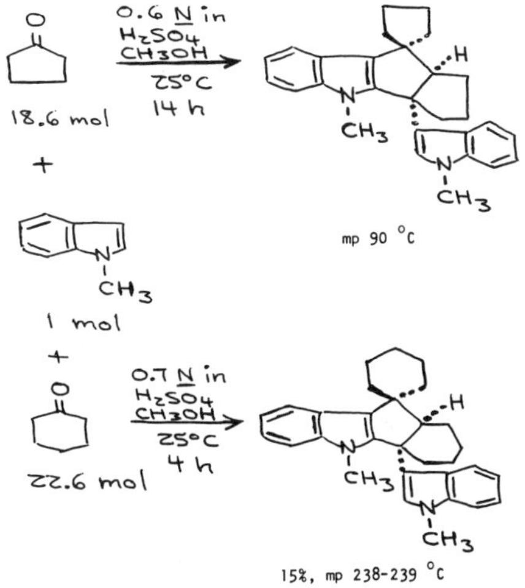

mp 90 °C

(13)

15%, mp 238-239 °C

In addition to the 2:2 five-membered ring product, higher condensation products are obtained, eg, 2:3 product of the indole with acetone minus

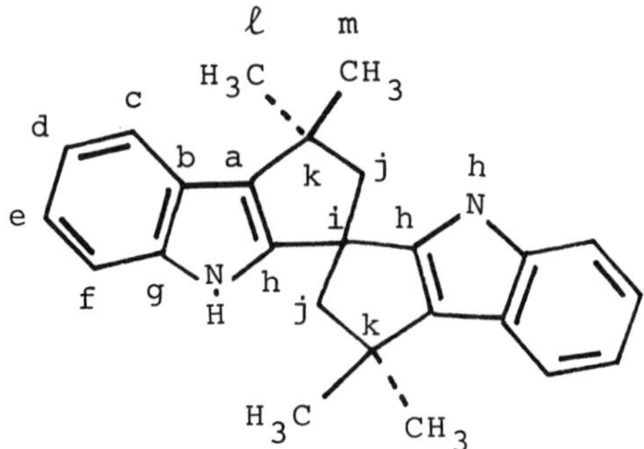

Figure 3. Product A of 2:3 indole-acetone reaction.

three molecules of water (Figure 3) [1a]. It is a beautifully symmetric molecule which we obtained in two crystalline forms. Actually, the lower melting form is a hemiethanolate of the compound. We determined the structure and were somewhat crestfallen to find we had competition from Mrs. Chatterjee and her associates. They reported isolation of the same compound from the reaction of boron fluoride etherate with indole and acetone in aprotic media [5]. They obtained a number of other products as well. We isolated and reported the compound in 1961, but we did not make a structural assignment [3c]. The NMR spectrum shows two equivalent AB patterns and two kinds of methyl groups, times two in each case. Similarly, the C-13 spectrum (Table 3) contains 25 carbon atoms but shows only 13 lines. The UV spectrum [λm,log ; 292(infl),4.180; 285,4.214; 234, 4.793; 225(infl),4.735] is that of a simple indole showing no additional conjugation. The probable mechanism for formation is shown in eq. (14). The formation of the 1:1 condensation product is shown in electrophilic form, but we need to insert another acetone equivalent into the center of the molecule. The product contains the framework of phorone; however, it is doubtful that phorone is a precursor of the product. It seems much more likely that the 1:1 condensation product is attacked by the enol of acetone giving in a series of steps what is formally the analog of the addition of the indole to mesityl oxide. If this central ketone unit in enolic form attacks another molecule of the 1:1 product in the electrophilic form, the skeleton can be put together. It is a matter of wrapping it together across the 2-position of the two indole

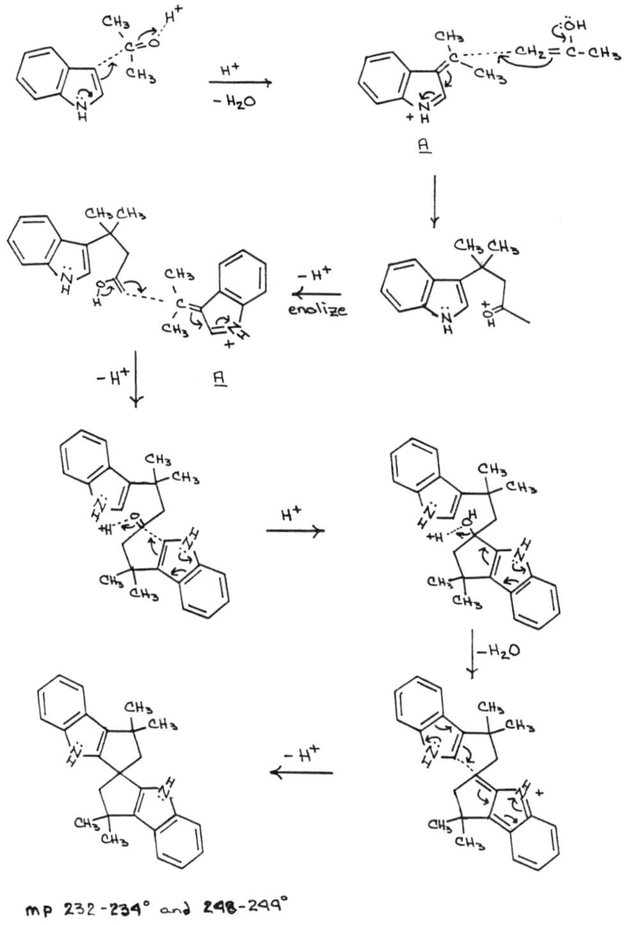

mp 232-234° and 248-249°

(14)

nuclei and the ketone carbonyl by a series of condensations using the 2-position of the carbonyl as the nucleophilic center. One would expect this reaction to be favored by an excess of acetone. All reactions that we carried out in this area involve an excess of acetone, but a particularly large excess seems to be associated with formation of significant amounts of the 2:3 product.

The 2:3 condensation product must have more strain than usual. It is very susceptible to autoxidative attack (eq. 15). Once this autoxidative attack has broken up one ring system, the molecule seems to revert to a more normal behavior. We ob-

Table 3. C-13 Assignment of the Spiro Compound "A"

Peak No.	Assignment	ppm	Type of Carbon
1	g	145.65	Q
2	b	142.13	Q
3	h	127.00	Q
4	a	124.05	Q
5	c	120.80	O
6	e	119.31	O
7	f	118.46	O
8	d	112.37	O
9	j	62.71	N
10	i	49.61	Q
11	k	39.39	Q
12	l,m	30.77	M
13	l,m	30.12	M

Q = quaternary, O = methine, N = methylene, M = methyl. Assignment letters are given in Figure 3.

+ $CH_3\overset{O}{C}CH_3$ EtOH / HCl → 2:3 and 2:2 Indole:Acetone condensation products

A

O_2 / $CHCl_3$ → dioxide 1 $C_{25}H_{26}N_2O_2$ mp 262-263°

2:3 spiro compound $C_{25}H_{26}N_2$

EtOH / O_2,hν → dioxide 2 $C_{25}H_{26}N_2O_2$ mp 155-160

(15)

mp 232-234° & 248-249° dimorphic forms

B

2:2 monobase $C_{22}H_{22}N_2$ mp 165-166°

served that in the presence of air and in chloroform solution with or without light the compound undergoes autoxidative attack to give dioxide 1. The carbon skeleton of the original material is retained, and all that we have done formally is to introduce an ether oxygen, to move a bond from the 2- to the 3-position, and to introduce a carbonyl oxygen at the 2-position creating an oxindole. This product also was reported by Mrs. Chatterjee and her co-workers [5]. They obtained it from an unstable and not well-defined product of their primary reaction. It appears to be derived via an autoxidation mechanism in their system as well.

We obtained dioxide 2 in ethanol solution in the presence of oxygen and filtered sunlight through window glass. The compound is the result of cleavage of a 2,3-double bond in one of the indole nuclei. This gives rise to the large ketolactam ring system. It is tempting to speculate that this was formed by a self-sensitized singlet oxygen reaction in which a dioxetane was formed across the 2,3-positions of one of the indole nuclei and then decomposed to the carbonyl groups. This would be reminiscent of reactions that Paul Bartlett and Leo Paquette described in connection with the chemistry of sesquinorbornene.

A suggested autoxidation mechanism which proceeds via the classical hydroperoxide approach can also be used to derive the product (eq. 16). This mechanism leads to a hypothetical oxonium ion that can give rise to cleavage of a 2,3-carbon-carbon bond. This species is followed by a series of hypothetical intermediates which afford a hemiketal that then could open to give the final product.

We also have obtained spiro-type five-membered ring compounds from 5-bromoindole and from 1-methylindole (eq. 17) [1e]. This appears to be a fairly general phenomenon. These compounds appear in up to 20% yield in some of our reactions. The major fragmentation pattern involves the molecular ion as the base peak in the mass spectrum and loss of a methyl group of which it has many to give. Alternatively, loss of one of the two 3-carbon *gem*-dimethyl groups may occur under electron impact.

RH

ROO•
(− H•)

R•

O_2

ROO•

RH
(+ H•)

heterolytic fission

^{-}OH

H_2O

(16)

mp 155-160°

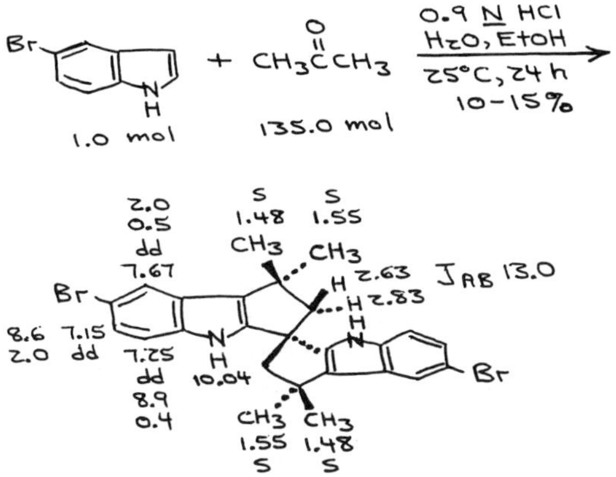

dimorphic forms

mp 247-248 °C

262-263

δ ppm in CD_3COCD_3

(17)

MS (70 eV, 70-80 °C, relative intensity) m/e

M	M-15 M-CH_3	M-43 M-$(CH_3)_2CH$
514 (50)	499 (31)	471 (20)
512 (100)	497 (60)	469 (43)
510 (52)	495 (31)	467 (23)

When you have a 2-alkyl substituent and a ketone which is more complex than methyl ethyl ketone, the reaction follows yet another pathway (eq. 18). The product that we expected was the 2:2 six-membered-ring cyclization product with the 2-alkyl substituent and a monobasic indole ring on the left hand side. This time the products involve only one molecule of ketone and 2 molecules of indole, but they are monobasic [1h,i,3b]. We obtained these from a large variety of ketones. The breakpoint seems to

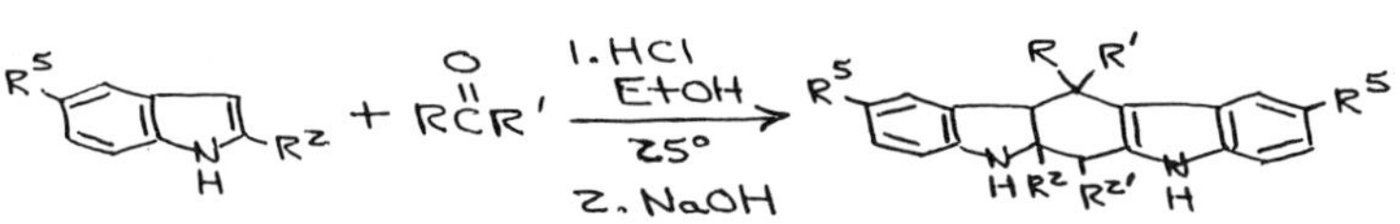

16 examples:

R = Et → neoPent → n-Dec

R' = Me (R ≠ Et), Et, n-Pr

R^2 = Me, Et

$R^{2'}$ = H, Me (2 diastereoisomers)

R^5 = H, Me

Yields of the precipitated hydrochlorides, 2-35%.

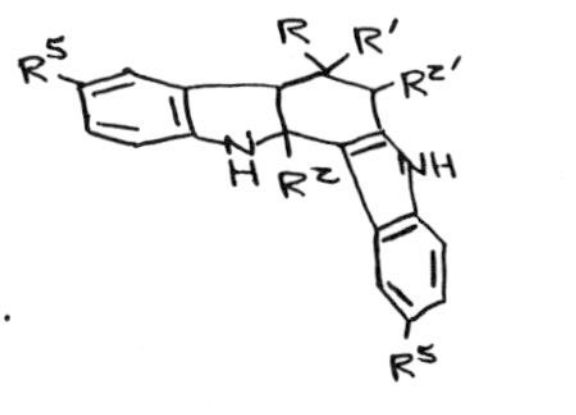

(18)

lie between methyl ethyl ketone (six-membered ring products [3a]) and anything with one more carbon atom in an alkyl group. Methyl n-propyl ketone and diethyl ketone give this type of product. We may have found a fairly fine dividing line.

From the NMR spectrum we could not differentiate between the two possible regioisomers. I believe we should be able to settle the question by NOE experiments. I have the data for the methyl isobutyl ketone product, but I am not prepared to make a statement on the basis of what I have found. It does not seem to rule out the lower structure in eq. (18) which I do not like. That structure still must be considered due to the inconclusive nature of the results. If we take a less substituted example with a little less complexity, we may be able to more cleanly resolve the dilemma. The problem can be analyzed mechanistically (eq. 19) [1h]. It seems very reasonable that the bisindoles from the ketones that give this type of reaction may be unstable intermediates, perhaps because of greater

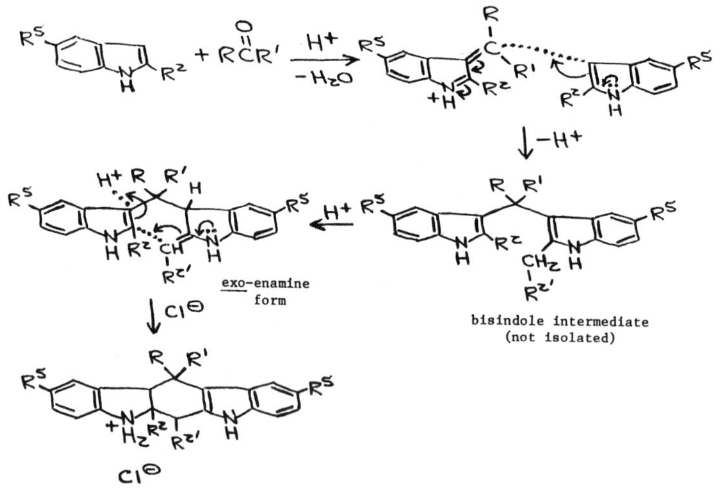

exo-enamine form

bisindole intermediate (not isolated)

(19)

steric hindrance around the bis (or central) carbon atom. The 2-methyl or 2-alkyl substituent participates in the ring closure and is incorporated into the ring that is formed. The difference in these two structures lies in the orientation of the right side of the molecule (eq. 18). It is either as shown in the upper structure or flipped over as shown in the lower structure. The proton NMR requirements would be the same in the two cases, so the answer cannot be resolved by regular NMR analyses.

Mechanistically, I have more trouble rationalizing the formation of the lower product. It requires that there be a nucleophilic attack by the right hand indole 2-methyl group (which would have to have tautomerized into the methylene enamine form) on the electrophilic carbon (which was originally the ketone carbonyl carbon) of the 1:1 ($-H_2O$ condensation product (portion on the left). This is hard to accept because 2-methylindole is highly nucleophilic at its 3-position and would be expected to react there first. If the unexpected should happen, then the subsequent ring closure step is easier to rationalize. The ring closure would involve nucleophilic attack by the 3-position of the right hand indole (after it has tautomerized back to the indole configuration) on the 2-position of the left hand indole. The molecule could become electrophilic when protonated at its 3-position. The alternate sequence would require initial attack by the right hand indole from its nucleophilic 3-position on the sterically hindered 2-position of the 1:1 ($-H_2O$) condensation product on the left. This breaks the conjugated diene system in the middle and leaves it as an _o_-aminostyrene system. This system would have to be attacked in the ring closure step nucleophilically at the wrong end by the 2-methylene enamine derived by tautomerization of the right hand indole nucleus. For these reasons, formation of the lower product seems mechanistically unpalatable. However, based on the proton NMR spectrum we cannot eliminate this possibility. This reaction has been examined with numerous precursors (eq. 20). The reaction was run with a variety of normal R^1 alkyl groups with R^2

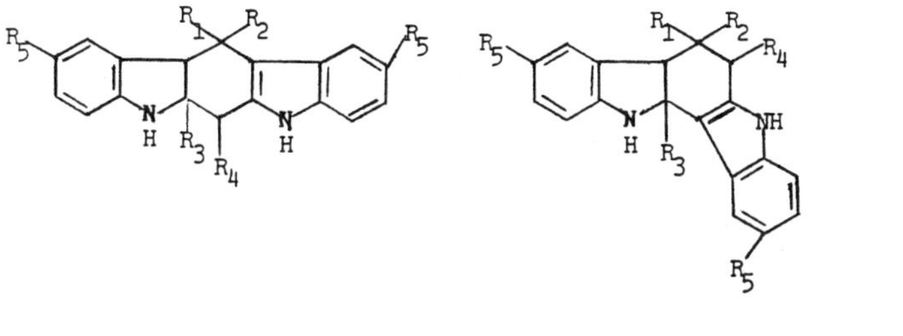

(20)

generally being methyl; however, we have several examples where R^2 was ethyl or n-propyl. The yields are modest and become quite small when both of the alkyl groups are larger than methyl. Interestingly, none of the bisindole is obtained in these cases. We are forced to conclude that the bisindole is unstable when alkyl (methyl or larger) substituents are present in R^1 and relatively large substituents are present in R^2.

We have seen a number of examples of cyclization reactions that involve intermediate 1:1 condensation products which tautomerize to vinylindoles which are the nucleophilic forms of such products. We have been able to utilize vinylindoles in what I call the "vinylindole synthesis of carbazoles" [6]. It is simply a Diels-Alder reaction utilizing the pendent vinyl group and the double bond of the aromatic indole nucleus as the diene (eq. 21). With vinylindole it gives relatively high

25° → [ ] −2H₂ → 91% (21)

mp 255–258°

yields of a product where the excess of naphthoquinone has oxidized the product to a fully aromatic derivative. Unfortunately, it is a lot of work to make 3-vinylindole. It is really a "super styrene" in terms of its reactivity. Most synthetically useful vinylindoles which are isolated contain electron-withdrawing substituents on the β carbon that stabilizes them. This increases the complexity of the product, but it certainly doesn't prevent the reaction from proceeding (eq. 22). We have used substituents such as methoxycarbonyl and cyano [7]. When you get to the dihydrocarbazole stage at these high reaction temperatures, aromatization occurs right in the reaction medium. We would like to be able to isolate simple vinylindoles by the condensation of the indoles with ketones since they probably are there as intermediates. 2-Methylindole with its blocking 2-substituents is the best bet for making simple vinylindoles from ketones, but it introduces a complexity. You may not want the 2-alkyl substituent at a later stage; it is hard to get rid of the 2-alkyl substituent. We have made a number of 3-vinylindoles by mild condensation of the indoles with relatively simple ketones (eq. 23) [1h,8]. 2-Indenone works relatively well but gives a more complex structure.

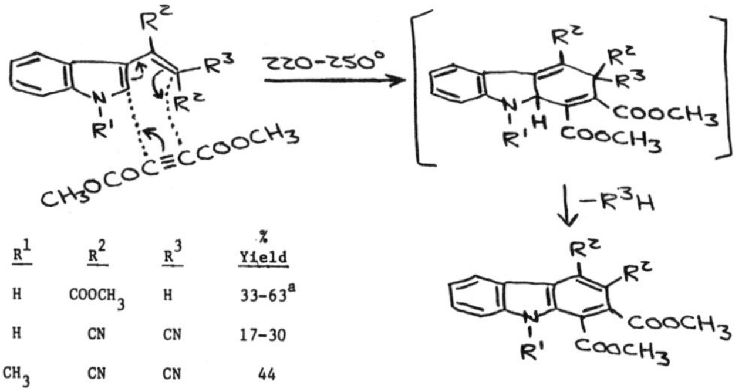

(22)

R^1	R^2	R^3	% Yield
H	$COOCH_3$	H	33-63[a]
H	CN	CN	17-30
CH_3	CN	CN	44

(a) From indole at 25°.

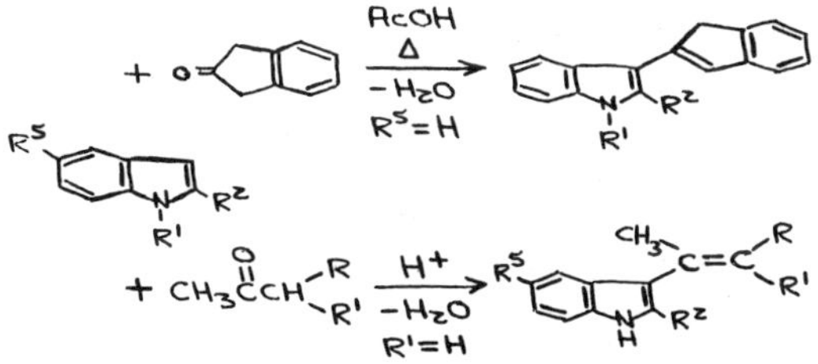

(23)

We have succeeded in combining the condensation of the indole with a ketone to form the vinylindole with a dienophile ready to capture it in a Diels-Alder reaction. Furthermore, you need an acid catalyst for this reaction, and in the examples in eq. (24) we have used maleic acid as the catalyst and as the dienophile. With a number of simple ketones we can get the corresponding tetrahydrocarbazolecarboxylic acids in reasonable though not oustanding yields. Also, it has been possible to build

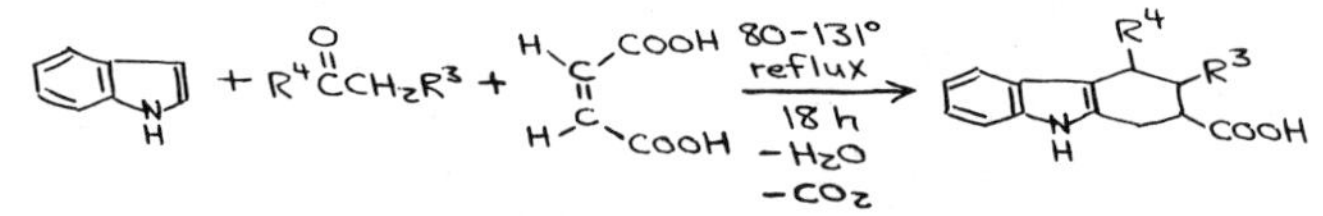

(24)

additional rings by using cyclic ketones at this point. The tetrahydrocarbazolecarboxylic acids can be made more soluble by esterification, and they can be dehydrogenated by standard methods to give the corresponding fully aromatic carbazoles (eq. 25). Thus, this sequence constitutes what we call "in situ vinylindole synthesis of carbazoles" [9].

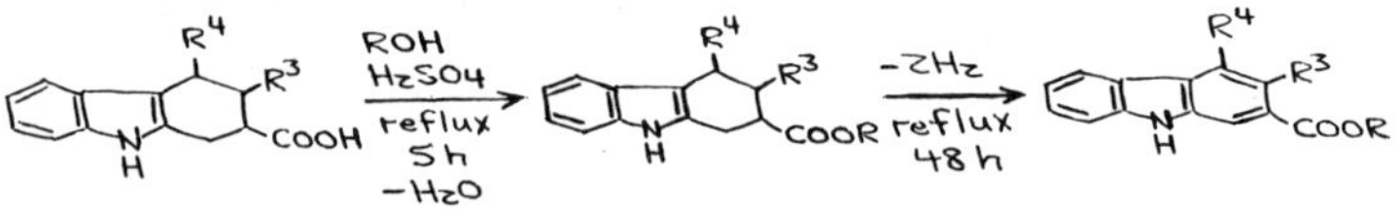

(25)

One of the interesting features of the mass spectra of tetrahydrocarbazole derivatives is that the retro Diels-Alder reaction is one of the major fragmentation pathways (eq. 26). The tetrahydrocarbazole derivatives were not formed by the same Diels-Alder reaction, but the molecules tend to split nicely at the point indicated to afford the corresponding dimethylideneindoline-type fragments. The common major fragmentation pathway is by

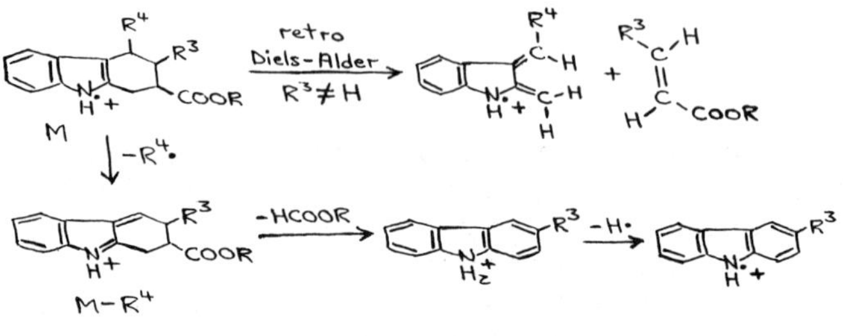

(26)

loss of the R^4-substituent. The R^4 substituent is well supplied with electrons to assist its departure by the indole nitrogen. Loss of R^4 followed by loss of the ester group or acid group gives a carbazole derivative.

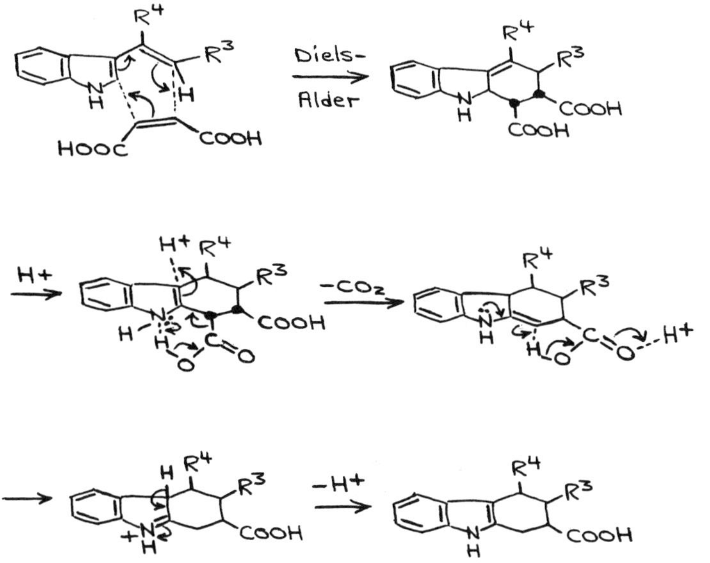

(27)

The probable mechanism for this sequence is shown in eq. (27). There are two things that have not been covered. One is the isomerization from the double bond initially formed at the R^4-bearing carbon in the Diels-Alder reaction to the indole configuration. This could be a simple acid-catalyzed allylic rearrangement. The other is the selective loss of the carbazole 1-carboxy group from the indoleacetic acid functionality. There is literature precedent for the loss of the carboxy group from 2-indoleacetic acids. This situation is analogous to a β-keto acid where a proton attaching at the 3-position increases the electron withdrawal through the indole double bond. That leaves the remaining carboxy group in the carbazole 2-position. A situation somewhat analogous to the decarboxylation is a reaction we observed some time ago (eq. 28) [1k]. Indoles can be polymerized to dimers or trimers in the presence of acid. When maleic acid is used as the acid, the indole in the trimer which has opened to form a primary amine is then converted to the amide. This gives rise to a malelyl derivative of triindole. Once that molecule eliminates an indole, we have a vinylindole with built in dienophile which permits an intramolecular Diels-Alder reaction. This gives rise to the cyclization product. Again we have an indole-2-acetic acid functionality which appears to decarboxylate spontaneously in the reaction medium to give the final product.

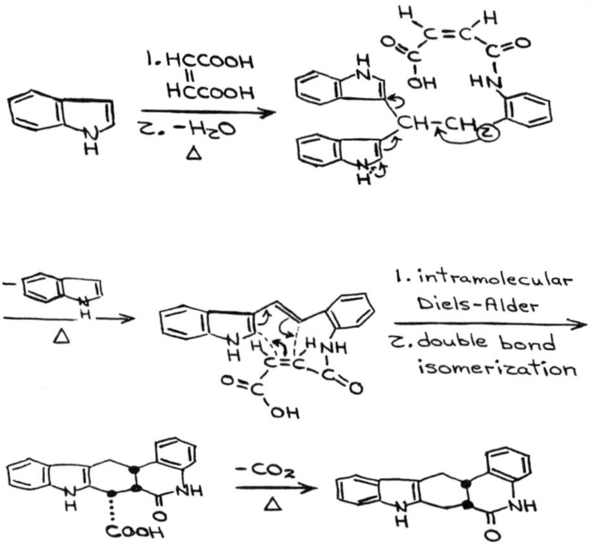

(28)

We have been able to utilize the <u>in situ</u> vinylindole synthesis of carbazoles to introduce with cyclic ketones additional rings. This includes five-membered and larger rings (eq. 29). In the 1-methylindole series we have been able to make derivatives of this type and even one with an eight-membered ring derived from cyclooctanone. When the dienophile being utilized is a strong dehydrogenating agent, the fully aromatic carbazole may emerge directly from the reaction. We saw this in a previous example using naphthoquinone, but it occurs even with maleic anhydride and maleimide [1e].

COOH / COOH 1.0 mol, reflux 24 h
1.0 mol   4.0 mol   14%, mp 269-272 °C

COOH / COOH 1.0 mol, reflux 22 h
CH_3 ϕCCH_3 (C=O) CH_3 COOH
1.0 mol   2.5 mol   21%, mp 263-264 °C   (29)

COOH / COOH 2.1 mol, reflux 5 h 16% CH_3OH, H_2SO_4, reflux 4.5 h 68%
1.7 mol CH_3 $COOCH_3$ 11%, mp 128 °C

That is the end of the story as far as indoles are concerned. The remainder of the discussion relates to indene chemistry. Our interest in this area was aroused by a reaction of 2-methylindole with indenecarboxylic acid. This was intended simply as a Michael-type addition across the double bond of indenecarboxylic acid. We ran into a situation where the product was actually the dimer of indenecarboxylic acid. This caused us to look at the chemistry of indenecarboxylic acid in some detail (eq. 30) [10]. We are going through an isoindene mechanism in which tautomerization is required before dienophiles such as maleic anhydride, maleimide, or dimethyl fumarate will react with the compound to give Diels-Alder adducts. These adducts are derived from the isoindene-type of intermediates. This behavior establishes a precedent for what we believe is happening with indenecarboxylic acid reacting with itself. The compound isomerizes into the isoindene form (a diene) at elevated temperatures and reacts with a molecule of the unisomerized acid which acts as a dienophile. This gives an adduct which has a head-to-head structure [11]. It is tempting to suggest the reason for the head-to-head structure is that the carboxylic acid is in its normal dimeric form that holds the two carboxyls together. The rest of the molecule simply undergoes the Diels-Alder addition after one of the ring systems has equilibrated over into the highly reactive isoindene state. This dimer was first reported in 1911 [12], but the structure was not elucidated.

	R	%	mp °C
2a	H	71	227-228
b	CH_3	43	168-169

	R	%	mp °C
3a	H	46	263-265
b	CH_3	0	—

	R	R'	%	mp °C
4a	H	CH_3	19	192.5-193.5
4b	H	H	*	272-274

*(69% from 2a; 11% from 3a; 68% from 4a)

(30)

The head-to-head structure has been established by a number of chemical methods and is consistent with all of the spectral data that are available. The fact that this molecule forms an anhydride rules out head-to-tail orientation. It undergoes selective oxidation at the benzyl methylene group, and we have carried out a whole series of derivatizations involving the oxidized series. Further evidence bearing on the manner of cyclization is provided by the results of reduction to the diol which was then converted to the tetrahydrofuran derivative either by a direct reduction of the acid to the diol or via its ester and reduction of the ester to the diol. Conversion to the diacid chloride, treatment with sodium azide (Curtius reaction), and refluxing with water gives the corresponding imidazolone. A similar reaction, in which we apparently did not completely convert the diacid to its diacid chloride led to an oxazinedione-type of cyclized product. Thus, the chemical evidence seems quite convincing for head-to-head dimerization [11].

Eq. (31) shows a generalized mechanism summarizing what we know from the work of others and ourselves about the cycloaddition reaction of indenes with dimethyl acetylenedicarboxylate (DMAD) [13-15]. The first reaction appears to be a Diels-Alder cycloaddition involving a double bond in the aromatic ring system. This is an interesting reaction since it occurs readily at room temperature. The resulting 1:1 product is highly reactive since it no longer has an aromatic system. It has a reactive double bond system at the R^3-bearing carbon. If R^3 is hydrogen, DMAD quickly captures the intermediate, so it has not been isolated. The reaction results in the formation of a 1:2 adduct (A). The product is formally the result of [2+2] cycloaddition. It has been suggested by Strachan and co-workers [14], that a reasonable mechanism would be a radical-type addition of the DMAD to the R^3-bearing carbon. It would certainly lead to a nicely stabilized allylic tertiary radical at the adjacent ring-junction carbon. I find that explanation attractive; at least, it is consistent with the facts shown here. If one pushes the thermal addition a little bit further, a third molecule of DMAD captures the diene system. It appears to do this on the side away from the blocking four-membered ring system. Maleic anhydride can do the

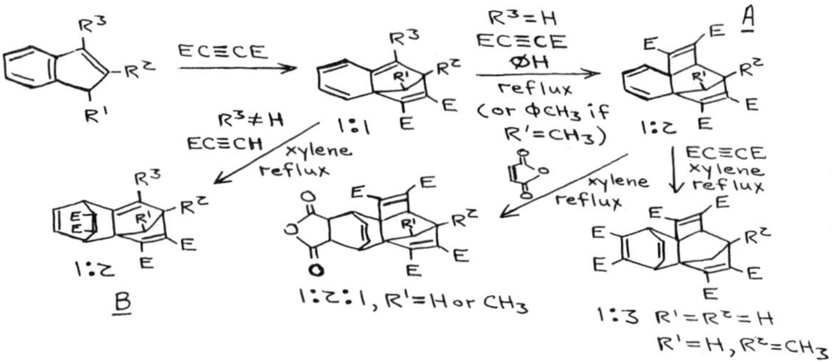

(31)

same thing. If R^3 is larger than hydrogen (eg, R^3 is a methyl, carboxy, or an ester group) this blocks the addition of DMAD at the R^3-bearing carbon. However, the diene system remains as the available site for further reaction, and 1:2 products are obtained. They are of a different type (B) from that first mentioned (A). The 1:2 product we observed with indenecarboxylic acid and DMAD is consistent with structure B [11]. Similarly, when you have a mixture of DMAD and maleic anhydride, the maleic anhydride does not do anything at first. It requires the high-temperature formation of an isoindene before it reacts. The DMAD initiates the addition across the diene system of the five-membered ring in the usual way. Then the maleic anhydride is quite able to capture the normal diene system that results from the 1:1 intermediate.

DMAD=EC≡CE, E=COOCH$_3$

Diels-Alder → electro-cyclic → ; homolytic ; homolytic ; (32)

(a) 3°, doubly allylic (a) 3°, triply allylic

(b) vinyl (b) 3°

Surprisingly, the product from 1,1-dimethylindene and DMAD was a 1:1 addition product involving a cyclopropane ring system (eq. 32) [14]. There is no doubt about the structure. It seems to be well established by all of the normal means, but we did confirm it via X-ray crystallography. The mechanism of this reaction is quite different from that of the other examples and is of interest to us. The most reasonable assumption seems to be that we get the normal Diels-Alder reaction, although there is some steric hindrance with the gem-dimethyl group. This gives us an intermediate 1:1 adduct (upper center). It can either undergo electrocyclic reaction of some sort directly to the cyclopropane, or radical fission could occur (lower right). The other end of the system would be a tertiary radical which would be relatively stable. Such a fragmentation does not seem beyond the realm of possibility. To try to break the bond which would lead to a vinyl-type radical (lower center) seems much less probable. Such a vinyl radical would be expected to close to a cyclobutene-type product (lower left), which was not observed. However, closure of the lower right structure would give the cyclopropane structure that is observed. This is the only example of this type of addition that we have observed in the indene series.

Discovery of this new type of structure caused us to look hard for other examples with 1,1-dimethylindene. A variety of dienophiles were tried in the hopes of obtaining crystalline adducts. We were only successful using azo compounds (dimethyl azodicarboxylate, diethyl azodicarboxylate, and the very familiar N-phenyltriazolinedione) for this purpose (eq. 33) [1j]. Fred Greene has pointed out that indene itself has been reported to form an adduct having structure (1) which would have a CH_2 in

ROOC–N=N–COOR + 1,1-dimethylindene (CH_3 CH_3) → (benzene, reflux, 8 h) → 1 → (xylene, reflux; 1a : 18 h, 1b : 11 h) → ROOC–NH–NH–COOR 2

	R	%	mp °C		R	%
1a	CH_3	74	118-118.5	2a	CH_3	54
			pale yellow		mp 126.5-129 °C	
1b	CH_3CH_2	14*	94- 95.5	2b	CH_3CH_2	36
			white		mp 132-134 °C	
		*69	(when run neat)			

+ N-phenyltriazolinedione → (CH_2Cl_2, 18° C, 0.25 h) → 3

quant., pale yellow
mp 164-166 °C
(turns green at 141 °C)
IR (KBr) cm^{-1}
1775 m, 1723 s (imide C=0) (33)

place of the gem-dimethyl in this case. Its formation was reported to be reversible, so we have a complete analog as far as we can see. If we push the reaction too far, by heating for too long in refluxing xylene, we do not get any of the adduct.

The only crystalline product isolated is the 1,2-hydrazinedicarboxylate (2). We interpret this as probably meaning that the formation of our adduct is reversible; this situation is analogous to the indene case. Additionally, this result suggests that decomposition has taken place to form 2 as the ultimate hydrogen acceptor. The N-phenyltriazolinedione adduct (3) is of the corresponding type and was formed in the usual high yields. The yields of the diesters (1a and 1b) were not as high, (ie, 74%, or a maximum of 69% when run neat). Comparison of the spectroscopic data for the diester adducts of type (1) with corresponding spectra obtained for the cyclopropyl case (2) (eq. 34) suggest that a benzenoid chromophore is undoubtedly present. Laser Raman shows that no double bond is present. The NMR spectrum confirms the presence of a gem-dimethyl group [16]. A significant feature of the NMR spectrum of the dimethyl azodicarboxylate adduct is the relative broadening of the H_b methine doublet at δ 4.68 relative to the H_a methine doublet at δ 5.26. This suggests the possibility of some conformational floppiness or rocking at the

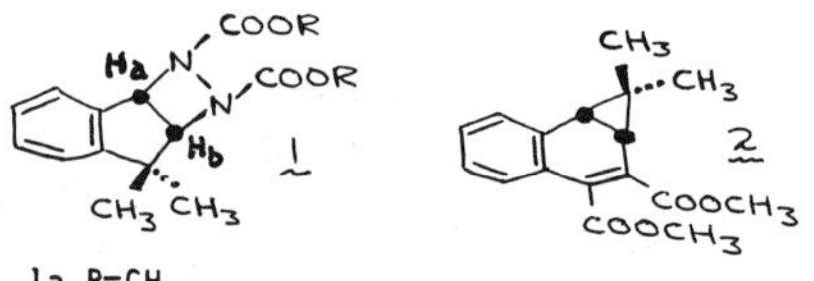

COOR, Ha, N, N, COOR, Hb, CH_3 CH_3 — 1 CH_3, CH_3, $COOCH_3$, $COOCH_3$ — 2   (34)

1a R=CH_3

1b R=CH_2CH_3

IR (KBr) cm^{-1} (C=O)

1a 1672 vs, 1b 1667 vs

IR (Nujol) cm^{-1}

1737 s, 1718 s (C=O)

1611 ms (C=C)

UV (95% EtOH) λ max (log ε)

1a	2
	240 (4.17)
257 (2.47)	296 (3.89)
264 (2.55)	308 (3.90)
271 (2.54)	325 sh (3.61)

ring carbon bearing H_b. The mass spectra of the dimethyl [m/e(rel ab): 290(10); 144(100); 129(45)] and diethyl [318(14); 140(100); 129(44)] azodicarboxylate adducts show that the adducts readily undergo cycloreversion in the mass spectrometer, giving the 1,1-dimethyl-1H-indene molecular ion as the base peak.

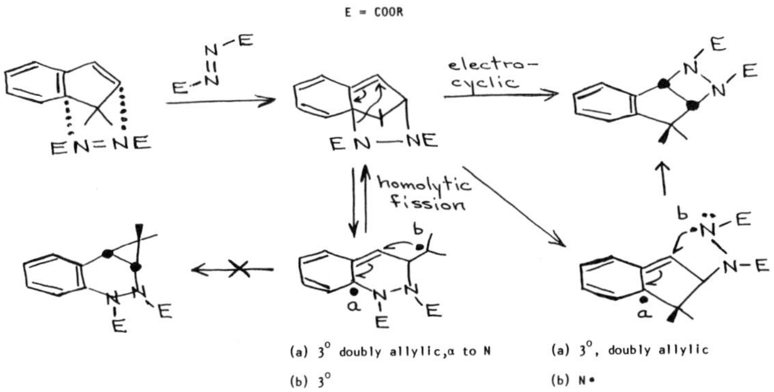

E = COOR

electro-cyclic

homolytic fission

(a) 3° doubly allylic,α to N

(b) 3°

(a) 3°, doubly allylic

(b) N•

(35)

Possible mechanisms for the addition of azo compounds to 1,1-dimethyl-1H-indene are shown in eq. (35). The mechanisms go through the Diels-Alder adduct (upper center) formed across the diene system of the five-membered ring of the indene as with DMAD. A 1,3-sigmatropic rearrangement of the ring-junction carbon (a) to nitrogen (b) bond would give a tertiary doubly allylic radical at a (lower right structure). Ring closure of the b nitrogen radical at the other end of the allylic system would aromatize the benzene ring and give the observed adduct (upper right). An alternative homolytic fission (lower center structure) of the bond from the same ring junction carbon (a) to the gem-dimethyl carbon (b) would give the tertiary doubly allylic radical a, which is now also α to a nitrogen, and a tertiary radical at b. Ring closure of the b radical at the other end of the allylic system would again aromatize the benzene ring and give an adduct (lower left), which was not observed, containing a cyclopropane ring analogous to the DMAD adduct.

Discussion

House: Isoindenes are often written as intermediates that precede many Diels-Alder reactions, but I never see anyone comment on what they see spectroscopically when indenes are heated to higher and higher temperatures.

Noland: I am not aware of any spectroscopic evidence, which does not mean there isn't any. The evidence is derived from products; it is derived from deuterium isomerization. Jerry Berson did a lot of good work on indene isomerization which certainly demonstrates mechanistically that it must have happened.

House: What has always puzzled me is why in some cases one seems to get Diels-Alder reaction across isoindene and in other cases it seems to involve the parent indene structure. I don't understand why sometimes it is one way and sometimes the other. Can you rationalize that in any way?

Noland: I would try and rationalize it on the basis of the rapid rate of interaction of the dienophile with this system, ie, ones that add across to give a normal Diels-Alder but involving the aromatic ring are DMAD and the azo compounds. Those are the only ones that we have noted that do that. All the ethylenic dienophiles seem incapable of doing that and must wait for the indene molecule to isomerize into the isoindene form. Now, that is not the order of reactivity of Diels-Alder dienophiles that has been reported in some other reactions but it seems to apply to this situation.

Foote: This is really a strange question. There are really two classes of dienophiles: those that react with indene in the ordinary cyclopentadiene dienophile form and those that are totally incapable of doing that. Singlet oxygen reacts at -80 .

Traylor: Addition products of indene and the azodicarboxylic acid esters seem to me to be compounds that would be subject to easy dehydrogenation. Is it possible that your hydrazine products that you got out of that reaction result from the dehydrogenation reaction to form an unsaturated four-membered ring? This would have some stability, I think.

Noland: Are you referring to the formation of the 1,2-hydrazinedicarboxylate ester being formed by some sort of dehydrogenation process?

Traylor: Yes, it is a good dehydrogenating agent.

Noland: Yes, it is.

Traylor: You just pluck off those two hydrogens. Doesn't that form a stable compound?

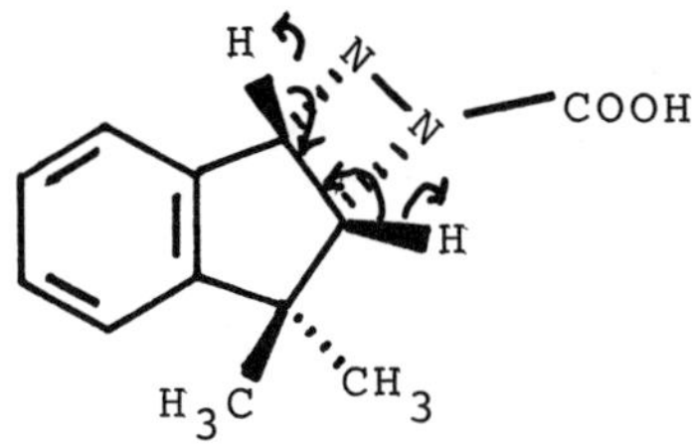

Noland: Oh, I see you are going to do this intermolecularly and remove these two hydrogens by another molecule. All right.

Traylor: Do you have excess diazoester?

Noland: Yes, a 1.5:1 molar ratio. This would certainly be a way of accounting for the formation of the 1,2-hydrazinedicarboxylate. We were not able to isolate any other crystalline products. If it were crystalline, I would guess it was not there. If it were crystalline, I would guarantee we would have found it.

Perrin: In order to get your 5-ring product, the 2:2 product, you use milder acidic conditions.

Noland: Not always, but some of the time you are correct. I used to think we could sort of see the difference there. The amount of sulfuric acid in methanol is small, in the range of 0.5 to 2 N.

Perrin: Well, I am no longer sure the question has much significance, but in the mechanism you presented, the branch point involved the deprotonation. If the branch point leading to the six-membered ring involved the deprotonation, where do you preserve the proton in order to go to the five-membered ring? I am curious if you have some reason or explanation why sometimes you get the 5- and sometimes the 6-membered rings. What is going on?

Noland: I wish I had a better one than I do. One of the problems is that these reaction mixtures often resemble primordial soups. They're not clean-cut reactions. I can show you a case where we isolated both products from the same reaction and that was in my first example. I have a suspicion that we do have an equilibrium situation and most of the time the precipitate of the hydrochloride of the 6-membered ring product is the first to fall out. But the preferred product is the five-membered ring product.

Gleiter: Do you see any spiro effect in the electronic absorption spectrum of your spiro compound? A long wavelength shift or a new band?

Noland: No, we do not. The spectrum is simply that of two indoles with 2,3-disubstitution which are simply added together. I was thinking I might find something of that sort, but there doesn't seem to be enough strain or perturbation to cause any change in the UV chromophore. And yet, there is a high reactivity toward oxidation.

References

[1] The following coworkers must be recognized for their contributions to this work:
(a) Brad C. Pearce. Senior thesis research 1978 and postgraduate research 1979; presented as a paper entitled "A Symmetrical Spiro 2:3 Cyclization product from the Condensation of Indole with Acetone" at the 22nd Annual (Minnesota) Undergraduate Symposium in Chemistry at the University of Wisconsin at River Falls, River Falls, WI, April 28, 1979.
(b) Alison J. Coulter-Knoff. Undergraduate directed study research 1980-81; presented as a paper entitled "Condensation Products of Indole with Acetone" at the 24th Annual (Minnesota) Undergraduate Symposium in Chemistry at St. Olaf College, Northfield, MN, April 25,

1981.
(c) Carl A. Houtman. Undergraduate directed study research 1981-82; presented as a paper entitled "The Condensation of 1-Ethylindole with Acetone" at the 25th Annual (Minnesota) Undergraduate Symposium in Chemistry at the University of Minnesota, Minneapolis, MN, May 1, 1982.
(d) Chanita Wongkrajang. Research for the M.S. degree 1974-75.
(e) Michael S. Tempesta. Senior thesis research 1977-78; NSF Undergraduate Research Participant, summer 1977; presented as papers entitled "2:2 Condensations of Indoles with Ketones" and "Reactions of Indoles with Ketones: The in situ Vinylindole Synthesis of Carbazoles" at the 21st Annual (Minnesota) Undergraduate Symposium in Chemistry at Bethel College, St. Paul, MN, April 22, 1978.
(f) Jay S. Simonson. NSF Undergraduate Research Participant, summer 1976; presented as papers entitled "Condensations of Indoles with Ketones" at the Undergraduate (Chemistry) Symposium at North Dakota State University, Fargo, ND, November 6, 1976, and at the 20th Annual (Minnesota) Undergraduate Symposium in Chemistry at Carlton College, Northfield, MN, April 23, 1977.
(g) Rodney D. DeKruif. M.S. Thesis, University of Minnesota, 1982.
(h) Roger C. Simurdak. Ph.D. Thesis, University of Minnesota, June 1972; Diss Abstr. Int. B., 34, 3175 (1974).
(i) Robert E. Krause. Graduate research 1963-65.
(j) Jay F. Schulz. Undergraduate directed study research 1980-82; presented as a paper entitled "Cycloaddition of 1,1-Dimethyl-1H-indene and Dimethyl and Diethyl Azodicarboxylate" at the 25th Annual (Minnesota) Undergraduate Symposium in Chemistry at the University of Minnesota, Minneapolis, MN, May 1, 1982.
(k) J. H. Sellstedt, Ph.D. thesis, University of Minnesota, November 1965; Diss. Abstr., 27B, 2300 (1967).

[2] M. Scholtz, Arch. Pharm. 253, 629 (1915).

[3] (a) W.E. Noland, M.R. Venkiteswaran, and C.G. Richards, J. Org. Chem. 26, 4241 (1961).

(b) W.E. Noland, M.R. Venkiteswaran, and R.A. Lovald, J. Org. Chem. 26, 4249 (1961). (c) W.E. Noland, C.G. Richards, H.S. Desai, and M.R. Venkiteswaran, J. Org. Chem. 26, 4254 (1961).

[4] W.E. Noland and M.R. Venkiteswaran, J. Org. Chem. 26, 4263 (1961).

[5] (a) A. Chatterjee, S. Manna, J. Banerji, C. Pascard, T. Prange, and J. Shoolery, J. Chem. Soc. Perkin Trans. I, 553 (1980). (b) J. Banerji, A. Chatterjee, S. Manna, C. Pascard, T. Prange, and J. Shoolery, Heterocycles 15, 325 (1981).

[6] W.E. Noland and R.J. Sundberg, J. Org. Chem. 28, 884 (1963).

[7] W.E. Noland, W.C. Kuryla and R.F. Lange, J. Am. Chem. Soc. 81, 6010 (1959).

[8] A.-M.M.Makky, M.S. Thesis, University of Minnesota, July 1961.

[9] W.E. Noland and S.R. Wann, J. Org. Chem. 44, 4402 (1979).

[10] W.E. Noland, L.L. Landucci and V. Kameswaran, J. Org. Chem. 45, 3456 (1980).

[11] Lawrence L. Landucci, Ph.D. Thesis, University of Minnesota, March 1967; Diss. Abstr. B, 28, 3223 (1968).

[12] R. Weissgerber, Ber. Dtsch. Chem. Ges. 44, 1436 (1911).

[13] W.E. Noland, V. Kameswaran, and L.L. Landucci, J. Org. Chem. 45, 4564 (1980).

[14] C.F. Huebner, P.L. Strachan, E.M. Donoghue, N. Cahoon, L. Dorfman, R. Margerison and E. Wenkert, J. Org. Chem. 32, 1126 (1967).

[15] D.W. Jones, J. Chem. Soc. Perkin I, 673 (1979).

[16] W.E. Noland and V. Kameswaran, J. Org. Chem. 46, 1318 (1981).

SUBJECT INDEX